Working with DNA

Working with DNA

Stan Metzenberg
California State University
Northridge, USA

Taylor & Francis
Taylor & Francis Group

Published by:
Taylor & Francis Group

In US: 270 Madison Avenue
           New York, N Y 10016
In UK: 4 Park Square, Milton Park
           Abingdon, OX14 4RN

© 2007 by Taylor & Francis Group

First published 2007

ISBN: 978 0 4153 7464 4 or 0 4153 7464 2

Library of Congress Cataloguing-in-Publication data

Metzenberg, Stan.
   Working with DNA / Stan Metzenberg.
         p. ; cm.
   Includes bibliographical references and index.
   ISBN 978-0-415-37464-4 (pbk. : alk. paper)
1. Recombinant DNA--Research--Methodology. I. Title.
   [DNLM: 1. DNA, Recombinant. 2. Genetic Techniques. QU 58.5 M596w 2007]

QH442.M485 2007
660.6'5--dc22

                                                                2006034192

Editor: Elizabeth Owen
Editorial Assistant: Kirsty Lyons
Production Editor: Georgina Lucas
Typeset by: Phoenix Photosetting, Chatham, Kent, UK
Printed by: Cromwell Press Ltd

Printed on acid-free paper

10 9 8 7 6 5 4 3 2 1

Taylor & Francis Group, an Informa business   Visit our website at http://www.garlandscience.com

Contents

Abbreviations

AAV	adeno-associated viruses	KLH	keyhole limpet hemocyanin
AD	activation domain	MLV	murine leukemia virus
AMV	avian myeloblastosis virus	MMLV	Moloney murine leukemia virus
ARS	autonomously replicating sequence	MOI	multiplicity of infection
BAC	bacterial artificial chromosome	MSDS	material safety data sheets
BSA	bovine serum albumin	PAC	P1 artificial chromosome
CCD	charge coupled device	PCR	polymerase chain reaction
CEN	centromere	PEG	polyethylene glycol
CMV	cytomegalovirus	qPCR	quantitative PCR
DBD	DNA-binding domain	RACE	rapid amplification of cDNA ends
ddATP	dideoxyadenosine triphosphate		
DEAE	diethylaminoethyl	RBS	ribosome binding site
DEPC	diethylpyrocarbonate	RF	replicative form
DGGE	denaturing-gradient gel electrophoresis	RFLP	restriction fragment length polymorphism
DMSO	dimethylsulfoxide	RISC	RNA-induced silencing complex
dNTPs	deoxynucleoside triphosphates	RNAi	RNA interference
dpm	disintegrations per minute	RT-PCR	reverse transcription-polymerase chain reaction
DTT	dithiothreitol		
dUTP	deoxyuridine triphosphate	rtTA	reverse Tet-controlled transactivator
EDTA	ethylenediaminetetraacetic acid		
EEO	electroendosmosis	SSCP	single-strand conformational polymorphism
ELISA	enzyme linked immunosorbent assay		
		SDS	sodium dodecyl sulfate
EMSA	electrophoretic mobility shift assay	siRNA	short interfering RNA molecules
		SSC	standardized saline citrate
EST	expressed sequence tag	TAE	Tris-acetate-EDTA
FACS	fluorescence-activated cell sorter	TBE	Tris-borate-EDTA
F factor	fertility factor	TdT	terminal deoxynucleotidyl transferases
FISH	fluorescent *in situ* hybridization		
FRET	fluorescence resonance energy transfer	TE	Tris-EDTA
		TEMED	tetramethylethylenediamine
GFP	green fluorescent protein	Ti	tumor inducing plasmid
GST	glutathione-S-transferase	TRE	tetracycline response element
HEK	human embryonic kidney	Tris	Tris [hydroxymethyl] aminomethane
HPLC	high performance (or high pressure) liquid chromatography		
		tTA	Tet-controlled transactivator
HSV	herpes simplex virus	TUNEL	TdT-mediated dUTP nick end labeling
IPTG	isopropyl-β-D-thiogalacto-pyranoside		
		UAS	upstream activation sequence
		YAC	yeast artificial chromosome

Preface

This is a book about the materials and methods that are used in the recombinant DNA laboratory. It is designed to help the novice understand how specific laboratory techniques work, but it also has sections that may stimulate the thinking of experienced researchers. This isn't a book of step-by-step protocols or recipes; there are many fine books of that kind already in existence. It is intended to broaden and strengthen the foundation of the reader so that working with DNA becomes more natural.

For example, it is easy to become addicted to using commercial kits for standard techniques. I enjoy using kits from time to time, but I can see that there is a hidden cost when they disconnect the hands from the brain (figuratively). This book is intended to help researchers think about their work, and be aware of the many ways that a molecular research question may be addressed.

I am grateful to the people who have taught me science over the years; my professors and lab-mates, my wife Aida, my father Bob, and now my daughter Gretchen. My own students have been the best teachers of all. I am particularly grateful to Louise Scharf for corrections and helpful suggestions during the writing.

Stan Metzenberg, Ph.D.
California State University, Northridge

Understanding recombinant DNA techniques

1

1.1 The *in vitro* workbench

Working with DNA is a very straightforward process. Most recombinant DNA techniques are performed on samples *in vitro*, using commercially available enzymes that are harnessed for specific tasks. Sometimes the goal of a procedure is to collect diagnostic information, for example regarding the length or base sequence of a nucleic acid. Other times the goal is one of engineering, in the sense of changing or rearranging the base sequence of a nucleic acid and constructing a novel sequence.

Cutting and pasting

The principal enzymes that are used in recombinant DNA techniques are the restriction endonucleases, polymerases, and ligases. There are hundreds of different restriction endonucleases, and they are used to cut DNA sequences internally at specific recognition sequences (as described in Chapter 5). Restriction endonucleases are useful in many diagnostic methods because the digestion reveals the location of enzyme recognition sequences in a DNA. On the other hand they are also useful in genetic engineering because the ends produced by enzyme digestion are highly characteristic. Each end is like a shaped puzzle piece that can fit only with certain matching pieces of DNA, usually by a specific pattern of hydrogen bonding. The DNA polymerases are particularly useful for changing and reshaping the ends of DNA molecules, and in adjusting the 'fit' of two ends of DNA (as described in Chapter 6). Polymerases are also used to copy portions of a nucleic acid into DNA or RNA, and to make planned changes to the sequence of a molecule through site-specific mutagenesis. Ligase enzymes are used for some of the finishing work in genetic engineering, because they build a covalent connection between two segments of DNA. There's a bit more to the story than just that, of course, but this book is intended to start with the basics and teach how each component of the system works. More sophisticated diagnostic methods and engineering approaches are presented in Chapters 7 and 8.

Getting things to bump into each other

At a very simple level, when we are thinking about reactions with DNA we have to worry a lot about how often the components are going to bump into each other. That's fundamentally a question of concentration; dilute solutions have solutes that bump into each other less frequently than concentrated solutions (as described in Chapter 3). If we think of DNA molecules as being like puzzle pieces, then each collision between two DNA molecule ends is another chance to try to fit them together. It takes longer to assemble a jigsaw puzzle if the pieces are spread around a room, and in a similar way it takes longer for a reaction between two DNA molecules to reach completion if the substrates are dilute.

Two DNA molecules bump into each other at a frequency proportional to their molar concentrations, but we sometimes have DNA concentrations that are widely different. For example in the polymerase chain reaction (PCR), described in Chapter 7, we may bring together a short DNA molecule and a long one, under conditions in which the short DNA is more than a billion times more concentrated. Using this high concentration gives us assurance that it will bump into the longer DNA so frequently that it will find the place that it 'fits' in 15 s or less.

Other times we increase the frequency with which two molecules collide by tethering them to each other, making sure that they don't diffuse away. If a topoisomerase enzyme is tethered to a DNA molecule, this permits very rapid DNA joining reactions (for example the TOPO® system described in Chapter 4). Furthermore, a single linear DNA molecule has two ends that can get no further apart than the DNA sequence connecting them, and so may be joined into a circle if the ends have a compatible fit.

There are electrostatic effects at the molecular level that sometimes cause repulsion between macromolecules, and shielding these effects is one way that the frequency of collisions between nucleic acids can be controlled (as described in Chapter 3). For example, DNA hybridization in Southern blotting is one way that the fit between two molecules can be checked. We agitate a probe incubating with a Southern blot to overcome the limitations of diffusion, and increase the likelihood that a probe and a target molecule collide at the surface of the blot. We also conduct the hybridization in a high ionic strength solution, so that the electrostatic repulsion between DNA molecules is partially relieved. If the probe and target have compatible base-pairing, their 'stickiness' for each other may exceed their mutual electrostatic repulsion, and the probe may bind to the target.

Engineering considerations

The design of a genetic engineering process is very much dependent on the experimental need. For example, suppose we want to express a eukaryotic protein in a prokaryotic host cell. The transcriptional and translational machinery will be prokaryotic, and so the regulatory sequences that direct those processes will have to be brought together with the coding sequence of the eukaryotic protein. If the gene for the protein has introns, these will have to be omitted from the constructed DNA that is introduced into the prokaryote, because prokaryotes do not perform pre-mRNA splicing. Other questions must also be posed before a plan can be made: Will the inserted gene in the prokaryote be part of an extrachromosomal element such as a plasmid, or will it be integrated into the host chromosome? Will the sequence need to be regulated so that it is sometimes transcriptionally active and other times silent, or will it be transcribed constitutively (i.e., all the time)?

In many cases, a plasmid vector may already be available that carries features that we want to use: prokaryotic or eukaryotic regulatory sequences, origins of DNA replication, and selectable marker genes to help us maintain the plasmid after introduction in the cell (described in Chapter 4). The genetic engineering project may be reduced to a simple matter of opening the plasmid sequence at a cloning site, inserting the novel sequence and closing the plasmid into a circle, and introducing the plasmid into a host cell.

Planning ahead

In recombinant DNA research we want to use approaches that will facilitate the construction of a sequence, and the analysis of its content. It wouldn't be of much help, for example, to build an elaborate DNA plasmid that brought together a prokaryotic transcriptional promoter with a eukaryotic gene sequence, if the ribosome could not bind to the mRNA and initiate translation.

We can exercise some control over the cloning process by preventing or eliminating side reaction products, but this takes some forethought in engineering. For example, the presence or absence of restriction endonuclease sites can help to control the outcome of a ligation reaction. If a DNA fragment 'A' has an end created with the enzyme *Bam*HI, and a

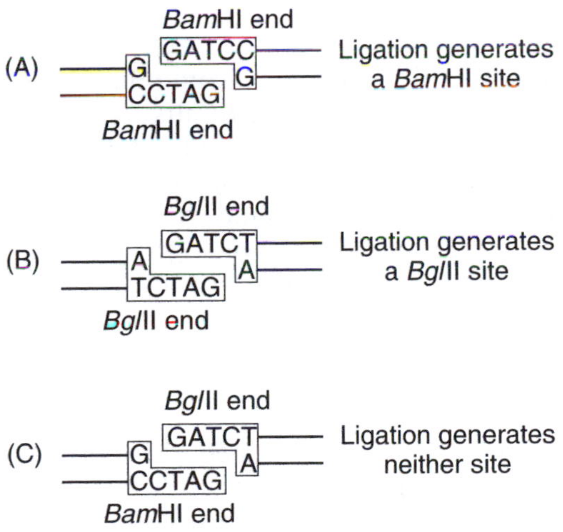

Figure 1.1

A simple genetic engineering trick: the restriction endonucleases *Bam*HI and *Bgl*II leave identical overhanging ends, which may be ligated together. (A) If ends left by *Bam*HI are joined, they generate the GGATCC sequence of a *Bam*HI restriction endonuclease site. (B) If ends left by *Bgl*II are joined, they generate the AGATCT sequence of a *Bgl*II restriction endonuclease site. (C) If an end left by *Bam*HI and an end left by *Bgl*II are joined, they generate a GGATCT or AGATCC sequence that is not a *Bam*HI or a *Bgl*II restriction endonuclease site.

DNA fragment 'B' has an end created with the enzyme *Bgl*II, then all of the pair-wise combinations of ends are compatible; A with A, B with B, and A with B (see *Figure 1.1*). These A and B ends can be joined by base-pairing and ligation because both *Bam*HI and *Bgl*II leave the same 5'-GATC overhanging end. However, we can select against the 'A with A' joints by digesting the ligated products with the original enzyme *Bam*HI. The 'A with A' connections will be separated again because the *Bam*HI recognition sequence GGATCC is regenerated when A joins with A. The 'A with B' connections are resistant to *Bam*HI because the joined fragments do not form a recognition sequence for *Bam*HI.

We also do careful planning when we think about how we might more easily distinguish successful cloning experiments from failed ones. When we insert a DNA fragment into a plasmid vector, we sometimes add enough new material that we can distinguish the product from the original plasmid by its increase in size. On the other hand, there are times when a change in size will not be easily discerned, and we need to use diagnostic restriction endonuclease mapping or DNA sequencing to characterize the cloned product. By planning a cloning experiment so that restriction sites are characteristically preserved or destroyed, we make it easier to locate the successful candidate plasmids among a collection of individual clones. We may also do our cloning work within easy reach of a universal primer binding site, to allow for definitive DNA characterization by sequencing.

1.2 When the goal is a clone

Why are clones so important?

One of the amazing and beautiful things about recombinant DNA techniques is that cloning work leads to unambiguous results. A plasmid structure is either right or it isn't, and there is a process for telling the difference. Furthermore, a correctly built cloned DNA can be copied and purified, and, at least ostensibly, each copy is the same as the original. Recombinant DNA techniques have revolutionized the biological sciences, in large measure because of the advantages that are brought to the laboratory bench and field by having genetically pure materials.

Climbing mountains

Building a clone takes a number of steps; however, these steps do not always work the first time, or every time. Achieving a successful result in the end is like climbing to the top of a mountain, and an important lesson can be learned from the methods of real mountain climbers. When climbing Mount Everest, for example, a person does not simply pack light and dash up the 3 500 m from base to summit in one day. That's a sure way to get a fast ride down Lhotse Face! Instead, the climber establishes a base camp and a series of intermediate camps that are separated by a few hours of climbing (Camp 1, Camp 2, Camp 3, etc.). Each camp up the side of the mountain is supplied with provisions, in case the weather turns bad and the climber has to fall back to make another attempt later. Without fallback camps, a mountaineer would have to return to the base of the mountain after each failed attempt, and would lose progress that had been made.

Genetic engineers also endure 'heavy weather' while making clones, in the form of nonfunctioning enzymes, unwanted side reactions, and lost reaction products. The researcher needs to have a series of fallback positions so that a failed attempt does not lead to too much lost time. For example, a cloning experiment might involve the preparation of a plasmid vector (discussed in Chapter 4), and a DNA fragment that is to be inserted into it. Both plasmid vector and DNA fragment might be digested with specific restriction enzymes (discussed in Chapter 5), and then added together in a ligation reaction. The ligated DNA products would then be introduced into competent *Escherichia coli* cells by transformation (discussed in Chapter 4).

However, bad things can happen to good researchers: What happens if one of the restriction enzymes doesn't work, either because the DNA is impure, or the enzyme is dead? The researcher needs to have some way of assessing the problem before committing time and resources to the DNA ligation and transformation steps. One way of doing this would be to compare the DNA samples before and after restriction digestion by gel electrophoresis. That is, the early step of restriction digestion is monitored carefully to make sure it is adequately complete. If the researcher intended to transform the *E. coli* with 50 ng of ligated plasmid, he or she might start the preparation by using 20 times that amount, or 1 µg of plasmid DNA in the restriction digestion (see *Figure 1.2*).

At each stage, a portion of the DNA is held in reserve for further analysis or repeated work. If the transformation fails to yield colonies (*Figure 1.2F*), the researcher can return to the ligated DNA held in reserve (*Figure 1.2D*) and either repeat the transformation, or analyze a portion of it by gel electrophoresis to see if there is evidence of ligation. A wise researcher develops cloning work like a mountaineer planning an ascent; making sure that the foothold at each step is checked, and that each 'camp' is adequately provisioned in case heavy weather closes in.

Courage and diligence

It is often the case that a recombinant clone becomes the foundation for a large volume of research. For example, a cloned DNA might be used to make a labeled probe for analysis, or might carry a recombinant gene that is to be expressed *in vitro* or *in vivo*. If the clone is incorrect in some way and this defect is not uncovered at an early stage, the research that is based on the faulty clone may need to be repeated, or if already published, then retracted. If the faulty clone has been sent to other laboratories, then apologies may also be necessary. Verifying a clone, making sure that its structure is exactly what was planned, requires attention to detail and diligence in performing a sufficient number of tests.

However, diligence also requires courage; the courage to face the truth. We may avoid doing crucial experiments or overlook conflicting bits of data because our minds do not readily entertain the possibility of failure. Facing the truth is a psychological issue that may be noted in many human endeavors. An airplane pilot stuck in a fog, with no visual cues

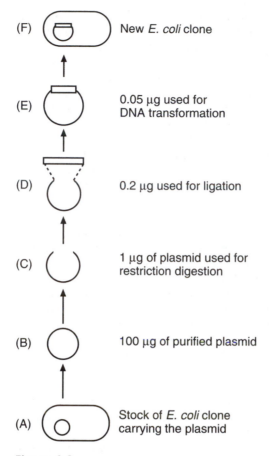

(F) New *E. coli* clone

(E) 0.05 µg used for DNA transformation

(D) 0.2 µg used for ligation

(C) 1 µg of plasmid used for restriction digestion

(B) 100 µg of purified plasmid

(A) Stock of *E. coli* clone carrying the plasmid

Figure 1.2

A genetic engineer keeps a reserve of material at each step before pushing an experiment to completion; the reserves are a resource if a procedure is repeated, or a remnant for analysis if troubleshooting work must be done. (A) The most important reserve is a frozen stock of the clone, from which more DNA can be generated. (B) A purified plasmid DNA will keep for years. (C) Plasmid DNA treated with a restriction endonuclease, and linearized. (D) Plasmid used in a ligation, and made into a circular molecule. (E) Plasmid introduced into *E. coli* by transformation. (F) A colony of cells representing the new clone, from which a frozen stock is made.

out of the window, may drift into a gradual turn that leads to a crash. The problem is that the turn causes enough centrifugal force to make the pilot feel that he is flying level, even though the compass, gyroscopic artificial horizon, and altimeter all indicate otherwise. It takes courage for a pilot to trust the instrument data, and accept the evidence of pilot error.

Contaminants

One thing that can go wrong during a cloning process is that a contaminant can slip into a culture and pretend to be the right clone. For example, if a DNA sample that was used in a previous experiment somehow contaminates a pipetting device or laboratory solution, it might inadvertently become an interloper in a transformation experiment. Small amounts of supercoiled plasmid contaminant are very efficient in DNA transformation of *E. coli*, and the researcher can accidentally pick a spurious colony from a bacteriological plate and not realize that it carries a plasmid unrelated to the work at hand. This type of contamination problem can usually be avoided by careful handling of materials and decontamination of wastes; however, filamentous bacteriophage vectors such as M13 are an added challenge in that they can spread from culture to culture through the air, in aerosol droplets.

We use antibiotics in bacteriological media as a means of selecting for plasmid maintenance, but finding a colony on an antibiotic-containing plate after a transformation is not *prima facie* evidence that the colony carries a plasmid, or is even *E. coli* for that matter. Transformation protocols typically call for a period of growth of the transformed cells in antibiotic-free medium, prior to spreading on an antibiotic-containing plate. If the antibiotic-free medium is handled with poor laboratory technique and becomes contaminated with an antibiotic-resistant bacterium, such as *Staphylococcus*, the plate carrying the transformed cells may be covered with spurious colonies.

Even if a colony on a plate is made up of the species *E. coli*, it may still have grown up without carrying a plasmid. This can happen in cases where the true antibiotic-resistant cells in a mixed population secrete an enzyme that destroys the antibiotic outside of the cell, for example β-lactamase which destroys ampicillin. Cells will grow up to form a colony in a region of the plate in which they have destroyed the antibiotic, but unfortunately, that does not preclude plasmid-free cells from freeloading and making their home around plasmid-carrying cells. This problem can be aggravated if the transformed cells are plated at high density, or if the bacteriological plate is old and no longer carries enough antibiotic activity to be selective. If the bacteriological plate was 'spread' with the antibiotic, but spread unevenly, there may also be regions of the plate in which antibiotic-sensitive cells can survive.

We may be hopeful, or guardedly optimistic about colonies that appear on a bacteriological plate after DNA transformation; however, some diligent work needs to be done to make sure they are not contaminants. Contaminating colonies that are not *E. coli* will not look or smell right, and will probably not produce any result if a plasmid extraction is attempted. Colonies that are plasmid-free *E. coli* 'freeloaders' will not propagate if streaked to a fresh antibiotic-containing plate, or transferred to an antibiotic-containing liquid culture. Colonies that are *E. coli* carrying a contaminant or 'interloper' plasmid will have to be sorted out by means of DNA diagnostics, for example restriction analysis.

Carryover

Plasmids can carry over into a transformation, not because they are contaminants *per se*, but because they were not properly processed or eliminated during earlier steps in an experiment. For example, suppose a researcher wants to excise a small piece of DNA from plasmid A and transfer it to plasmid B. Plasmid A might be digested with a restriction enzyme or two, to physically separate the small DNA fragment from the larger remnant of plasmid A. Plasmid B might be digested with compatible enzymes to make it into a 'landing site' for the small DNA fragment from the other plasmid. What happens if either of the two parental plasmids, A or B, are not completely digested by their restriction enzymes? There may be some carryover of the undigested plasmid into the ligation reaction, where nothing will happen to it since it is already covalently closed, and then into the transformation reaction where it will create havoc!

This is just one example of a cloning artifact that can plague a researcher; more examples of common cloning problems and their solutions are discussed in Chapter 4. A critical problem for a researcher is to devise a DNA diagnostic approach to distinguish between the 'right' plasmid clone, and the plasmid clones that may result from plasmid carry-over. In a genetic engineering method that is carefully planned, a researcher thinks about all of the side reactions that could be troublesome, and tries to incorporate a means of testing into the structure of the plasmid.

Rearrangements and unexpected side reactions

A plasmid may have an incorrect structure, even if some of the diagnostic tests suggest that the cloning procedure worked. For example, suppose a researcher wishes to insert a 200-bp DNA fragment into a 5-kbp plasmid, at a single restriction endonuclease site such as *Eco*RI. Would the researcher be able to tell if several copies of the DNA fragment were inserted in tandem? If the resulting plasmid were re-digested with *Eco*RI to check for the presence of a 200-bp DNA insert, then the two copies that were excised from each plasmid would co-migrate during gel electrophoresis and the error might be overlooked. The researcher must consider this possibility when designing diagnostic tests of plasmid structure.

A common site for trouble during DNA cloning work is at the ends that are being joined by ligation. A small extra fragment of DNA may sneak in during the ligation reaction and be nearly invisible in most diagnostic tests. Or, an end may be 'nibbled' by a contaminating exonuclease activity with the result that DNA is lost. The researcher must carefully plan a set of diagnostic tests to confirm that no mischief has occurred at a cloning site, and must maintain some level of skepticism about the clone until the evidence is unambiguous.

For example, suppose a researcher is joining two DNA ends that have been prepared using the restriction endonuclease *Sma*I, and predicts that the *Sma*I site will be regenerated as a result of cloning. What would happen if a 5-bp piece of 'junk' DNA is accidentally sandwiched between the *Sma*I ends during ligation? The junk DNA is probably too small to be discerned by restriction mapping across the junction, but the extra piece may disrupt the regeneration of the *Sma*I site. The absence of an expected *Sma*I site should tell the researcher that a potential problem exists. However, the lack of plasmid digestion with an enzyme is sometimes overlooked as a problem because enzymes can fail to function if the DNA is impure. Like the airplane pilot stuck in a fog and ignoring his instruments, the researcher may trust his own rationalizations more than the evidence to the contrary.

Modification of a plasmid by the host cell

Bacteria accept a plasmid during DNA transformation, but they do not always maintain its structure faithfully. For example plasmids can recombine with each other, particularly if they have repetitive sequences, or can recombine with the host DNA. The host cells can be party to all manner of mischief, and if DNA additions, deletions, or rearrangements lead to a faster-growing clone then the variant may overtake the original, correct cell clone.

Variants can develop in a cell culture at any point, and particularly annoying are the ones that develop after a plasmid clone has been thoroughly characterized and shown to be correct. A researcher cannot take anything for granted, including the purity of a clonal population. Fortunately, good microbiological practice can be used to limit the spread of variants. When a culture is being transferred from one plate to another, or when a liquid culture is being established, a researcher should always pick a single colony from a freshly streaked plate. This practice is like an artificial genetic bottleneck that makes a culture more genetically uniform. A single colony on a plate may still contain a few variant cells, but the problem can at least be minimized.

Postmortem or Champagne?

A researcher may be inclined to stuff bad results into a notebook and forget about them, but a lot can be learned from experiments that don't work. Trying to find out exactly what went wrong can help a person avoid the same situation in the future, and postmortem analyses of experiments can be the best teacher. For example, a researcher might ask: If an enzyme didn't work, which one was it? Was the enzyme really inactive, or was the reaction simply conducted under non-optimal conditions? If the reaction has worked previously but didn't work this time, what was different? Was a different reagent solution introduced; one that might have caused the failure? Were all manipulations of the sample performed just as they should have been? Without this type of careful consideration of the work, the researcher might just chalk up the failure to bad luck and do it exactly the same way again, with the same unhappy result.

On the other hand, if all of the appropriate diagnostic analysis has been conducted, and all evidence points to success, the researcher might consider breaking out the Champagne; a recombinant DNA clone has been made!

The recombinant DNA laboratory

2

2.1 Supplies

One of the beautiful things about recombinant DNA techniques is that a single molecule of plasmid DNA transformed into a bacterial cell can make the cell antibiotic-resistant, the cell can give rise to a colony on a plate, the colony on a plate can be used to start a liquid culture, and the liquid culture can give rise to many hundreds of micrograms of the original plasmid. While we can put that process to good use, our ability to make large amounts of a single type of DNA also means that we risk cross-contamination of our experiments if we have to wash and reuse all of the tubes and pipette tips. It is for this reason that recombinant DNA laboratories are notable for their consumption of large amounts of disposable plasticware.

Pipette tips

The plastic pipette tip probably heads the list of all disposable plasticware consumed by laboratories. Every time a micropipette is used to transfer a volume of liquid from one bottle or tube to another, the action very likely involves a plastic tip that will be ejected and discarded after that one use. An enzymatic reaction in a tube may be assembled using four or five disposable tips. Why are these tips not washed and reused? The primary reason is that tip re-use would risk cross-contamination of reagent stocks and experiments, which is an expensive problem. Even if washed and sterilized, the plastic tips are precision parts that are not meant to be re-used. The plastic inside the tip may hold a residue of reagent or cleaning solution that affects its wettability and accuracy of liquid delivery, and the plastic collar that seals the tip to the micropipette shaft may become fatigued with re-use and begin to leak.

Pipette tips are available in a variety of sizes, but two brands of tips that appear to be similar may still not fit the same pipetting instruments. Commercial distributors of pipette tips provide compatibility tables that indicate which tips can be used with each pipette model. Generally speaking, tips are designed for use in volumetric ranges of 0.1 to 2 µl, 1 to 200 µl, or 200 to 1000 µl. There are larger disposable pipette tips, but they are generally designed for use with a single model of pipette. Some tips have graduated markings on the side, to assist the user in deciding if the correct volume is being dispensed, and others have filter plugs to prevent cross-contamination of samples through the air forced into the top of the tip. There are tips that have long needlelike extensions on their ends that are specifically designed for loading gels. These tips may be rounded or flattened, and fit between the glass plates of a vertical gel to allow injection of a sample. Most tips hold and deliver their liquid volumes by the application of mild air pressure from the shaft of the micropipette. However, many pipette tips and specialized pipetting devices employ a solid plunger that enters the tip and provides positive displacement of the sample volume, like a syringe. These positive displacement pipettes are very useful with viscous samples, because the plunger sweeps the tip walls clean when the sample is expelled.

The quality of plastic tips can be a variable of some concern, depending on the manufacturer, with the less expensive tips occasionally having problems with plastic flashing residue and molding errors. Some brands of tips may be contaminated with heavy metals, pyrogens, or nucleases that interfere with certain laboratory applications. Tips can be purchased pre-sterilized, or without any guarantee of sterility. The tips can be purchased loaded in racks with a standard 96-well format, a different nonstandard format, or loose in bags. Some manufacturers provide racked tips in towers, where a layer of rack is removed when it is empty, exposing a new layer of fresh tips. There are also systems that allow a researcher to refill a rack from a dispenser, saving on storage space and plastic waste.

Plastic 'bullet' tubes of 0.2 to 2 ml

Much of the work in a typical recombinant DNA laboratory is conducted in small, bullet-shaped plastic tubes that have a volume of approximately 1.5 to 2.0 ml. These tubes have a snap cap that is tethered to the main body of the tube, and are sometimes called microfuge tubes because they fit into small microcentrifuges. Some tubes have volumetric markings on the side, or have a frosted section of plastic that allows for the direct writing of a label using a permanent marker. Depending on the brand, these tubes can often be obtained in a variety of colors to help in tube identification. The cap may be flat, making for easy writing of information on the top of the tube, or indented to add strength to the sealing connection between tube and lid.

Tubes with large caps are difficult to pack into close proximity in a holding rack, but they may be stronger than tubes with smaller caps. The hardness of the plastic varies between brands, with softer plastic tubes requiring less force to open by hand but being more susceptible to popping open accidentally when pressurized in a boiling water bath. Repeated opening and closing of a tube fatigues the cap seal, and may also lead to accidental tube 'popping' problems.

A laboratory may also use smaller bullet-shaped plastic tubes with a capacity of approximately 0.5 to 0.65 ml. These have less space for writing labels and may be more difficult to handle if a researcher has large fingers, but a collection of them can be stored in a smaller volume in a freezer. They typically fit inside the larger 1.5-ml tubes, with the small tube rims resting on the lip of the larger tube. The 1.5-ml tubes can sometimes be used as microcentrifuge tube adaptors, provided that the microfuge does not operate at a very high g-force. Some high-speed microcentrifuges will shatter the 0.5-ml tubes if they are not used with a specially made tube adaptor to support them, or are not used with a special rotor designed for small tubes. Most tubes of this size are designed for high-speed centrifugation, and are accordingly very strong. Others are 'thin-walled' and designed for use in a thermocycler where heat transfer through the plastic is important. It may be necessary to transfer samples out of these thin-walled tubes and into centrifuge tubes before the samples are subjected to centrifugation.

There are also smaller 0.2-ml tubes that are widely used for polymerase chain reactions and have roughly the same dimensions as a well from a 96-well plate. These tubes are thin-walled and may break during high-speed centrifugation. Some 0.2-ml tubes may be obtained in a 'tube strip' format, in which the tubes are connected and can be inserted as a unit into a standard 96-well format rack or heating block. Tubes of this size do not have much room for writing identifying marks, but tube strips and their corresponding cap strips may have numbers pre-printed on them to help with identification and orientation of the strips.

Plastic vials

Plastic vials with O-ring seals and capacities ranging from 2 to 5 ml are widely used for freezing biological specimens at −70°C or −196°C (liquid nitrogen temperature). These tubes usually do not fit in a microcentrifuge, but are made of a plastic that is less susceptible to shattering at extremely low temperatures. The tubes can be obtained in several styles, with the cap being either threaded internally or externally. The O-ring seal is particularly important for samples stored in liquid nitrogen, because if a tube contains residual liquid nitrogen when it is removed from a storage container then it may pressurize and explode as it warms. Furthermore, the liquid nitrogen can carry contaminants such as mycoplasma infections between cultures that are stored in the liquid phase in the freezer.

Plastic sample and centrifuge tubes of 5 to 50 ml

There are several types of plastic tubes with soft snap caps that are useful for storage or centrifugation of samples in the 1 to 10 ml range. Similar tubes are also available without a snap cap, and these may be sealed with a plastic film such as ParaFilm®. These tubes may have round or V-shaped bottoms, the shape of which affects the ease with which liquid is collected and the type of pellet obtained by centrifugation. The ability of these tubes to withstand autoclaving, centrifugation, and organic solvents such as chloroform must be considered before they are put into use. A tube may appear to withstand an organic solvent such as chloroform in the short term, but long-term exposure may cause the plastic to dissolve and soften.

Screw-top plastic tubes of 15 and 50 ml are widely used for sample storage and centrifugation. These can be made out of several different types of plastic, some being more suitable for organic solvent extractions and others being used for cell culture applications. The hardness of the plastic in the tube varies, but most will tend to become more brittle at −70°C and susceptible to shattering.

Microtiter plate formats

Plastic microtiter plates have many applications in molecular biology, including cell culture, sample storage, and diagnostic tests. These plates may be obtained in a standard 96-well format (8 rows × 12 columns), or 384-well and 1536-well formats that fit 4 and 16 times as many wells into the same plate area, respectively. The latter are particularly useful for large collections of small-volume samples for robotic handling, such as chip production or recombinant library storage.

Some microtiter plates are pre-sterilized and made of a pyrogen-free plastic that is appropriate for cell culture. These plates are typically packaged with a lid to maintain sterility in an incubator. Other microtiter plates are lidless but may be sealed with an adhesive strip if necessary. The plastic may be UV-opaque or transparent, and vary in optical purity. For applications involving thermocycling, thin-walled 96- or 384-well plates are used because the samples can change temperature rapidly. There are 96-well plastic tube blocks that are much taller than standard 96-well plates, used for storing several milliliters of solution in each deep well.

Ninety-six-well plates have flat, U-shaped, or V-shaped bottoms. The flat-bottomed wells are used for ELISA (enzyme linked immunosorbent assay) tests, and other applications involving optical readings or microscopy, but it is difficult to collect all the liquid from them without tipping the plate at an angle. U-shaped bottoms and V-shaped bottoms allow for easy collection of liquid, which is important for robotic liquid handling, and there are cell culture applications in which it is helpful to have a plate that allows for gravitational or centrifugal accumulation of sediment and cells in a small area of the well. On the other hand, these plates are difficult to use in microscopy and may give inconsistent readings in an ELISA plate reader.

Glass tubes

Borosilicate glass tubes come in many sizes, but they are generally not appropriate for storage of small amounts of nucleic acids because DNA sticks to glass. Conversely, small samples of protein are often stored in glass instead of plastic, because protein sticks to new plastic. Glass is compatible with organic solvents, but not with strongly alkaline solutions such as sodium hydroxide. Over time, a sodium hydroxide solution will dissolve the walls of a glass container, contaminating the solution with silicates and possibly weakening the bottle or tube. Disposable glass tubes rarely have capping systems, and they are not suitable for high-speed centrifugation.

There are stronger types of glass tubes made out of Corex® or Pyrex® that are designed for centrifugation, but these are not disposable labware. These glass centrifuge tubes generally require a rubber adaptor or cushion when they are used in a centrifuge, and will occasionally shatter in a centrifuge if they are weakened through etching or scoring of the glass.

Laboratories sometimes use glass tubes for growing bacterial cultures, and these may have a loose-fitting metal or plastic cap that allows gas exchange, or a tight-fitting airtight cap. If the tubes are not sterile when purchased, they should be autoclaved before use or re-use in growing cultures. Some cultures are grown in glass tubes with cotton or fiber plugs to allow for gas exchange. Fiber plugs are generally not used if the tube contains liquid medium and needs to be shaken in an incubator, because splashing of medium onto the fiber plug may invite the invasion of contaminating fungi and bacteria.

Glass and plastic pipettes

Glass and plastic pipettes are widely used in laboratories to transfer volumes of 1 to 25 ml. These pipettes can be disposable or reusable, and there are slight differences in handling for each. Disposable pipettes may be either plastic or glass, stuffed at the end with cotton or not stuffed, individually wrapped or taken from a bulk pack, and sterile or nonsterile. Serological pipettes are typically of the blow-out type, in that the last drops of draining fluid need to be expelled by air pressure from the pipetting device to deliver the true volume. Volumetric pipettes may not be blow out, in which case the researcher does not expel the remaining volume. Pipettes are marked 'TD', meaning 'to deliver,' or 'TC', meaning 'to contain.' The TD pipettes should not be washed out using the diluent, because there is an anticipated amount of solution volume that adheres to the inside of the pipette and this residue is accounted for in the calibration. Some pipetting devices use disposable tips with volumes in the range of 1 to 5 ml, and these may also substitute for serological pipettes.

Reusable pipettes can be washed and prepared in any way that the laboratory cares to have them, but after each use they should be completely immersed in a pipette jar filled with a disinfectant soaking solution, so that residues do not become dried onto the inside. Periodically these pipette jars are emptied and the pipettes washed with a cleaning solution such as Alconox®, followed by a thorough tap-water rinse, and several rinses with deionized or distilled water. This is a general approach to cleaning any piece of laboratory glassware, but it can lead to some carry-over of detergent residue. If the pipettes are used for a specialized purpose such as mammalian tissue culture, or if the available water is particularly mineral rich, then the detergent selected and method of rinsing may need to be altered accordingly. Once washed and dried, and possibly stuffed with cotton, glass pipettes can be sorted by size into metal pipette cans that are then autoclaved. Cotton-stuffed pipettes are used if there is concern that a solution might be contaminated by spores or debris entering the top of the pipette. However, if the pipettes are baked in an oven to make them RNase free, then cotton plugs will turn to ash during the baking process.

Disposable pipettes may be made from borosilicate glass or plastic, and some consideration needs to be given to chemical compatibility with plastic pipettes. If pipettes are flamed before use, for example in transfers of bacteriological medium, then plastic pipettes may soften if the pipette lingers too long in the burner. Disposable pipettes have a distinct advantage over reusable glass pipettes, in that they are unlikely to cause any cross-contamination between experiments or introduce contaminating enzymes such as RNases. For the transfer of cultures of sensitive cells, disposable pipettes also have an advantage in that they do not carry detergent residues that might inhibit cell growth.

Graduated cylinders

Volumes in the range of 10 to 2000 ml are usually measured in reusable graduated cylinders, made either of glass or plastic. A solute can be dissolved into solution in a graduated cylinder, and the solution then brought to its final volume. Graduated cylinders can be covered with foil and autoclaved, if sterility is important during solution making. The cylinders are necessarily long so that volumes can be measured accurately; however, this also causes them to be unstable and easily knocked over. Some glass cylinders have plastic rings that protect the rim of the cylinder from breakage, and these rings can withstand autoclaving. The plastic rings cannot withstand high-temperature baking for RNase-free applications, nor can cylinders that are made of plastic.

Beakers, flasks, and bottles

Reusable beakers, flasks, and bottles are made of glass or plastic, and are used for solution preparation and storage. Volume markings on bottles are usually not sufficiently accurate to use for volume measurements and a solution should be brought to volume in a graduated cylinder and then transferred to a bottle for storage. The exceptions are the volumetric flasks used in analytical chemistry laboratories. These flasks are designed to bring a solution to a single volume, for example 100 ml, and are not used for storage of reagents.

Beakers are not favored for solution storage because they are not easily sealed, and because they have a wide open top that might allow liquid or fumes to escape when the solution is swirled, or allow admittance of contaminants. However, beakers are often used if a large volume of solution is being prepared because the wide top allows reagent powders to be added without a funnel. They also allow the insertion of pH meter probes and other instruments through the top, and the sides of a beaker can be readily washed with a diluent and scrubbed with a rubber policeman if particulates are being collected for filtration.

Erlenmeyer flasks have a flat bottom and sloped sides that rise to a narrow opening at the top. While this design makes them more difficult to clean than an open-top beaker, it provides flasks with a stable base that is resistant to tipping. The narrow top of an Erlenmeyer flask makes it easy to maintain the sterility of the contents, and many flasks also have loose- or tight-fitting capping systems. As a solution is swirled in a flask the sides tend to keep the liquid contained, and this makes flasks particularly useful for the growth of liquid microbial cultures. Flasks that are used for microbial cultures may have internal baffles that increase the oxygenation of the solution during agitation. Some Erlenmeyer flasks have a side arm and are sufficiently strong that they can be sealed with a stopper and attached to a vacuum line. However, as discussed in Section 2.5, glass flasks that are designed for general use should not be attached to a vacuum line through a stopper, because their glass is not thick enough to withstand the vacuum pressure. Glass Erlenmeyer flasks can be used to warm solutions on a hot plate, but for vigorous boiling a rounded Florence flask may do a better job of preventing eruptions of boiling liquid through the top of the flask.

Gloves

Disposable latex, nitrile, or vinyl gloves are a staple in recombinant DNA laboratories, particularly in those that work with RNA. There are certainly times in the laboratory when the gloves are necessary to protect the researcher from a toxin, mutagen, infectious agent, or radioactive compound in the sample, but gloves are also worn to protect the integrity of the sample from contaminating hands. Flaking skin can contribute nucleases and microbial contaminants to a sample, and oily fingerprints can cause high background on a protein or nucleic acid blot. Furthermore, if a researcher contaminates a sample with his or her DNA, it may become an active template in a polymerase chain reaction if the oligonucleotide primers base-pair to a complementary sequence. Gloves are therefore usually not washed and re-used in the recombinant DNA laboratory; in fact, they are stripped off and discarded whenever the researcher has any doubt about whether they are still clean.

2.2 Reagents

Purity

Water

Water is a reagent that is used for nearly every application in the laboratory, and contributes to the success or failure of nearly every experiment. The hot and cold water taps that produce 'tap water' for a laboratory sink may be rich in dissolved salts and minerals, depending on the supply, and flecks of rust and lime. It is generally unfit for any use except a preliminary washing of equipment and glassware. Depending on building construction, laboratory taps and icemakers may even deliver nonpotable water that is not suitable for human consumption. That should not interfere with an experiment, unless a residue of the tap water is being carried through the washing of glassware; however, it may cause a researcher to pause before using tap water to make Monday morning coffee or tea, or using laboratory ice for Friday afternoon daiquiris.

In regions where water is collected from a limestone-rich ground, the water may carry a high concentration of dissolved calcium carbonate. This can cause the buildup of lime scaling on parts of the building infrastructure such as the insides of pipes and hot water heaters, and in laboratory equipment such as water baths and water distillation apparatuses. Buildings may have a water softening unit to protect the pipes and hot water heater, and these consist of an ion exchange column that replaces the calcium ions with the more soluble sodium ions. Periodically, the water softener column is recycled by application of salt brine that relieves the column of its accumulated load of calcium ions and delivers it to the sewer. During the recycling process, the building taps may deliver untreated hard water instead of softened water.

For water that is used as a reagent in experiments, laboratories usually have highly purified water that is either deionized or distilled, or both. Distillation of water separates dissolved solutes from the pure water, provided that the solutes are not themselves volatile. Deionization columns capture both cations and anions in a mixed bed resin, and these columns must be periodically changed to maintain a high level of water purity. Reverse osmosis systems may further separate solutes from water using a pressurized semipermeable membrane. Highly purified water is used for solution preparation in a laboratory, and for rinsing of glassware before drying. Depending on the hardness of tap water it may also be a convenience to add purified water to water baths, as they gradually evaporate their water and would otherwise develop lime scaling.

A common method for determining the purity of the water is to measure its conductivity. As the concentration of dissolved ions increases, the conductivity increases and the electrical resistance decreases reciprocally. Water that has a resistance of 18 MΩ, for example, would be considered highly pure for most laboratory applications. Deionization columns may have an electrical probe that continually measures conductivity, and a lamp indicator or digital readout to tell the researcher when the quality of water has declined and the columns need to be replaced.

Dissolved ions are easy to detect because they affect the electrical conductivity, but what about nonionic contaminants? Plastic tubing may leach compounds such as plasticizers that are not easily detectable, and as they age between periodic changings the deionization columns can become sites of microbial growth. Some water systems may incorporate an ultraviolet lamp to kill microorganisms in water passing through an irradiation cell, but even with this measure taken the bacteria and fungi may secrete enzymes such as RNase and DNase. Water systems may have an activated charcoal column to capture these organic substances, and this part of the unit also may require periodic maintenance. There are various tests that can be conducted on water to determine its level of purity, including measuring organic contaminants, and test kits are available that may be helpful in sorting out a problem.

Unlike wine, fine water does not improve with age and storage. Some laboratories store water in large plastic or glass carboys, and these containers can leach substances into water that was delivered to them clean. Furthermore, they can be sites of microbial growth, accumulating biofilms of algae, bacteria, and water molds that add organic contaminants to the stored water. Many researchers sterilize samples of water that are intended for solution preparation. If bulk water storage is necessary in a laboratory setting, the carboys should be capped with a sterile cotton-stuffed thistle tube, or dry 0.2–0.45 µm filter, to allow air pressure equalization. The carboy should be periodically emptied and flushed, and checked for signs of microbial growth. Water can be tested for some types of microorganisms by filtration of a large volume through a sterile nylon filter, which is then applied to a microbiological plate and analyzed for colony forming units. Where water is purified by a centralized system that serves an entire building, the researcher may want to take note of whether a cistern is filled on the top floor or roof, and water delivered by gravity to the labs underneath. A pipe that delivers deionized water to a laboratory may actually be delivering water that has been deionized and then stored for an indeterminate amount of time, in a cistern of doubtful sterility.

Chemicals

Many of the reagents that we use in molecular research can be purchased in a variety of different chemical grades. The names given to the grades of purity are often trademarks and do not readily admit comparison (e.g., is 'Ultrapure grade' NaCl from company X superior to 'HPLC grade' NaCl from company Y?). One exception is that the American Chemical Society has standards for purity that, if met, allow a reagent to be labeled 'A.C.S. grade' and accompanied by a certificate of analysis. Another is the United States Pharmacopeia that sets a standard for 'USP grade' substances used in pharmaceuticals. Certificates of analysis usually define a level of purity and list impurities by percentage or parts per million, for example 'purity: ≥99.5% by titration; insoluble material: ≤0.005%; heavy metals: ≤5 ppm'. A researcher should be aware of the grade of a reagent that is needed for a specific application, and should keep careful records of which reagents (by supplier and lot numbers) were used in experimental work. The materials and methods sections of published papers often list the commercial sources of unusual reagents, to help other researchers reproduce the findings if necessary.

It is difficult to make a broad statement about reagent purity that will fit every circumstance. For example, bacteriological media that are used to grow cells for DNA extraction may not need to be fancy, but if the bacteria are being studied for their uptake of trace metals then a crude reagent may be so contaminated with trace metals that it confounds any interpretation of results. As a second example, sodium dodecyl sulfate (SDS) that is used to lyse cells may not need to be particularly pure, but the SDS that is used in polyacrylamide gel electrophoresis of proteins may need to be exceptionally pure if the researcher wants uniform results. Some substances can be cleaned up by recrystallization, provided that the solvents used to precipitate and wash the crystals are of high quality themselves.

Enzymes

Enzymes used in molecular biology work also have problems of purity. For example, restriction endonucleases may be contaminated with small amounts of exonucleases that 'nibble' at overhanging ends. While this might be inconsequential if the enzyme was being used for mapping, where the size change would not be detected on a gel, it could be very serious if cohesive ends need to be brought together during a cloning process. Distributors of restriction enzymes will often indicate what fraction of digested DNA fragments can be re-ligated after cutting, to show whether overhanging ends remain intact.

Commercially available RNase A enzyme is often contaminated with DNase, and it is a common practice to add ethylenediaminetetraacetic acid (EDTA) and heat treat the RNase at 80°C to inactivate the contaminating

DNase. Furthermore, DNase can have RNase as a contaminant. While this would not matter if both RNase and DNase were being added to a sample anyway, for example during a protein extraction, it is a serious problem if the researcher is preparing RNA for a cDNA preparation and wants to rid the RNA of any residual genomic DNA. For these applications, researchers obtain a more expensive RNase-free DNase and use additional RNase inhibitors to prevent residual RNase activities.

Reagent forms

Reagents used in the laboratory are often available in different chemical forms or preparations. Sometimes these chemical forms can be used interchangeably, or with adjustment of a recipe, and other times not. If an old bottle of reagent is used up and the 'wrong' form is ordered to replace it, will the recipe still work? It depends on the situation.

Hydration

For example, $CaCl_2$ is available in anhydrous, dihydrate, and hexahydrate forms, representing different amounts of water that are co-crystallized. Different gram amounts of each would be used in a solution recipe, because the water is accounted for in the formula weight. Magnesium salts such as $MgCl_2$ and $MgSO_4$ are also available in anhydrous and hydrated forms, and gram weights in recipes need to be altered if the reagent form changes.

Other than correcting the gram weights in a recipe, does it matter which form of hydration is used? Yes, it could make a difference. Anhydrous salts are heated during their manufacture to drive off the extra water, and this heating may result in decomposition. For example, different samples of anhydrous $CaCl_2$ may vary in their alkalinity depending on how much calcium oxide was generated from calcium carbonate during heating.

Different types of scientists take different approaches to the issue of hydrated chemicals. Analytical chemists generally prefer anhydrous chemicals for gravimetric analysis because any residual moisture content leads to uncertainty in measurement. In the molecular biology lab, bottles of hydrated salts are common and the co-crystallized water is simply weighed along with the salt. Perhaps the molecular biologists prefer to have granulated reagents that do not harden into a solid brick if a bit of atmospheric moisture enters the bottle. Perhaps they don't want to have to store reagent bottles under desiccation. Or, perhaps, they hated the gravimetric analysis lab that was part of their analytical chemistry course. Whatever the reason, the hydrated chemicals work just fine for molecular biology, except when taken to excess: a bottle of $MgCl_2 \cdot 6H_2O$ that has been left with a loose lid may become so waterlogged that the bottom of the bottle is half crystalline and half saturated solution. There are obviously times when a reagent bottle needs to be retired and replaced!

Acid and base forms

The nature of acid dissociation is discussed in Chapter 3, but suffice it to say that a reagent such as monobasic sodium phosphate (NaH_2PO_4) has a different formula weight than dibasic sodium phosphate (Na_2HPO_4), and yields a different pH when put into solution. When ordering replacements for bottles of reagent that have been used up, it is important not to ignore the information about acid and base forms in the chemical description because an experiment may easily fail if the wrong form is used.

When a laboratory recipe depends on a specific acid or base form of a reagent, substituting a different form could require an adjustment in the way the solution is made. For example, a 0.5 M solution of EDTA at pH 8.0 could be made using either disodium EDTA, or EDTA in the free acid form, but the amount of NaOH needed to bring the pH to 8.0 would be considerably greater if the free acid is used. Using a different acid or base form of a reagent may lead to problems that are not so easily correctible; for example, the sodium salt of ampicillin is soluble in water, but the trihydrate form is only sparingly soluble.

The free acid form of EDTA has no sodium ion, and is not a salt because all four acetate groups carry protons. The free base form of Tris (tris [hydroxymethyl] aminomethane) is similarly neutral, and not a salt, because its amine group is not charged with an extra proton. On the other hand, a chemical reagent such as Tris-HCl is a salt because it represents the acid form of Tris, with chloride as a counterion.

RNase-free reagents

Bottles of chemical reagents such as Tris, EDTA, and sodium acetate would not be expected to have a significant amount of contaminating enzymes, particularly in a newly opened bottle, but might accumulate human RNase contamination from day-to-day handling. There is usually no need to purchase a special RNase-free grade of these types of reagents, but successful work with RNA may require buying a new bottle and labeling it so that researchers take appropriate precautions to avoid contamination. Reagent solutions cannot be made RNase-free by autoclaving, because the RNase enzyme refolds after the autoclave temperature drops, but sterilizing a solution will prevent growth of microorganisms that secrete additional RNases. Many salts such as NaCl that are inorganic can be made RNase-free by baking, a process that turns organic contaminants to ash, but organic reagents such as Tris will decompose at high temperatures and cannot be made RNase-free in this way.

Some researchers treat water and other solutions with diethylpyrocarbonate (DEPC), an acylating compound that was once used as a preservative by the soft drink industry. DEPC can be used to destroy RNase in water, but is reactive with amines and therefore cannot be used to make RNase-free solutions of Tris, proteins, or ammonium salts. DEPC should decompose in an aqueous solution to yield carbon dioxide and ethanol, but traces of DEPC may remain and chemically damage the RNA in an experiment. Furthermore, there may be a health risk to researchers who breathe in the sweet-smelling fumes that emanate from DEPC-treated solutions.

RNase A activity is a common contaminant of reagents that are crude natural products, for example enzymes and other proteins extracted from organisms. Sometimes an enzyme is supplied in a highly purified form so that it is largely free of RNase A activities, for example with commercially available RNase H and reverse transcriptase enzymes. Other times a supplier may inactivate the RNase, along with other contaminating enzymatic activities, by a chemical treatment such as acetylation. Bovine serum albumin (BSA) is an example of a commercially available protein that is likely to carry endogenous RNase activity as a contaminant; however, it is available in an acetylated form that can be used as a protein carrier in enzymatic reactions where endogenous activities would interfere with the experiment.

Storage, stability, and expiration dates

Most reagent bottles in a molecular biology laboratory need to be stored with tightly sealed lids, at a controlled environmental temperature between about 15°C and 30°C. Some reagents are refrigerated at a temperature of 4°C to 10°C, and others are stored frozen at −20°C or −80°C. The labels of reagent bottles should be consulted for specific instructions about storage and handling of each compound.

Safety is an important concern when storing reagents because improper storage or inadequate labeling can bring risks to researchers and fire fighters who may need to enter a lab during an emergency (see Section 2.5). While it may seem sensible to store reagents alphabetically, there are reasons to not want to store incompatible chemicals near to each other if breakage of the bottles during a natural disaster (e.g., earthquake) or lab accident (e.g., collapsing shelf) could result in explosion, fire, or generation of toxic fumes.

Tight capping of reagent bottles is important for keeping atmospheric moisture and other gases out of the bottle. Humidity can be a particular problem with hygroscopic compounds such as anhydrous $CaCl_2$ and NaOH, or compounds susceptible to decomposition such as tetramethylethylenediamine (TEMED) and ammonium persulfate. To prevent water damage, small bottles of moisture-sensitive compounds might be stored in a desiccator with a pan of moisture adsorbent such as Drierite®. Atmospheric gases other than water vapor can also damage reagents, for example some compounds may become oxidized by atmospheric oxygen, and NaOH may absorb carbon dioxide and water. Light can damage reagents, and many are supplied in brown or opaque bottles to prevent decomposition or chemical change.

Reagents are refrigerated or frozen if they are at risk of heat decomposition or enzymatic attack. It is often the case that reagents are stable when dry, but susceptible to damage when moist or in solution. For example, nucleic acids and proteins that are sold in a desiccated form are typically stored at 4°C, or −20°C. Before samples are taken from these reagent bottles, it is usually good form to allow them to warm to room temperature before opening the lids so as to prevent condensation inside the bottle. Solutions of high-molecular-weight genomic DNA are typically stored at 4°C with EDTA added as a preservative, bacterial plasmids and other 'short' DNA samples are often stored at −20°C. Storage of a sample at 4°C will retard the growth of microorganisms, but certainly not prevent it if the reagent is contaminated and has no preservative.

Freezing and thawing can damage some macromolecules, for example proteins, but it is their exclusion from the water crystal that causes the damage and not the low temperature, *per se*. Protein solutions such as antibodies and enzymes are often provided by companies in an aqueous solution of glycerol to prevent solidification at $-20°C$, or sold in a highly purified form that may be stored at $4°C$ without fear of proteolysis. Enzymes may be shipped with dry ice, frozen solid, but this is only a temporary condition; the intent is that these reagents should be thawed to a liquid state in a $-20°C$ freezer for storage.

When reagent solutions are stored frozen, for example stocks of $MgCl_2$ or $10 \times$ reaction buffers that are used during assembly of an enzymatic reaction, they need to be thoroughly mixed each time they are thawed. Freezing un-mixes aqueous solutions because pure water crystals form without solute. The solute is concentrated in a higher density liquid phase that drops to the bottom of the tube by gravity, and upon thawing the solution is not homogeneous until it is remixed. This un-mixing during freezing can have a damaging effect on nucleotide solutions that are prepared in water without any pH buffering component added, perhaps because the pH varies locally during freezing. Freezing may also lead to damage of high specific activity ^{32}P-labeled DNA and RNA probes if the high local concentration of the probe leads to collateral radiation damage. Sometimes damage from freezing can be minimized by 'flash freezing' a sample in a dry ice and ethanol bath, or in liquid nitrogen. This rapid freezing may result in a more amorphous solid that incorporates solutes in inclusions, rather than excluding them and concentrating them. It is an approach that is often used for freezing solutions of proteins, which might otherwise aggregate and precipitate during slow freezing.

Reagents that are part of a kit may need to be separated after arrival at a laboratory, and the kit components stored differently. For example, a kit might include buffers and solutions that need to be stored at room temperature, other solutions and DNA samples that need to be stored at $-20°C$, and frozen cells or biological extracts that need to be stored at $-80°C$. If the kit components are not stored properly they may lose activity quickly or become inactivated. If components of a kit do not arrive at a laboratory in good condition, for example if the dry ice added to the shipment is completely sublimated or the wet ice packs are entirely thawed, then the kit provider and shipping firm should be contacted immediately. One or the other might take responsibility for damage due to delayed arrival and replace the components of the kit that are ruined.

Biological reagents, such as enzymes, frozen cells, and extracts, are usually provided with an expiration date printed on the label. There can be a reasonable expectation that the reagent will remain functioning until this date, provided that it has been properly stored and handled. If a reagent ceases to work before the expiration date the researcher may wish to contact the distributor to negotiate a replacement component. After the expiration date the reagent may continue to function, without warranty. Some adjustment may need to be made in a protocol if a reagent has decreased in activity, for example if a restriction enzyme has passed its expiration date and dropped in unit activity concentration. However, with proper storage and handling of components a reagent kit may have an extended lifetime before it ceases to function properly.

2.3 Technique

Measuring things

We use many different measuring devices in the laboratory, including pipetting devices and other instruments that dispense precise amounts of reagents, and analytical devices that make physical measurements such as optical density. While every instrument is different there are a number of common features that may be discussed, such as calibration, zeroing, and scaling.

Do I know how to use the device?

A key to successful technique with laboratory instrumentation is the knowledge of how to use it properly. Many months of experimentation, and thousands of dollars of laboratory resources can be wasted because a researcher has a simple unasked question about a piece of equipment. Sometimes laboratories have well-trained personnel that can explain the intricacies of a piece of equipment to a novice. However, in a laboratory with a high turnover of personnel the knowledge base may gradually erode to the point where instruments are either misused or not used. At this point a researcher might be best advised to look for an instruction manual in a laboratory file cabinet, or copy down the model number and search for a manual on the company website.

Micropipettes

Micropipetting devices are the principal tools of the molecular biologist, and it is important that they be understood and used properly. In brief, a micropipetting device uses changes in internal air pressure to control the flow of liquid in the pipette tip. A moveable stainless steel cylinder, which is controlled by the user's thumb, causes air displacement in the pipette tip (see *Figure 2.1*). Some micropipettes deliver a fixed volume, for example 100 µl, but most can be adjusted to deliver variable measures in a defined range of volumes. A researcher first selects the appropriate size of micropipette for the job at hand, and sets the measure on the instrument. The size of the micropipette is important because each device is designed to work in a specific range: 0.5 to 10 µl, 5 to 40 µl, 20 to 200 µl, or 200 to 1000 µl. If a micropipette is adjusted out of its range, for example if a 20 to 200 µl pipette is set at 5 µl, it may not perform reliably. Next, the researcher applies a clean pipette tip to the end of the shaft. This pipette tip has to be compatible with the micropipette chosen, or it may not form an airtight seal with the shaft of the micropipette. Some researchers pick up the pipette tip by making a stabbing motion into a rack of clean tips. Others seat each pipette tip by hand, by giving it a slight twist on the shaft. The important principal is that the tip has to form an airtight seal when it is seated (see *Figure 2.1B*).

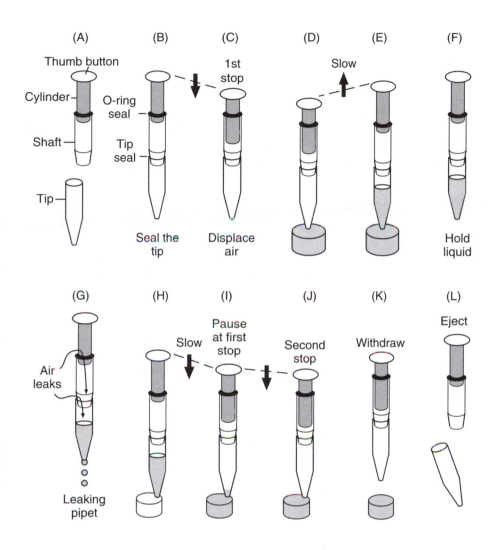

Figure 2.1

Operation of a simple pipetting device. (A) The pipetting device has a cylinder that is driven into a sealed shaft through action of a thumb button (top). A disposable tip is applied to the end of the shaft before use (bottom). (B) For correct operation the tip must be firmly seated and sealed to the shaft, and the O-ring seal must prevent leakage around the cylinder. (C) Pushing the thumb button to the first stop displaces air through the tip. (D) The tip is placed in a solution, with the thumb button held at the first stop. (E) Allowing the thumb button to return to its resting position causes liquid to be drawn up into the tip. (F) The pipette is withdrawn from the solution, and should hold its volume by suction. (G) If air leaks into the shaft or the tip, the suction will be broken and the pipette will leak. (H) The pipette tip is lowered into the receiving tube. (I) Liquid is expelled from the tip by pushing the thumb button to the first stop. (J) After the tip has drained, the thumb button can be pushed beyond the first stop to expel the last drop. (K) The pipette is withdrawn from the solution, with the thumb button depressed. (L) The tip is ejected and discarded.

The researcher next depresses the thumb button to its first stopping point, which displaces an amount of air equal to the volume setting on the pipette (*Figure 2.1C*). Holding the pipette vertically, the researcher inserts the tip a short distance into the solution (*Figure 2.1D*) and gradually allows the thumb button to rise to its starting position (*Figure 2.1E*). This allows the pipette tip to draw up the set amount of liquid from the solution, as negative air pressure from the rising stainless steel cylinder is relieved. If the thumb button is allowed to rise too quickly, then the liquid measure may be inaccurate. In particular, viscous solutions climb into the pipette tip more slowly, and require more patience during pipetting. If the researcher pipettes too quickly with a nonviscous solution, it may leap into the tip so quickly that it splashes the shaft of the micropipette, which may then require cleaning. In any case, it is good technique to pipette solutions slowly, and to pause at the point when the thumb button is fully raised, so that the pressure in the tip can equalize (*Figure 2.1E*).

The researcher then withdraws the pipette tip from the solution (*Figure 2.1F*). At this point the liquid is held in the tip by the suction of air in the shaft and tip. A pipette may leak fluid (*Figure 2.1G*) if the tip is not seated properly and leaks air, or if the O-ring that seals the cylinder is in need of replacement. If there is solution adhering to the outside of the tip, it may be touched to the inside of the bottle or tube so that it is not carried out with the measured solution. However, if the researcher tries to dry the outside surface of the tip by wiping in a circular stirring motion, all around the inside of a tube or bottle, the pipette tip may become unseated and drop off the end of the shaft. Some applications may require wiping the outside of the tip with a sterile adsorbent wiper, but this can lead to loss of a few microliters of the measured sample through the end of the tip. If the micropipette is carried between solutions, it must never be turned horizontally because the liquid in the tip might then drain into the shaft of the micropipette. If that happens, the pipette may need to be disassembled and cleaned immediately to prevent corrosion of the stainless steel cylinder, or damage to the O-ring that seals the cylinder. Furthermore, a micropipette shaft that has been contaminated with a solution may cross-contaminate samples taken in the future, so this is a problem that needs to be taken care of promptly.

Holding the micropipette vertically, the researcher places the tip into the solution into which the volume is to be delivered, or places it against the inside wall of the vessel (*Figure 2.1H*). The thumb button is gradually depressed until the first stop point is felt, and the liquid in the pipette tip is expelled (*Figure 2.1I*). During solution delivery it is important that the tip be touching either the vessel wall or the solution, because if the pipette tip is suspended in the air a small droplet may adhere to the tip and not be counted in the measured volume. Once again, viscous solutions require the most patience. The solution will drain down the inner walls of the tip as the thumb button is being depressed, and the rate of squeezing of the thumb button should be commensurate with that drainage rate. If the solution is expelled too quickly, before the solution has drained off the walls, then the tip will deliver a blast of air at the end instead of measured solution volume.

It is good technique to pause at this point, with the thumb button held at its first stop point, to allow the draining liquid to collect at the bottom of the tip (*Figure 2.1I*). This collected liquid can then be expelled by pushing the thumb button beyond its first stop point, while still holding the tip in the solution or against the inner wall of the vessel (*Figure 2.1J*). One reason that it is important to maintain this connection between the end of the tip and the wall of the vessel is to draw away the adherent liquid as it is delivered; in some experimental techniques it is also important to avoid creating aerosol droplets by having an air bubble form and pop during liquid delivery. Following delivery of the last drop, the pipette is withdrawn from the solution (*Figure 2.1K*) and the tip ejected and discarded (*Figure 2.1L*).

Micropipette tips are usually 'TD', meaning that they should not be washed out with the diluent after the volume is delivered. However, a researcher might decide to disobey that directive when pipetting small volumes of a viscous and expensive reagent, such as an enzyme that is stored in glycerol solution. As discussed earlier, some micropipettes are positive displacement pipettes because the fluid is controlled by a plunger that fits in the tip, similar to a syringe. These do a better job of delivering viscous solutions because they do not rely on air pressure (see Section 2.1).

For most applications, a researcher changes the pipette tip each time the micropipette is used to deliver a solution. This is necessary to prevent contamination of stock reagents, and cross-contamination of samples. Pipette tips are made out of plastic and become wetted with the solution the first time they are used, and some pipetting devices are calibrated to give the most accurate results if the tip is pre-rinsed in the solution before use. A small amount of liquid is expected to stay behind with the tip when it delivers its volume, and this may not be accounted for in the 'TD' calibration of the device. If a pipette tip is pre-rinsed, the correct volume will be delivered the second time because no part of the solution needs to be retained for wetting. However, the calibration method may vary with each micropipette maker, so the instruction manual should be consulted for best results.

There is a method called 'reverse pipetting' that can sometimes be employed when the researcher does not wish to change tips but nonetheless wants a reproducible delivery. It is also useful in working with samples that are viscous. In this method, the thumb button is depressed past the first stop point and the researcher withdraws an amount of solution that is greater than the set amount. A portion of the solution is then delivered to a vessel by depressing the thumb button to the first stop point, but not beyond it. The tip is then withdrawn from the delivery vessel, with the thumb position still held at the first stop point, and returned to the original solution for another filling. This method may be used for repetitive additions of a solution, but more typically, concerns about cross-contamination will prevent reverse pipetting from being useful.

Some solutions are more challenging to handle with a micropipette than others, and the problem of viscosity has already been discussed as one example. A second type of problematic solution is one that has a high vapor pressure, because this can lead to an unstable air pressure in the pipette tip and the shaft of the micropipette. For example, if we try to pipette an aqueous solution that is about 90°C, close to its boiling point, the vapor pressure of the water and the increase in temperature of the air column above it will pressurize the shaft of the micropipette and cause the solution to begin to expel itself without any depression of the thumb button. Chloroform will also self-expel at room temperature, probably as a result of its vapor pressure and low solution cohesiveness. These are problems to keep in mind because an unanticipated 'eruption' of a micropipette during transfer of solutions can make the protocol unsuccessful and dangerous to the researcher.

Graduated cylindrical pipettes

Graduated glass or plastic cylindrical pipettes are typically used for transfer of solution volumes between 1 and 25 ml. The graduated markings are used as a guide during pipetting, and the amount of solution delivered is the difference between the starting mark and the finishing mark. Some pipettes have graduated markings that are printed so that they increase from bottom to top, and others are printed in the reverse direction. A researcher selects a pipette that is appropriate for the job at hand; for example, a 5-ml pipette would be much more accurate at delivering a volume of 2.0 ml than a 25-ml pipette. If the pipette is in a canister or a bulk pack, it must be removed carefully so that it does not touch a dirty part of the package. The other pipettes should also not be contaminated by the researcher's fingers.

Once a pipette is selected, the proper technique is to place the tip of the pipette just below the solution surface, and draw in liquid using a pipette bulb or other air control device so that the liquid has risen above the top graduation being used. Then, air is allowed to leak into the pipette slowly until the bottom of the meniscus drops to the desired graduated marking when the pipette is held vertically. Some pipetting devices have a means for allowing this slow leak of air, and others require that the researcher cover the top of the pipette with an index finger to control the leak. It is generally considered poor form to use a thumb to cover the top of the pipette, because the thumb is usually more difficult to control than a forefinger. Reading the meniscus properly requires that the researcher hold the pipette at eye level to avoid inaccuracies due to parallax. The researcher may touch off, or wipe off drops that are on the outside of the cylindrical pipette so that they do not get delivered along with the planned volume.

Then, the researcher places the tip of the pipette in the delivery vessel, and holds the tip against the side of the vessel or in the solution. The volume is delivered by allowing air to leak into the top of the pipette, until the meniscus has dropped to the desired level when the pipette is held in a vertical position. The solution will be draining down the inside of the pipette during this delivery, and must be allowed to catch up with the meniscus before a reading is taken. The tip of the pipette should be touched to the side of the vessel so that drops do not adhere to the outside of the pipette during delivery. Some serological pipettes have graduated markings that extend all the way down to the tip, and must be blown out to deliver the last drop of solution. Volumetric and Mohr pipettes are not blow-out pipettes, and should simply be touched to the side of the vessel when the volume is dispensed. Despite the evocative name, blow-out pipettes do not involve mouth pipetting; a researcher should always use a pipetting device during liquid handling.

Pan balances

Pan balances are designed to cover specific ranges of weights, just as micropipettes are designed to work within specific volumetric ranges. Some balances are 'analytical' and work best in the range of 0.001 to 10 g, while others are used to weigh larger amounts such as 0.01 to 1000 g. The researcher must first select the appropriate level of precision needed for the job at hand, but it is generally inadvisable to use a balance at the lower end of its range. For example, a digital balance that gives weight increments of 0.01 g would not be very accurate in measuring 0.02 g. For that, an analytical balance that gives weight increments of 0.0001 g would be best.

Next, the researcher must zero the balance with a weighing dish or piece of weighing paper. Some researchers like to make a crease in the weighing paper to help funnel or collect the powder in the center. The reagent is added to the pan balance (more is said about this aspect of technique in Chapter 3) until the final weight is correct. Modern pan balances are electronic and capable of zeroing the tare weight automatically, but on older instruments the tare weight must be recorded and subtracted from the final weight.

Pan balances must be placed on a level surface, or leveled themselves using set screws at the corners, so that the Earth's gravitational field is perpendicular to the pan. In addition, they must be used in a part of the lab that is reasonably free of vibrations and air currents. This is a significant problem with analytical balances because they are very sensitive. A researcher may find that leaning on the countertop or opening a drawer underneath the balance changes the measured weight of a sample, or that whistling a tune while weighing out a sample leads to air currents that depress or lift the sample slightly and change its apparent weight. Most analytical balances have shielded chambers with side doors to prevent air currents, and these should be closed when a measurement is being taken.

Measuring temperature

Glass thermometers filled with alcohol or mercury are commonly used to make temperature determinations, and they may be obtained in a wide variety of ranges and levels of precision. The typical laboratory thermometer has Celsius gradations printed on it, and the researcher can interpolate by eye to determine a temperature, plus or minus 0.2°C. This level of precision requires that the researcher sight across the thermometer face perpendicularly, to avoid parallax effects.

Thermometers may have a printed line that is circumscribed around the shaft, and this line indicates the proper immersion level; the researcher dips the thermometer into a water bath only up to that immersion line. The researcher must allow time for the thermometer to equilibrate before taking a reading, and the amount of time will depend on the type of thermometer and whether it is being used in air or fluid. If the thermometer is giving consistent readings, it has probably equilibrated.

Incubators and other pieces of electronic equipment may have digital thermometers that give a read-out of the internal conditions, but the calibration of these thermometers cannot always be trusted. A researcher should regularly check the inside temperature on each shelf of an incubator for uniformity, using a full immersion glass thermometer. It may seem like a small thing to worry about, but many a laboratory has had trouble growing cells in culture because the digital readout on the incubator assured everyone that the temperature was 37°C, but the true temperature inside was closer to 35°C or 39°C.

How do electronic instruments work?

On that note of mistrust for all things electronic, we move on to a discussion of modern instrumentation. Many electronic instruments gather their signals from a probe that changes in electrical resistance, for example a thermistor that varies in resistance in response to temperature changes. Other electronic probes generate a small current or voltage change, for example a photomultiplier tube in a spectrophotometer. Modern chip probes, for example charge coupled device (CCD) chips that are used in light-gathering instruments, rely on the measuring a change in capacitance in each pixel element.

When a researcher uses a device to measure something in the laboratory, the measurement is often based on a fixed reference point. For example, when weighing a reagent on a pan balance the tare weight is subtracted from the total weight, or the scale of the balance is zeroed physically or electronically which amounts to the same thing. This zeroing process is the first step of calibration; it establishes a reference line and permits a researcher to make a relative measurement. A useful analogy for zeroing an instrument might be a symphony orchestra tuning itself to a single reference note, a 440 Hz 'A' provided by an oboe. With strings tightened or loosened, and pipes shortened or lengthened, the players have calibrated their instruments and ears to the reference vibration, and all of the other notes can be played in tune by reference to that 'A'.

Analog electronic instruments may have a 'zero' knob that must be adjusted by hand, while keeping an eye on an analog or digital readout, until the reading is 0.000. Old spectrophotometers require the user to calibrate the zero transmission point, after blocking the path of light with a solid object. This establishes the background signal or 'noise' from the photomultiplier tube when no light is striking it. On newer spectrophotometers this aspect of calibration may be performed automatically and without the user's knowledge.

When a detector or probe in a lab instrument produces an electrical signal, that signal is amplified and scaled by electronic components on an internal circuit board. On old instruments, the signal might cause deflection of a needle or gauge that can be interpreted by where it points on a printed scale. The user might calibrate the zero point on these old instruments by zeroing the needle against the very same scale. On newer instruments, an analog signal might be converted to a digital readout and scaled so that the numbers reflect meaningful units such as grams, °C, or absorbance.

The magnitude of response of an instrument to a signal is a second factor that requires calibration. For example, a pan balance needs to be calibrated using a set of standard masses, usually by a service technician. Fortunately, pan balances need this recalibration infrequently. On the other hand, some instruments need constant recalibration. For example, a user will 'blank' a spectrophotometer using a cuvette that is empty or contains only pure diluent, and must perform this calibration each time because the cuvettes and diluents may vary. The photomultiplier tube or light detecting chip in the instrument generates an electrical signal that represents the amount of light passing through the blank cuvette; that information is used in conjunction with the first calibration point, in which no light hits the photomultiplier tube, to establish the scaling of the signal. If a user blanked the instrument with one type of cuvette and measured the samples with an unmatched one, the results would not be interpretable.

Some spectrophotometers have programs that allow the fitting of data to a standard curve, for example in a Bradford protein assay, or calculate concentrations of nucleic acids using pre-set extinction coefficients. These programs may be helpful in performing repetitive calculations, but researchers must be cautious about accepting their conclusions at face value. In particular, spectrophotometer programs that state the concentrations of nucleic acids and proteins in a sample from absorbance readings at 260 and 280 nm are notoriously inaccurate. When a sample is impure, a calculation based on an assumed extinction coefficient is not meaningful. That's just a fancy way of restating an old adage of computer scientists: 'Garbage in, garbage out.'

Is the instrument really working?

Sometimes when we're using or collecting data from an instrument, we get a feeling that something just isn't right. For example, we might have just placed a pH probe in a solution to measure its pH, but after obtaining an initial reading of pH 7.20 the readings drift steadily down: 7.15, 7.10, 7.05, and so forth. Should we believe any of these values? Is the pH meter or probe in need of repair? Or, to take another example, what if we measure out 100 µl of a liquid using a pipetting device, and the liquid begins leaking out of the tip immediately (*Figure 2.1G*). Is an O-ring seal in the pipetting device in need of replacement or lubrication? It is hard to say categorically when an instrument isn't working correctly, but whenever a researcher sees unusual behavior it is worth noting and reporting. With some instruments, for example centrifuges, an unusual vibration or sound may indicate a problem that could make the device unsafe. In laboratories with shared equipment, it is important to post a note so that others in the laboratory will not use the possibly faulty piece of equipment unawares.

When an electronic instrument is first turned on, it may give readings that are different than when it is fully warmed up. In the case of a spectrophotometer the bulb flickers and gradually brightens as it warms up, and calibration is usually delayed for about 10 min until its light output stabilizes. This is not an effect that is limited to spectrophotometers. Any piece of electronic equipment may change subtly as it warms up because electronic components are temperature sensitive. Sometimes when an instrument has an electronic circuit component that has gone bad, it will produce electrical noise after the instrument starts to get hot. Electronic equipment may also malfunction if the cooling vents or air circulation fans are blocked.

Even when working properly, electronic measuring instruments have a usable range in which they deliver a linear response and they may be unreliable at the extremes of that range. For example, a spectrophotometer may give accurate absorbance measurements between 0.010 and 1.500, but give inaccurate ones outside of that range. This is because weak light signals may not be above the limit of detection of a probe, or may not be distinguishable from instrument noise. Strong signals may overwhelm the detection system and be misreported.

Handling tubes

Keeping things clean and organized

In the molecular biology lab, reactions are conducted and samples are stored in small tubes or microtiter plates. These pieces of plasticware have been described earlier in this chapter, and they are available in many sizes and formats. Plastic microcentrifuge tubes are usually sold in bulk, in plastic bags. The tubes are not typically biologically sterile, but an unopened bag may be free of most biological contaminants such as RNase and DNase. Some researchers autoclave microfuge tubes to make them biologically sterile; however, other researchers are more concerned that the steam in the autoclave might leave deposits on the plastic surfaces. In either case, a difficulty with handling tubes is finding a way to extract a single tube from a bag or autoclave canister without contaminating the other tubes. It is bad technique to reach into a bag with a bare hand, and rustle around for a tube to grab, because then the other tubes will be contaminated with skin cells, microorganisms, and enzymes such as RNase. It is better for a researcher to use a gloved hand to grab a clean tube, or to tap a tube out of a container onto the benchtop. The area of work on a benchtop must also be kept dust-free, so as to not introduce contaminants into the samples and reaction tubes.

If many tubes are being assembled for an experiment, then each tube needs to be individually labeled so that the researcher does not lose any work if the rack is upended. Organizing the tubes while making solution additions is a second concern; for example, how can we add 2 μl of a stock reagent to each of 30 tubes without losing our place? One technique is to leave gaps in the rack and move each tube after the solution addition, either side-to-side or forward and backward. Then, if the telephone rings in the middle of the set-up, a researcher can tell by visual cue which tube was the last one handled (*Figure 2.2*).

One difficulty with this approach is that with compact racks that hold the tubes closely together, extracting a single tube to move it may cause

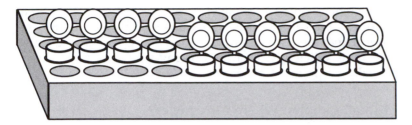

Figure 2.2

A visual cue for remembering which tube was last handled. A solution was added to the first four tubes on the left, and each tube was shifted one space back in the rack as the addition was made. It is easy to see by this visual cue that the fifth tube will be the next in line to receive the solution.

other nearby tubes to flip out of the rack and spill their contents. The tubes can have their tops closed between each solution addition, but this causes extra wear and tear on the thumbs and increases the risk of flipping over the entire rack if it is done one-handed. Furthermore, repeated opening and closing of a tube weakens the seal and makes the top more susceptible to popping open in a boiling water bath. A second problem with moving tubes after each solution addition is that this technique of staying organized may be more difficult to employ when the tubes are immersed in an ice bucket, and not neatly arrayed in a grid.

Keeping solution additions organized in a microtiter dish may be made easier by the use of multi-channel pipetting devices, or repeater pipettes, or, better still, robotic liquid handlers. If these are not available, the surface of a plate can be partially covered by a plastic wrap or film to keep the researcher focused on adding solutions to a single row or column at a time, or the researcher can keep track of the additions by making check-marks in a laboratory notebook. A researcher may also chant the name of the microtiter well last used, to keep it in short term memory; however, if the telephone rings the memory of place may be lost. Experiments with 96-well or 384-well plates are probably best conducted by researchers with very simple lives, the better to not lose their places during experimental set-ups.

What can I handle?

There is a limit to the size of an experiment that each researcher can conduct without making a mistake through mental exhaustion, and it is important to know that limit and not cross it. For example, if we have 100 DNA extractions to conduct, we might discover that we can perform the 100 extractions in 5 hours straight, but in doing so we become exhausted and mix up two or three of the samples along the way. However, suppose we could have performed 50 extractions in 2 hours, without making any mistakes. Then it would have been a better plan to split the 100 extractions into two days' work. A similar argument might be made for the time of day or night when work is done. Our biological clocks do not always allow us to do our best work at 2:00 AM, and sometimes a researcher would be well advised to get some rest. This is not to advocate laziness, but rather wisdom in knowing how to do error-free work efficiently. Experiments often work better if they are scaled down and conducted more efficiently, at a time of day when the researcher is alert and focused.

Recovery of small samples: Where's the pellet?

Work can be done on samples that are very small, but some techniques are helpful in keeping the sample from being lost. For example, what do we do if we have a microfuge tube with only 5 μl of fluid in it, and the fluid is broken into little droplets that cover the walls and cap of the tube? That's a problem that is easily solved. We can place the tube in a micro-centrifuge, with a balance tube across from it in the rotor, and turn on the centrifuge for 5 to 10 seconds. Then, all the microdroplets of sample will be recollected at the bottom of the microfuge tube. This is a particu-larly useful technique for recollecting a sample after it has been shaken or vortexed. Very small volumes can also be corralled in a centrifuge tube by adding a drop of mineral oil, which is less dense than the aqueous solution and floats on top. Pure mineral oil does not mix with the aqueous phase and does not inhibit enzymatic reactions. An aqueous drop of 1 to 5 μl will tend to pull away from the sides of the tube and form a sphere, suspended on the bottom of the mineral-oil layer. This sphere can be collected in a micropipette tip efficiently, and residues of mineral oil can be extracted by rolling the droplet on a piece of ParaFilm®. The mineral oil, being nonpolar, sticks to the ParaFilm® and the aqueous solution forms a small bead.

A sample may be small in volume or small in concentration, and these are two separate problems. For example, if we want to apply a sample to the top of a chromatography column, but the sample only contains 1 pg of DNA, will the DNA stick nonspecifically to the column matrix or the glass column cylinder? Perhaps it will, and this is a problem of scaling of the experiment. If the sample had had 10 μg of DNA there might have been a sufficient amount for a few nanograms of it to be lost.

If our 1 pg of DNA were radiolabeled and we only needed to separate it from unincorporated radiolabeled substrate, we might add a nonspecific carrier DNA from salmon sperm, herring sperm, calf thymus, or some other commercially available carrier. It doesn't matter what the carrier DNA is, since it is only being used to protect our radiolabeled sample from adsorption and loss. On the other hand, if our 1 pg of DNA is a fragment that we intend to use in ligation for cloning work, the addition of a nonspecific carrier DNA would increase the background in our ligation substantially. Even an RNA carrier might be sufficiently contaminated with DNA to cause problems. We would probably have to devise a different approach to work with our 1 pg sample, one that did not involve adding nucleic acid carriers.

Small samples of DNA or RNA can be preserved and collected by ethanol precipitation, using 10 to 20 μg of highly purified glycogen as a carrier. This polysaccharide does not interfere with subsequent enzymatic reactions, and does not contain contaminating nucleic acids that might interfere with cloning work. The length of time of centrifugation is a critical factor in recovering small amounts of nucleic acid by ethanol precipitation; centrifugation for 10 to 15 min at $10\,000 \times g$ is usually adequate. If the microcentrifuge tubes are loaded into the instrument in an orderly way, for example with all the hinge sides of the tubes facing the spindle of the centrifuge, then the location of a nucleic acid pellet can be predicted even if the pellet is too small to be seen. Then, the ethanolic supernatant can be withdrawn from the tube with a pipette tip while avoiding disturbing the probable region of the pellet. A magnifying lens can be very useful at this stage, and can reveal a pellet that is not visible with the unaided eye.

The best advice for working with small, precious samples of nucleic acids that cannot be tracked by visual inspection of a pellet is to not discard any fractions until the sample is safely recovered. For example, the ethanolic liquid lying over an invisible pellet ought to be saved in a second clean tube, and labeled. If the pellet has accidentally become dislodged and lost in the ethanol, the researcher can backtrack and find it. The second tube can be subjected to centrifugation and the pellet recovered. However, if the ethanolic supernatant is routinely and immediately discarded down the drain, then there is no possible recovery of the lost sample. Making a habit of saving 'garbage' fractions for a few days can save a researcher many weeks or months of work, over the long run of a career, by allowing the complete recovery of a sample after accidental loss.

2.4 Looking at DNA on a gel

Gel electrophoresis separates nucleic acids based on their ability to move through the gel, and size and shape are both factors that affect gel mobility. Mobility also depends on the type of gel, whether it is agarose or polyacrylamide, and whether the gel matrix is prepared at a low percentage or high percentage. During electrophoresis, DNA moves towards the '+ terminal', or the anode. It moves in that direction because the phosphodiester backbone of DNA is negatively charged; DNA is a polyanion and the number of charges it carries is proportional to length. As the DNA moves through the gel matrix, it has to fit through tight openings and narrow passageways because the gel acts as a sieve, and this slows down the DNA migration. With the smaller DNA molecules, the gel sieve is less of an impediment to movement than it is for the larger DNA molecules.

Size is important

In gel electrophoresis, the DNA samples are loaded at the origin of the gel, in wells that are cast to hold the sample before the voltage is applied. The DNA moves in straight lines, or lanes, across the gel, in the direction of the anode (see arrow, *Figure 2.3A*). The individual bands on the gel have the same shape as the well, and each band simply represents DNA molecules that happen to migrate together in the gel lane because of their size and shape. Often, a single band on a gel represents a collection of DNA molecules that are all the same. However, a single band can also be made up of DNA molecules that are different but co-migrate.

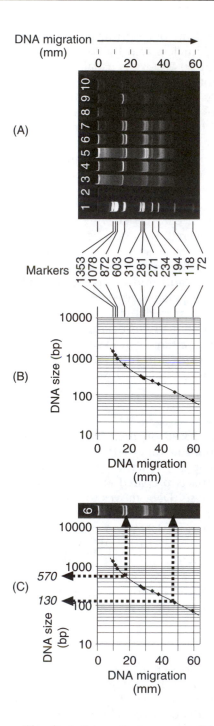

The first thing that we usually look at on a gel is a marker lane into which we have loaded a collection of DNA fragments of known size: a set of size standards (see lane 1, *Figure 2.3A*). There are many ranges of molecular weight standards, and we pick a standard that will help provide information about our 'unknown' samples that have been run in adjacent

Figure 2.3

Electrophoresis of DNA in a gel. (A) Fluorescent staining (ethidium bromide) of DNA in a 10% polyacrylamide gel. Lane 1: φX174 DNA digested with *Hae*III. Lanes 3–10: various DNA fragments generated by a polymerase chain reaction. The direction of electrophoresis is left to right. A millimeter scale is shown (top) and the corresponding lengths of the φX174 DNA fragments are indicated in base pairs (bottom). (B) The sizes of the φX174 DNA fragments are graphed as a function of electrophoretic migration, using a semi-logarithmic scale. The scaling on the x-axis matches the gel. (C) The apparent sizes of two DNA bands in lane 6 are determined from the graph correlating φX174 DNA migration and size (in bp). As in the previous part of the figure, the scaling on the x-axis matches the gel.

lanes. For example, if we are using a gel to look at fragments of DNA in the range of 1 to 10 kbp, we might load a marker lane with size standards that range from 0.5 to 20 kbp. In the example shown in *Figure 2.3A*, the size markers are derived from bacteriophage φX174 digested with *Hae*III, and yield 11 bands ranging in size from 72 to 1353 bp. These sizes are indicated at the bottom of *Figure 2.3A*. We can estimate the sizes of the 'unknown' bands by comparison with the known sizes of the standards.

The DNA fragments in the marker lane of *Figure 2.3A* are separated by size with the smaller fragments migrating a greater distance (to the right) than the larger fragments. The amount of migration of linear fragments of DNA is not proportional to the sizes of the fragments, but is roughly proportional to the logarithm of their size. This relationship is demonstrated in the graph *Figure 2.3B*, where the migration of the size standards in lane 1 of *Figure 2.3A* is indicated along the horizontal axis in millimeters. The figure is constructed so that the position of the bands on the gel (*Figure 2.3A*) is aligned vertically with markings in the graph (*Figure 2.3B*); however, many researchers take a picture of a gel alongside a ruler with millimeter markings so that information can be collected from the picture and transferred to a graph. The curve in the graph links together the ordered pairs (migration, log(size)) for the 11 marker fragments and can be used in interpolation to estimate the sizes of unknown linear fragments. An example of this is provided in *Figure 2.3C*, where lane 6 from *Figure 2.3A* is reproduced and the migrations of two bands are correlated with their apparent fragment sizes. The size of a fragment can be estimated by reference to the vertical axis of the graph, which is labeled using logarithmic spacing.

Size isn't everything

DNA molecules that have a compact shape, such as supercoiled DNA molecules, can migrate through a gel more quickly than relaxed DNA molecules with the same number of base pairs, but a less compact shape. Most of the plasmid molecules we extract from *E. coli* will be negatively supercoiled, a matter that is discussed in Chapter 4. However, some of the plasmid will have one or more breaks in the phosphodiester backbone so that the molecule is a relaxed circle. We understand that this relaxation is due to the fact that if either strand is not covalently closed all the way around the circle, then supercoiling is relieved. If some of the plasmid has breaks in the two strands, very close together or directly across from each other, then the circular plasmid may be broken into a linear form. Linear forms can also be generated by restriction endonuclease treatment of DNA, or polymerase chain reaction. Linear DNA has no supercoils in it, because its two ends are not constrained and can turn around themselves.

The migration of supercoiled plasmid DNA on a gel depends on the conditions under which the gel is run, so we cannot say that supercoiled plasmids always migrate further than linear molecules of the same size. DNA molecules can be supercoiled to different extents; for example, the amount of supercoiling in a plasmid depends on the temperature in which the host cells were grown before the plasmid was extracted. Furthermore, some DNA binding substances such as ethidium bromide have an effect on the structure of DNA, and can change the amount of supercoiling of a covalently closed circle. In terms of interpreting a gel migration pattern, this is problematic if we run a gel with ethidium bromide mixed into the gel or electrophoresis buffer.

Ethidium bromide is an intercalating agent, meaning that the ethidium fits between the bases in a DNA strand. We can imagine that ethidium binds to DNA in the way that a bookmark fits into a book, separating the 'pages' very slightly. We know that if we interleave too many bookmarks in a real book, the book pages swell out from the spine and the shape of the book is distorted towards trapezoidal. As ethidium binds to DNA, it acts like a little wedge or spacer between the DNA bases and increases the amount of twisting or winding of the double helix. By this, we mean that the average number of bases per turn decreases as the DNA becomes more tightly wound.

If a DNA molecule is negatively supercoiled, meaning that it is under-wound, the binding of a small amount of ethidium will decrease the amount of negative supercoiling and make it more relaxed. As more ethidium is added, a covalently closed circular molecule will become over-wound and positively supercoiled. Because of these shape changes, the migration of covalently closed circular DNA on a gel will also change if ethidium is bound to the DNA during the running of the gel. Linear molecules don't become supercoiled, because their ends are not constrained, so they will continue to migrate in much the same way, whether ethidium is present in the gel or not.

The information we gather

We have already seen an example of one thing that can be learned by gel electrophoresis, and that is the apparent sizes of linear fragments of DNA in comparison to standards (as presented in *Figure 2.3*). This type of information is very important in identifying DNA fragments, for example after a restriction endonuclease digestion or polymerase chain reaction. When the migration of DNA fragments is compared to size standards, the measurement of migration is made from the lanes in which the samples were run. If the picture of the gel is crooked, the measuring ruler or the angle of viewing must be adjusted so that the measurements of migration from the wells are accurate.

What happens to small fragments of DNA when they reach the end of the gel? If electrophoresis is continued, they run off the end of the gel into the electrophoresis buffer, and are lost! We can lose important information about a sample by running a gel for too long. For example, suppose that we digest a plasmid with a restriction enzyme such as *Eco*RI and we have predicted that the digestion will produce two bands on a gel: a vector band of size 3000 bp, and an inserted DNA fragment of size 1500 bp. What if there were a third fragment of 250 bp generated by the *Eco*RI digestion and we ran the gel for such a long time that this small fragment ran off the end? We would not detect the extra DNA fragment as a band, and so might assume that our results were consistent with our prediction. The unseen 250 bp *Eco*RI fragment might be a cloning artifact; perhaps two *Eco*RI fragments were cloned in tandem into an *Eco*RI site in the vector. It is easy to imagine a situation in which having the extra 250 bp might ruin an experiment, perhaps by separating two regulatory elements in a plasmid that are important for expression, or introducing an unintended sequence that might become part of a labeled probe and lead to spurious results.

We can sometimes extract information by comparing the relative migration of two bands on a gel, even without consulting known size standards. Two DNA fragments that are nearly the same size might be compared and distinguished if they are loaded next to each other on a gel, and that is a useful technique if we are trying to detect a small insertion or deletion in one of them. Furthermore, we can compare the migration of bands on a gel if we are trying to determine if a restriction digestion has been successful. A shift from supercoiled to linear DNA is usually easily discerned; however, there are problems in comparing lanes that have not been loaded and run under identical conditions. The salts that are part of the electrophoresis buffer are primarily responsible for electrical conductivity, but when DNA is migrating towards the anode during electrophoresis it also contributes, in a very small way, to the overall current on the gel. If the lanes on a gel do not all have the same salt conditions, for example if one sample is loaded in a high salt buffer and the others are not, the sample loaded in a higher salt buffer will be more electrically conductive and have slower DNA migration (see *Figure 2.4*). That is, the lanes will be different at least initially, and by the time the sample conditions become equilibrated to the electrophoresis buffer the damage may have already been done.

There is also a problem of greater power dissipation in the high salt lanes of a gel, and localized heating effects that may distort the migration of those DNA samples. When a gel is run very fast, the middle of the gel runs a bit faster than the edges and causes 'smiling' effects that can make comparison of lanes more challenging. Many vertical gel apparatuses have heat dissipation plates or aqueous cooling systems to keep the gel running straight, at high voltage and power levels.

When DNA fragments are run on a gel and stained with an agent such as ethidium bromide, or a SYBR® stain, the amount of fluorescent signal is roughly proportional to the mass of DNA. A band containing 10 ng of a 1-kbp DNA fragment will look about the same as a band containing 2 ng of a 5-kbp fragment. A corollary of this is that when we have equimolar

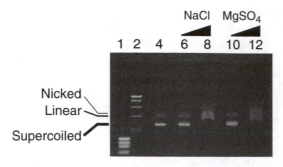

Figure 2.4

Gel electrophoresis is distorted by excess salt in the samples. The direction of migration is from the top of the gel to the bottom. Lane 1: 2 µg φX174 DNA digested with *Hae*III (only the largest four fragments are visible). Lane 2: 0.5 µg λ DNA digested with *Hind*III (only the largest six fragments are visible). Lane 4: 250 ng of a 4.5-kbp plasmid (the migration of nicked circles, linear fragments, and supercoiled DNA are indicated, at left). Lane 6: 250 ng of 4.5-kbp plasmid loaded in 0.5 M NaCl. Lane 8: 250 ng of 4.5-kbp plasmid loaded in 4.5 M NaCl (note severe distortion). Lane 10: 250 ng of 4.5-kbp plasmid loaded in 0.1 M MgSO₄. Lane 12: 250 ng of 4.5-kbp plasmid loaded in 0.95 M MgSO₄ (note severe distortion). A 0.7% agarose gel was loaded with DNA samples, immediately subjected to electrophoresis in a submarine gel apparatus (200 V, 0.25 to 0.30 amps, for 40 min in 1 × TAE buffer), and stained with 0.50 µg/ml ethidium bromide.

amounts of DNA of different sizes, the amount of staining of those DNA fragments is proportional to their size. A band containing 15 fmol of a 5-kbp DNA fragment will stain with approximately five times more intensity than a band containing 15 fmol of a 1-kbp fragment. This is because double-stranded DNA is approximately 660 kDa per 1 kbp, so 15 fmol of a 5-kbp DNA is approximately 50 ng, while 15 fmol of a 1-kbp DNA is approximately 10 ng (e.g., $(1.5 \times 10^{-14} \text{ mol})(6.6 \times 10^5 \text{ g mol}^{-1}) = 1 \times 10^{-8} \text{ g}$).

Let's consider the implications of this when we're looking at fragments of DNA cut from a single plasmid sample. In our previous example of a plasmid that is cut into three fragments with the enzyme *Eco*RI, the fragments are all present in equimolar amounts. We might think in particular about the difference in staining intensity between the smallest fragment of 250 bp, and the largest fragment of 3000 bp, because their relative sizes are in a ratio of 250:3000, or 1:12. If we digested 0.50 µg of the plasmid with *Eco*RI for analysis, and loaded it on a gel, the 3000-bp fragment would have 0.31 µg of DNA, the 1500-bp fragment would have 0.16 µg of DNA, and the 250-bp fragment would have 0.03 µg of DNA. While the staining in the smallest fragment would be well above the limit of detection of DNA using a gel stain such as ethidium bromide, it is easy to see that such a small fragment could have been easily overlooked if less of the overall plasmid had been loaded. The relationship between DNA size and molecular weight is shown in *Table 2.1*, and includes sample calculations of the gram amounts of 100 fmol of DNA of various sizes.

Table 2.1 The relationship between DNA size and molecular weight

DNA size	MW[a]	How much is 100 fmoles of DNA?[b]
20 000 bp	1.32×10^7	1.32×10^{-6} g = 1.32 µg
10 000 bp	6.60×10^6	6.60×10^{-7} g = 660 ng
5000 bp	3.30×10^6	3.30×10^{-7} g = 330 ng
2000 bp	1.32×10^6	1.32×10^{-7} g = 132 ng
1000 bp	6.60×10^5	6.60×10^{-8} g = 66 ng
500 bp	3.30×10^5	3.30×10^{-8} g = 33 ng
200 bp	1.32×10^5	1.32×10^{-8} g = 13.2 ng
100 bp	6.60×10^4	6.60×10^{-9} g = 6.6 ng
50 bp	3.30×10^4	3.30×10^{-9} g = 3.3 ng
20 bp	1.32×10^4	1.32×10^{-9} g = 1.32 ng

[a] The average molecular weight (MW) of a double-stranded DNA is 660 Da per bp.
[b] Sample calculation: $(1.00 \times 10^{-13}$ mol$) \times (1.32 \times 10^7$ g mol$^{-1}) = 1.32 \times 10^{-6}$ g.

We can use these facts about DNA staining to discover cases when restriction endonuclease digestions are incomplete, a matter that is discussed more thoroughly in Chapter 5. If in our previous example the enzyme *Eco*RI had not digested the DNA to completion, we might have not only the expected bands of 3000 bp, 1500 bp, and 250 bp, but also several bands representing partial digestion products such as 4500 bp (3000 joined with 1500), 1750 bp (1500 joined with 250), and 3250 bp (3000 joined with 250). It would certainly look strange to see a very weakly staining band at 1750 bp, just above a strongly staining band at 1500 bp; we would know that these were not present in equimolar amounts and that one or the other had to be a partial digestion product. That's important information, because if we simply catalogued all of the fragment sizes we saw and added them up, we might think that the size of the plasmid was 14 kbp instead of 4.75 kbp! To take a slightly different example, if we digested a plasmid with the restriction endonuclease *Eco*RI and saw on a gel that the 3000-bp and 1500-bp bands had equal staining intensity, we might consider that the digestion had released two fragments of length 1500 bp for every one that is 3000 bp. That is, the overall plasmid size might be closer to 6000 bp than 4500 bp. We can gather this information from the gel because the staining of DNA is proportional to mass; one word of caution is that when a labeled probe is being used in a Southern blot (discussed in Chapter 8), the signal is proportional to the overlap between target and probe and the interpretation of relative signals is completely different.

Smaller DNA fragments are not resolved into gel bands that are as 'tight' as the larger fragments. This is because there is more diffusion of the smaller DNA fragments during the course of running the gel. Once again, this is an opportunity to notice gel artifacts that don't make sense. If we see a band at 1 kbp that is more diffuse than a band at 0.5 kbp, what are we to make of that? The most likely explanation is that there is some true variation in size or shape of the 1-kbp band, and not that it somehow experienced a greater rate of diffusion. Perhaps the 1-kbp band is heterogeneous, made up of a 0.95-kbp and a 1.05-kbp fragment.

It is common for a band on a gel to leave a slight trailing smear behind it, as it migrates, particularly if the gel lane has a considerable amount of

DNA loaded and the voltage of electrophoresis is high. Many researchers use a lower voltage of electrophoresis when running gels that are loaded with a substantial amount of DNA, for example 10-20 µg of genomic DNA per lane, and simply run the gel for a longer time. A gel band that migrates 1 cm in 1 h at 100 V will migrate approximately the same distance in 10 h at 10 V. This rough proportionality between volt-hours and migration rate makes it easy to calculate a voltage that would allow a gel to be run overnight.

A gel band that shows smearing on its leading edge is unusual, and may suggest that the DNA fragment is being gradually degraded or 'nibbled' from its ends by an exonuclease. If a DNA has been purified and stored correctly, this type of shortening is not expected. However, if the DNA is genomic and the smeary band is near the origin of the gel, this smearing may simply indicate heterogeneity in sizes of sheared DNA. If there is staining material left in the well of the gel, perhaps in addition with some slight smearing from the well into the gel, this may indicate that the DNA sample was not entirely in solution when it was loaded on the gel. The solid DNA may rest on the top of the gel during electrophoresis, and gradually go into solution as electrophoresis is conducted, resulting in a stained well margin and a trailing smear.

When very high molecular weight DNA does enter a gel, it usually does not resolve well on a gel that is run under continuous electrophoresis conditions (i.e., not pulsed field), and large fragments may collect and form a spurious band that has an apparent size of about 25 to 30 kbp. It is not uncommon to see this DNA in plasmid extractions, if there is contamination of the sample with host genomic DNA. If the plasmid sample is digested with a restriction endonuclease, a genomic DNA contaminant may form a broad smear that extends over most of the gel lane as it produces restriction fragments of all sizes.

Not all diffuse bands on a gel are DNA. When plasmid DNA is extracted from a cell and placed on a gel, there may be some residual RNA that also runs in the same lane. This RNA is heterogeneous in size, and probably is largely degraded because of endogenous RNase enzymes. It forms a smear or glow on the gel that runs with an apparent size of approximately 50 to 500 bp, against the DNA size standards. If RNase is added during the course of a plasmid DNA extraction, this RNA is degraded to small enough sizes that it is usually run off of the end of most gels during electrophoresis.

Gel systems

Electrophoresis of nucleic acids is usually performed in gels composed of agarose or polyacrylamide. Agarose is a polysaccharide that is a natural product of kelp, and it is available in many grades of purity, melting temperature, and other characteristics such as mechanical strength. The agarose that we use for most applications is 'high melting point', meaning that it can be solubilized in a boiling aqueous solution and will solidify below about 50°C. Agarose used for nucleic acid gels is usually rated as having low EEO (electroendosmosis) or −mr. The amount of agarose used in preparing a gel varies, depending on the size range of DNA fragments that are being studied, but usually agarose is used at concentrations between 0.7% (w/v) and 2.0% (w/v). As explained in Chapter 3, a 1% (w/v) solution has a concentration of 1 g per 100 ml. The 0.7% (w/v) gel is good for resolving DNA fragments in the range of 0.5 to 25 kbp, but it may be lacking in mechanical strength and difficult to handle without breaking. Higher percentage gels, such as a 1.5% (w/v) agarose gel, is better for resolving DNA fragments in the range of 0.2 to 10 kbp. It is difficult to prepare a gel solution with more than about 2% (w/v) agarose, because of the low solubility of agarose; however, NuSieve® agarose can be used at much higher percentages for the resolution of very small DNA fragments.

'Low melting point' agaroses are available, and are particularly useful for applications in which it is desirable to isolate the DNA from a gel band without using NaI or an enzyme such as agarase to disrupt the gel matrix. The granulated low melting point agarose can be solubilized in a boiling aqueous solution, but it sets at a much lower temperature than high melting point agarose, usually between 25°C and 37°C. The agarose can be re-melted, for example after a gel band is excised, at 65°C. Most low melting agaroses are highly purified and do not inhibit enzymes used in molecular biology; so, for example, we can readily conduct a DNA ligation reaction while the DNA is embedded in low melting point agarose. In a related application, blocks of agarose containing high molecular weight DNA may be treated with restriction enzymes in advance of pulsed field gel electrophoresis, and the enzymes are not inhibited by the highly purified agarose.

When we cast an agarose gel, we prepare a solution of agarose by boiling the granules in an electrophoresis buffer, then cooling the mixture to approximately 50°C before pouring the gel into a mold. During the boiling step, we must inspect the solution and continue the boiling for a long enough period of time to make sure that all of the solid granules of agarose have gone into solution. We can prepare agarose solutions in an autoclave, a boiling water bath, or a microwave oven. In all of these cases there is danger in working with boiling-hot liquids, so caution is important. It is a good idea to swirl a flask of agarose a few times while it is equilibrating at 50°C, to make sure that the solution remains homogeneous. This cooling step reduces the vapor pressure of the liquid, which prevents excessive loss of moisture when the gel is poured. Furthermore, many acrylic glass gel apparatuses cannot easily handle the sudden temperature changes brought about when boiling hot agarose is poured into them, and may eventually fatigue and crack.

Agarose gels are cast on a flat glass or acrylic glass sheet, with sides provided by a molding unit of some kind. The mold needs to be level so that the gel is cast with uniform thickness. Many gel trays need to be sealed at their ends to hold the agarose during casting, using a sealant such as a water-resistant tape or Press'n Seal® film. A plastic comb is mounted on the gel before it has solidified, and this forms the wells that will be used for sample loading. The comb should not touch the glass or acrylic glass support for the gel during casting, so that the wells have a solid agarose bottom. The gels can be cast with variable thickness, but a 3- to 5-mm-thick gel is typical for most applications. Thicker gels have more mechanical strength, and they have deeper wells that can take larger sample volumes, but on the other hand they may take longer to stain to completion because the stain has to diffuse through a thicker layer.

Most laboratories use agarose gel systems in which the gel is run as a 'submarine gel', meaning that the gel is submerged in a buffer solution during electrophoresis. The DNA does not run out of the top during electrophoresis, because the anode and cathode are arrayed below the level of the gel tray and keep the DNA running low in the gel. There are several important advantages to running agarose gels as submarine gels: the gels are not susceptible to drying out, and tend to remain equilibrated with the gel electrophoresis buffer in which they are being bathed; the surrounding buffer helps to keep the gel cool and uniform in temperature; and the surface of the buffer solution is flat and level; this helps to make the flow of current through the gel uniform.

Agarose gels can be poured with additives such as formaldehyde, which is used as a denaturant for RNA electrophoresis, or NaOH, which is used for alkaline agarose gel electrophoresis of single-stranded DNA molecules. In either case the additive is added after the agarose is already boiled and equilibrated at 50°C. In the case of formaldehyde gels, the casting should be performed in a fume hood. Samples are usually denatured before loading on these types of gels, and the denaturant in the gel simply prevents the formation of base-paired regions.

Polyacrylamide is a polymer that can be chemically cross-linked, and when we use it to make a gel for electrophoresis it forms a very fine sieve that is useful for separating small DNA fragments in the range of about 10 to 1000 bp. The cross-linking agent for polyacrylamide is usually bis-acrylamide, used in a ratio of 19:1 or 29:1 acrylamide:bisacrylamide, or a proprietary mixture of polymer such as LongRanger® gel solution. Polyacrylamide gels can be made in a wide range of concentrations, but a typical gel is between 6% (w/v) and 20% (w/v). Once again, the lower percentage gels have less sieving and facilitate the migration of larger DNA fragments, while the higher percentage gels are better for separation of smaller DNA fragments. Polyacrylamide gels are cast between two glass plates, and usually run as vertical gels. The plates are separated during casting by spacers, and a comb that is the same thickness as the spacers is inserted into the top of the gel to mold the wells. A variation of this approach is sometimes used with very thin polyacrylamide gels, the kind used for DNA sequencing applications, and that is to cast the gel without wells, leaving a flat gel surface and a space of a few millimeters at the top of a gel for insertion of a sharks tooth comb. The sharks tooth comb is essentially an external well that protrudes from the top of the gel and allows samples to be loaded very close to each other.

To make a polyacrylamide gel for separation of nucleic acids, a solution is prepared with polyacrylamide:bisacrylamide, buffer, water, and the catalysts ammonium persulfate and N,N,N',N'-tetramethylethylenedi-amine (TEMED). Unpolymerized polyacrylamide is a neurotoxin, so gloves must be worn and care must be exercised during handling. Dissolved oxygen inhibits polymerization, so many researchers degas the gel solution using a vacuum pump before the catalysts are added. The ammonium persulfate catalyst is prepared as a fresh solution on the day of use, and the TEMED is kept tightly capped to prevent moisture from entering the bottle and spoiling the reagent. Once the catalysts are added, and the solution gently swirled, the gel may be cast and the comb set in place.

Casting vertical gels is challenging if the apparatus is difficult to seal and tends to leak. Often a gel leaks if the glass plates have been improperly assembled with spacers, for example if the bottoms of the plates are skewed at different angles. Leakage may be prevented if the bottom of the plates are sealed with a film such as ParaFilm® or Press'n Seal® before placing the plates in a casting stand. Alternatively, some researchers pour a short plug of acrylamide in an apparatus, using extra catalyst to ensure rapid polymerization, and this plug seals the margins of the plates so that the rest of the gel will not leak when it is finally cast. Vertical gels may develop bubbles during casting, and these may be released from the plate by gentle tapping. However, if bubbles come to rest on the bottom of the comb these should be removed, if polymerization is not imminent, because they will distort the shape of the well and lead to oddly shaped bands on the gel. Generally speaking, a researcher will have a much easier time casting a good-looking gel if the plates are clean and free of fingerprints.

Urea is sometimes added to polyacrylamide gels as a denaturing agent, in concentrations of about 6–8 M, and acts to prevent secondary structure formation in single-stranded nucleic acids. When urea gels are being used, it is necessary to flush a well with buffer immediately before loading a sample to clear out the dense solution of urea that can form at the bottom of the well. As with the previous example of formaldehyde-containing agarose gels, the sample is typically denatured prior to loading on a urea gel, and the urea in the gel simply maintains the denatured state while the sample is running. Denaturing polyacrylamide gels may also be run at temperatures of approximately 50–55°C to prevent secondary structure formation within sample molecules. The temperature rises automatically when gels are run at high power, and many researchers will pre-run a gel before loading samples so that the gel is warmed up to its target temperature when the samples are first loaded.

With both agarose and polyacrylamide gels, a sample is loaded in a 'loading buffer' that is denser than water because it includes glycerol or ficoll. This loading buffer approximates the electrophoresis buffer in ionic strength and usually contains one or more anionic dyes that will migrate towards the anode and show the rate and uniformity of electrophoresis. If loading is performed using a micropipette tip, any extra air in the tip that is ahead of the sample should be extruded before the tip is placed in the well. Otherwise, the pipetting device will dispense bubbles of air prior to the sample, and these may disrupt smooth loading as they rise to the surface. The volume capacity of a well can be estimated by multiplying its height, width, and thickness, and bearing in mind that 1 mm^3 is 1 μl. Samples should not be loaded beyond the capacity of a well, because if samples spill out of the well during loading it can result in a confusing set of cross-contaminated lanes. Where it is critically important to isolate lanes from each other to prevent cross-contamination, blank lanes can be left between each loaded sample on a gel. Gels are loaded with samples after the buffer chambers and wells are already filled with buffer. Having the entire gel apparatus ready to run before loading minimizes disruption afterwards; the electrodes and cover can simply be attached to the power supply, and the electrophoresis commenced.

When an agarose gel is being loaded with a sample, the pipette tip should be placed just inside the well, but the researcher must take care not to damage the surface of the agarose. The wells are very fragile, and not attached to the rest of the gel by very much material, so any roughness or jabbing of a well with the pipette tip could break the well and cause sample leakage. The sample should sink naturally to the bottom of a well because it has a higher density than the surrounding buffer; however, if the sample was previously precipitated with ethanol and still carries some residue of ethanol, it may rise to mix with the surface of the water. Once this mixing starts, the entire sample may come out of the well and disperse on the surface like an oil slick! There isn't much that can be done to save a sample that has been lost in this way, but similar samples might be rescued before they are loaded by adding loading buffer stock to increase their densities. Then, after loading, the gel might be allowed to rest for 5–10 min before electrophoresis is commenced, to allow the salts in the sample to equilibrate with the surrounding gel.

When polyacrylamide gels are being loaded with a sample, many researchers use Hamilton® syringes or 'loading tips' with long, thin extensions that allow the tip to reach deep into a well. Loading tips allow the well to be filled from the bottom to the top, with less disturbance of the sample. Some polyacrylamide gels may have a dense residue of unpolymerized gel solution or urea that needs to be flushed from the well surface before loading. If the wells are filled with sheets or threads of polymerized gel debris then this may indicate that the comb is ill fitting or mismatched with the thickness of the vertical spacers.

Polyacrylamide gels are mounted in a vertical stand so that an upper buffer chamber, containing the cathode, and a lower buffer chamber, containing the anode, are in contact with the gel. If the liquid levels in the upper buffer chamber drop during electrophoresis, perhaps due to a leak in the sealing gaskets of the apparatus or formation of a channel between the gel and the vertical spacer, the buffer may no longer make contact with the wells and electrophoresis will cease. The buffer chambers in the electrophoresis unit should be covered to prevent passersby from accidental electric shock, and the removal of the cover should automatically disconnect the gel from the power supply.

With both polyacrylamide and agarose gels, it is important to check the cabling of the gel apparatus and power supply to make sure that the current will be flowing in the right direction. The DNA migrates in the direction of the anode, which is usually color-coded red. If the cables are plugged into the power supply backwards, or the power supply has a polarity switch that is flipped the wrong way, or the gel tray has been placed into the apparatus backwards, the DNA will move out of the gel instead of into it. Once the DNA has been lost into the gel buffer, it is too late to fix the problem.

DNA gels are usually run in either a Tris-Acetate-EDTA (TAE) buffer, or a Tris-Borate-EDTA (TBE) buffer, at pH 8.3. The TBE electrophoresis buffer provides better pH stability than TAE buffer, because TBE is a combination of two buffers in equimolar amounts: Tris base, with a pKa of approximately 8.2, and boric acid, with a pKa of approximately 9.2. The Tris acid and borate base become each other's counterion in solution, and the pH is more stable with this type of buffer than it would be with only a single buffering species in solution. The matter of pH and pKa is discussed more thoroughly in Chapter 3.

The TAE buffer is a combination of Tris base and acetic acid, but the pKa of acetate is 2.5 units below the pH of the buffer so the acetate ion does not have much buffering capacity *per se*; it is acting as a counterion for the Tris acid, which is at its pKa. If a TAE buffered gel is run at very high voltage and without any exchange of buffer between the cathode and anode ends, the pH of the anode end of the gel can fall significantly and the gel can literally melt and disappear into the solution. Furthermore, the high pH at the cathode end can lead to a reduction of mobility of DNA, as the phosphodiester backbone becomes protonated and electrically neutral. However, a significant advantage of TAE as an electrophoresis buffer is that it is usually compatible with enzymes that might be used to modify DNA fragments purified from the gel. DNA isolated from TBE-buffered gels must be specially purified to remove traces of borate, before it can be used in applications such as restriction digestion and DNA ligation. TAE buffer can also be prepared as a $50\times$ stock solution, which is a great convenience for storage; TBE buffer has a limited shelf-life and is typically prepared as a $10\times$ stock solution because of solubility constraints.

When we have completed electrophoresis of DNA, the next step is usually to stain the sample with ethidium bromide or a SYBR® stain. It is possible to add some types of gel stain to the gel during electrophoresis; however, as discussed earlier, this can lead to changes in DNA mobility. The stain is diluted and mixed into either water or a fresh sample of electrophoresis buffer, and poured into a clean tray. The gel is then added to the tray for staining. The researcher wears gloves at this point to avoid exposure to mutagenic chemicals (e.g., ethidium) and to avoid making the gel dirty through skin contact. It is poor form to put the gel into a tray of water and then add a concentrated sample of stain while agitating the tray to mix in the stain. The mixing is not as rapid as it might seem, and it leads to localized over-staining of the gel and high background. In the case of SYBR® dyes, the researcher should follow the manufacturer's recommendations for dilution, storage, and use. Many SYBR® dyes will stick to glass and plastic bottles and trays, for example, limiting the shelf-life of diluted solutions. In the case of ethidium bromide, the stain can be stored at 10 mg ml^{-1} and used at a working concentration of 0.5 to 1 µg ml^{-1}. A dilute solution such as this has a faint orange color that is barely discernable, but stains a gel nicely without over-staining. If extremely high concentrations of ethidium bromide are used in staining, a gel may become over-stained, in which case the background fluorescence is notably higher. If this happens, the gel can be washed and destained in several changes of fresh water over a 30-min period, and the background fluorescence will drop.

Photographing a gel that has been stained with a SYBR® dye or ethidium bromide requires a UV light source, usually a flat transilluminator screen on which the gel can be placed, and a vertically mounted camera or gel documentation system. The camera needs to be fitted with a filter that will absorb the UV and violet light of the transilluminator, so that the background appears black in the photograph. The camera can be adjusted and focused on the gel with regular visible light illumination, so as not to expose the gel to UV light for any longer a time than is necessary. A camera is best focused with as much light coming through the lens as possible, so the f-stop should be lowered during focusing and then raised to its proper exposure level afterwards. Many researchers place a transparent ruler on the gel, or next to the gel on the transilluminator screen, so that it is included in the picture. This allows measurements of band migration to be extracted directly from the photograph.

When the transilluminator is turned on, the researcher must be protected from the UV light, which can cause sunburn of exposed skin and permanent damage to the eyes. Face shields and wrap-around safety goggles that are UV opaque are commercially available, and exposed skin can also be protected by a lab coat and gloves. The gel photograph must be taken with no visible light contamination from the room lights or window, and many systems for recording the results of gel electrophoresis are set up in darkrooms or enclosed in light-tight cabinets.

Problems

The solutions to these problems are provided at the end of the chapter.

(i) Suppose that you want to cast an agarose gel on a plate that has dimensions of 8×10 cm, and you want the gel to be 6 mm thick. What volume of molten gel solution do you need to prepare? If the comb has teeth that are 2 mm thick and 6 mm wide, and the comb rests 1 mm above the surface of the plate during gel casting, approximately what maximal volume of sample will the well hold?

(ii) Suppose that you loaded an agarose gel and started it running at a voltage of 100 V, at 9:30 AM. At 10:00 AM you get a call from the day-care center, telling you that your daughter has a fever and needs to be picked up right away. At this point the loading dye has migrated only 1.5 cm in the gel. You want the gel to stop after the dye has run for another 6 cm, but you won't be able to make it back to the lab until 5:00 PM to turn off the power supply. What new voltage should you set so that the gel is ready to shut off at 5:00 PM? Remember that your daughter isn't feeling well, so don't take too long with the calculation!

(iii) Suppose that you have completed electrophoresis on an agarose gel, and after staining with ethidium bromide you take a picture of the gel next to a ruler, on a UV transilluminator. The standard linear DNA markers have the following pattern of migration: 5000 bp, 2.7 cm; 4000 bp, 3.3 cm; 3000 bp, 4.4 cm; 2000 bp, 6.0 cm; and 1000 bp, 9.0 cm. What is the apparent size of an unknown DNA in an adjacent lane, that shows a migration of 6.5 cm?

(iv) Suppose that the standard DNA marker fragments on your gel (5000, 4000, 3000, 2000, and 1000 bp) have equal fluorescence intensity after staining with ethidium bromide. What are the molar ratios of these fragments?

2.5 Safety

Maintaining the health and safety of researchers and visitors in a biological laboratory is the most important principle of lab operation. In a properly run facility, new workers are trained in safety at the point when they commence their jobs, and re-trained at intervals to ensure that their safe practices continue. This section of the book is not intended to substitute for a lab safety course, but rather to discuss some of the thinking and common sense behind safe practices. Laboratory researchers are responsible for both their own safety, and the safety of the other researchers and laboratory personnel around them. For example, a researcher who drops a piece of sharp broken glass into a plastic-lined garbage container, instead of into a proper sharps receptacle, creates a dangerous situation for the custodian or janitor of the facility. A researcher who notices that a coworker has departed and accidentally forgotten to turn off a Bunsen burner is obliged to turn off the burner to make the lab safe. Furthermore, a researcher needs to help coworkers realize and overcome their unsafe behaviors. Laboratories cannot operate safely if the researchers and workers do not feel a mutual sense of responsibility for each other.

Thinking about what might go wrong

It is a skill to be able to look at a laboratory set-up and perceive the flaws in its safety, not in terms of what is an immediate hazard, but what represents a potential hazard in the future. We must try to think about how equipment might become fatigued and break, how reagents and solvents may become dangerous if they are mixed or heated, and how skin, eyes and lungs might be exposed to hazards.

For example, Bunsen burners are usually connected to a benchtop or wall outlet by a piece of rubber tubing connected to a hose barb, and the flow of gas is controlled by a large valve at the outlet. However, many burners also have a thumbscrew valve at their bases that allows the gas to be shut off at the burner. This feature is a convenience that allows a researcher to start and stop the flame by turning this small valve, while leaving the wall or benchtop outlet valve turned on at all times. Is shutting off the gas at the burner a safe practice? There is no immediate danger, but what would happen if the rubber hose gradually became fatigued over a period of months? The pressurized hose might burst, or pop off the hose barb, and this might happen at night when no one was in the lab to hear or smell the gas as it filled the room. A simple oversight, leaving a rubber hose under pressure, could lead to a massive explosion.

Another type of forward thinking is for a researcher to imagine what would happen if a container broke or spilled its contents. For example, suppose a laboratory stored a bottle of liquid bleach next to a bottle of HCl on a reagent shelf; what would happen if the two bottles fell and shattered, and mixed their contents? In the worst scenario, workers might be overcome by toxic chlorine gas. Many substances are incompatible when mixed, and can generate toxic fumes, fires, or explosions. Bottles can break during a natural disaster, such as an earthquake or fire, or something as simple as a shelf collapsing.

Problems with heating things up

We use fire in the laboratory to flame-sterilize equipment, and if handled carelessly a fire can ignite nearby reagents and other combustible materials. For example, when we spread bacteria on a plate we first dip a glass spreader in an open vessel containing ethanol and pass the spreader over a flame to ignite the ethanol. What could go wrong with that procedure? For one thing, we have to plan and think about our hand movements when working with flames, because a lab coat or shirt sleeve could catch on fire if passed over a Bunsen burner. Gloves may melt or burn, particularly if they are latex, and loose hair could ignite if it gets too close to the flame. While the ethanol on the glass spreader is in flames we have to hold it horizontally so that the burning liquid does not drain down the spreader onto our hands. Burning droplets of ethanol may fall onto the bench surface, igniting paper towels, notebook paper, and perhaps even the open vessel containing the ethanol, if that lies underneath. Did we think ahead to what we would do if the vessel of ethanol is ignited? If the vessel is made out of glass, we might have planned to cover it with a glass lid to smother the flame. On the other hand if the vessel is made out of plastic, it may melt in the flame and release a burning river of ethanol that spreads the fire to other combustibles. Laboratory accidents can escalate from a simple problem to a serious one, by lack of foresight.

Fires should not be left unattended because of the danger they pose in igniting other materials. If a Bunsen burner has been left on in an enclosed vertical space with no air flow, such as a laminar flow hood that has been turned off, the air column above the burner may become extremely hot and cause a fire, for example by ignition of the HEPA filter. Bunsen burners may also be extinguished in a current of air, and allow natural gas to build up in a container or room.

Electrical heating devices such as hotplates can cause fires if they are left unattended and without thermostatic control. Water baths set at a high temperature, for example, may lose their water by evaporation and then start a laboratory fire as the device becomes overworked trying to raise the temperature of water that is no longer there. Mechanical switches in these high wattage electrical devices can also generate a substantial spark when they are turned on, or when their thermostatic control makes a self-adjustment, and this spark may ignite combustible fumes that have collected from a spilled organic solvent or a gas leak.

Problems with combustion

Many substances used in the laboratory are combustible, and susceptible to ignition. The flash point of a liquid is the temperature at which its vapor pressure is sufficient to cause ignition. For example, the flash point of ethanol is approximately 12°C, and that of acetone is approximately −18°C. Of course, the amount of combustible vapor increases with temperature, and so does the risk of accidental ignition. That is why it would be extremely unsafe to heat an open beaker of ethanol on a hotplate; as the vapor concentration increased, the solvent could be accidentally ignited by a spark from the hotplate.

Solvents of this kind are also not stored in regular laboratory freezers or refrigerators. Some 'flash free' freezers and refrigerators are specifically constructed for solvent storage, and these have their electronic components placed outside the refrigerated compartment and in a location where their sparks could not ignite the contents.

Nitrocellulose, which is used in nucleic acid blotting methods, is an example of an easily combustible solid. It has a low flash point and can easily ignite from exposure to a spark or flame. When used in Southern blotting, DNA is often covalently bound to the nitrocellulose by baking the blot in an 80°C vacuum oven; the vacuum being particularly important to exclude oxygen from the chamber and prevent spontaneous ignition and burning of the nitrocellulose.

Problems with pressurization

Air expands as it is heated, and this is why we do not apply heat to most closed containers. If a tightly capped bottle is heated, the air pressure may increase in it to the point that the bottle cracks or explodes. Flasks and beakers are usually used for heating liquids, but one problem is that glass can become fatigued by use and weakened by etching, scratches, or pits. These imperfections in the glass can become cracks and breaks, as the flask or beaker is heated unevenly, leading to possible leakage of a hot solution. If this happens when a piece of glassware is on a hotplate, the hot liquid may leak onto the device and spatter as it touches the even hotter ceramic or metal surface. The liquid may flood the benchtop and leak onto the floor, possibly injuring the researcher. If the solution is electrically conductive, it may cause a short circuit in the device or expose laboratory personnel to a high voltage electrical shock from the surrounding fluids.

Glassware may crack and weaken during an autoclave cycle, or during heating in a microwave oven. If a bottle or flask is filled with a solution for autoclave sterilization it may break and release its boiling-hot contents at an inopportune moment, for example when the bottle or flask is being picked up and transported. Glassware is loosely capped in an autoclave to allow for expansion and contraction of the contents without pressurization, but this also means that the caps are not secured for lifting a bottle safely. Furthermore, bottles in an autoclave or microwave oven may contain superheated liquid with a temperature just over the boiling point. If the bottle is touched or moved in this state its contents may suddenly boil and erupt through the top, injuring the researcher. This sudden boiling of a superheated solution is called bumping, and can be avoided by letting solution bottles cool undisturbed until they have reached a safe temperature.

While nitrogen and carbon dioxide cannot be ignited, liquid nitrogen and dry ice produce substantial amounts of gas as they boil and sublime, respectively, and this can result in pressurization and explosion of a sealed vessel. For example, if a researcher decides to put liquid nitrogen into a test tube so that a tissue sample can be dropped in and flash-frozen, it is critically important that the tube not be capped. The tube could become pressurized in seconds and explode. As discussed in Section 2.1, plastic vials removed from the liquid phase of a liquid nitrogen tank may explode if liquid nitrogen has accumulated in the vial during storage.

Gas cylinders are widely used in laboratories to hold compressed gases, such as carbon dioxide and nitrogen. While these are safe to use, they must be handled properly during transportation and storage. The cylinders have a main shut-off valve and are attached to gas regulators to control the rate of release of gas in an experimental procedure. When the gas is not immediately needed, the main shut-off valve should be closed. The cylinders have an awkward shape and must always be fixed to a solid support by a chain or other restraint, to keep them from falling over. Similarly, during transport they must be restrained and handled carefully so they do not fall. When not attached to a regulator, the cylinder should be capped to protect the stem or neck from damage if a fall does occur. Breakage of a cylinder at the stem can lead to the explosive release of the compressed gas, and a rocket-like launching of the cylinder that can cause structural damage and bodily harm.

Just as positive pressurization of a vessel may lead to explosion, negative pressurization can lead to implosion. Unfortunately, both explosions and implosions can lead to flying glass shrapnel and injury. A solution bottle that is sealed or capped when it is hot may become negatively pressurized when it cools. If the bottle is made out of plastic, it may become simply deformed as the airspace contracts, but if it is made out of glass, and the glass is already fatigued, it may crack or implode. Some laboratory glassware is designed for use under vacuum conditions, for example a side-arm flask that is made with very thick glass. Even still, it is best to place shielding around any piece of evacuated glassware, and for researchers to wear eye protection during vacuum procedures. A side-arm flask that is cracked or badly scratched should be discarded. Standard laboratory flasks and bottles should not be attached to a vacuum line, for example by attaching a vacuum hose to a stopper, because their glass is not sufficiently strong to withstand the negative pressure.

Vacuum desiccators are designed for use with vacuum lines, and may be made out of plastic or thick glass. In either case, the desiccator should be regularly inspected for cracks or signs of fatigue, and should be shielded during use. Glass desiccators often have heavy lids that are readily dropped and chipped on their rims, and these devices should be retired if they incur this type of damage. While a desiccator is under partial vacuum, it may implode if it is hit or dropped. A good desiccator can hold its vacuum indefinitely, which means that it may be an implosion hazard if it is evacuated and then left in a place where it might accidentally be knocked off a bench onto the floor.

Problems with electricity

Special consideration must be given to laboratory devices that use electricity, because these may be a shock or electrocution hazard if they are misused, become damaged, or become wet. For example, electrophoresis is a common procedure in the molecular biology laboratory, and it carries many risks. If the electrophoresis apparatus is home-made, or jury-rigged from a variety of parts, it may not have been designed with safety in mind. Buffer chambers and gels must be shielded to prevent passersby from accidentally touching the contents. The installation of the shielding should be integrated into the process of attachment of the electrode leads, so that a researcher cannot, for example, pour additional electrophoresis buffer into the chamber without disconnecting the power. The electrode leads should be male to female, so that when the leads are connected to a power supply their tips are not exposed. The power supply should have a safety feature that allows it to shut off automatically if there is a break in the circuit. It is important that safety features of electrophoresis equipment not be bypassed or disabled.

Laboratory equipment should be properly grounded through the electrical receptacle, and ground fault detecting outlets should be used where there is a chance that the equipment could become wet. Researchers should be attuned to any signs of instrument damage that might lead to electrical fire or shock, for example burning smells, mild shocks or tingling when the chassis of a device is touched, or unusual heat. Electrical instruments that have any of these symptoms need to be inspected and serviced, before the problem worsens and leads to a fire or injury.

Problems with moving parts

Care needs to be taken around pieces of equipment that have moving parts. For example, vacuum pumps and other motorized pieces of equipment may have exposed belt drives or spindles that can grab and wind up a loose hank of hair, a necktie, or a lab coat sleeve that is inadvertently brought too close. A well-designed piece of equipment should have metal guards that protect the workers from injury, but many laboratories retain old pieces of equipment that have not been maintained or updated to modern safety standards.

Shaking incubators are usually designed to have a flat horizontal surface that oscillates a few centimeters around a rotating spindle. Fingers or hands may be accidentally crushed if they are inserted into the gap along the side of the shaking platform while it is in operation, because the oscillation closes the gap. If a glass flask breaks in a shaking incubator and spills its contents, the liquid may pool underneath the plate and drain down the spindle shaft into the electric motor. When an accident such as this is detected, the incubator may need to be shut down so the oscillating plate can be removed for inspection. The area underneath the plate may need to be dried and decontaminated, and if leakage into the underlying electrical parts has occurred the device is not safe to use and may need professional cleaning to prevent metal corrosion, electrical short circuiting, or other damage.

Centrifuges are also constructed on a motorized spindle, but the rotor rotates symmetrically around the spindle and the rate of rotation of the centrifuge motor is substantially higher. Centrifuges should never be opened while they are under rotation, except under controlled conditions such as zonal centrifugation protocols, because the centrifuge chamber and lid are designed to shield the researcher from injury. A hair clip or a pen that is accidentally dropped into an open centrifuge could ricochet off of the rotor and be launched into the air at high speed. The loose object could even break a centrifuge tube, destabilize the rotor, and lead to a catastrophic accident. If a centrifuge is unbalanced it will vibrate and

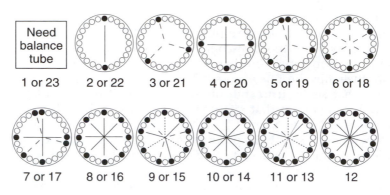

Figure 2.5

Even loading of a 24-place rotor with 2 to 22 balanced tubes. The diagrams show loading patterns, where lines of symmetry are indicated by solid, dashed, or dotted lines between groups of tubes. For example, the '5 or 19' pattern shows one pair of black circles symmetrically distributed (top and bottom) and three black circles evenly spaced, to make five in total. If five balanced tubes are distributed in a rotor in the manner of the black circles, the rotor will be balanced. The pattern can also be viewed as a loading pattern for 19 tubes, if the 19 white circles are taken to be the tubes and the black circles are unfilled rotor positions. An extra 'balance tube' is always required if 1 or 23 tubes are loaded, to bring the number up to 2 or 24, respectively.

make a substantial amount of noise as it is being started, and this is a cue to the researcher to shut off the centrifuge, let the rotor come to a rest, and then rebalance the tubes. *Figure 2.5* shows an example of a 24-position rotor, and how it can be evenly loaded to prevent imbalance. A small centrifuge that is operated with an imbalance may be damaged, and may become so destabilized that it 'walks' off the benchtop and crashes to the floor. Large ultracentrifuges may even 'walk' across a room if there is a rotor explosion during a high-speed centrifugation procedure. These catastrophes can be avoided by carefully balancing the tubes and adaptors that are inserted into a rotor. The tubes must be in good condition, free of cracks or scratches, and designed for centrifugation in the model of rotor at the intended speed. Furthermore, the rotor must be used within its rated rotation speed. If the spindle has a horizontal pin that fits a notch on the base of the rotor, these must be properly aligned. Some rotors have screws or knurled knobs to fix them to the centrifuge spindle, and these need to be firmly tightened before use so that the rotor does not become airborne during operation. Rotors may have a lid that is used to contain the sample, or reduce rotor air resistance during operation, and in this case the lid must always be used. If the rotor or its tubes just don't seem to 'fit right' or 'look right', then it is critically important that the researcher not operate the centrifuge until the problem is studied and resolved. It may be that the rotor has a crushed tube in the bottom that prevents tubes from fitting, or some other problem that could be readily fixed. If there are signs of damage to a centrifuge or rotor, the parts should be inspected by a qualified technician before they are used. Ultracentrifuge rotors and buckets, in particular, must be regularly inspected for hidden corrosion damage. The log books that are used to keep records on ultra-centrifuge rotor use are important because an ultracentrifuge rotor has a limited lifespan; metal fatigue sets in after extensive use, and a rotor may be retired or rated to a lower speed for reasons of safety.

Skin and eye exposure

The skin and eyes are susceptible to burns and flying shrapnel from all of the aforementioned types of laboratory accidents, but they can also be exposed to chemical burns or other forms of damage during routine experiments. Corrosive substances such as acids and bases can cause chemical burns to the skin that are extremely painful, and, in the case of the eyes, potentially blinding. Toxins may be absorbed through the skin, or through skin-to-mouth contact, and lead to acute or chronic poisoning. Ionizing radiation, such as that occurring during decay of ^{32}P or ^{125}I, may penetrate the skin and cause mutations. Other substances are chemically mutagenic, for example ethidium bromide, or carcinogenic.

Laboratory work must be conducted in a way that minimizes risks of skin and eye exposure. For example, a tube that is placed on a vortex device may launch some of its sample into the air, spattering the researcher with a chemical. Reagents should be combined and mixed in a way that minimizes the risk of splashing, or formation of aerosol droplets that might be carried in the air. Where toxic or irritating fumes will be generated during a procedure, for example in formaldehyde gel electrophoresis of RNA, or the decanting of a bottle of HCl, the procedure should be conducted in a working fume hood.

After good planning and technique, the next most important layer of protection is the physical barrier provided by a lab coat, a pair of gloves, and a pair of eye goggles. The lab coat should fit to allow free movement, and be buttoned to protect the skin of the torso. The fit of the coat around the forearms may be a concern if the cuffs do not button close. The gloves selected must be compatible with the chemicals being used and appropriate for the task at hand; some toxins will diffuse through latex gloves, for example. The protective eyewear must also be selected with care. The goggles or face shield should be impact resistant if an explosion or implosion is possible, sealed to the skin of the face if fume exposure is a problem, and UV-opaque if ultraviolet light sources are being used.

The researcher needs to foresee that these articles of protective equipment may prevent lasting damage to skin and eyes if something doesn't go quite right. Older researchers may not take all the necessary precautions, believing that their years of experience will keep them safe, but this is a flawed belief; accidents can happen to anyone. A wise researcher takes the same safety precautions that are expected of a novice.

Inhalation and ingestion exposure

Most laboratories have strictures against eating and drinking in the laboratory, or storing of food and drink in a refrigerator or freezer that also contains laboratory materials. This policy is intended to prevent the accidental ingestion of harmful chemicals or pathogenic organisms. There may be a specific area where food consumption is permitted, for example a lunch room or food service area, and researchers should not contaminate those areas with laboratory materials. Laboratory workers should remove gloves, lab coats, and other potentially contaminated apparel before entering an area where food and drink are consumed, and wash their hands to avoid contaminating countertops and utensils.

In the laboratory setting, researchers must avoid putting anything into their mouths, such as cigarettes, chewing gum, and pencils. It is unsafe to 'mouth pipette' liquids, a practice in which the mouth is used in place of a rubber bulb to apply suction. Even if the liquid is just water, the end of the pipette may carry residues from other experiments that could be harmful if ingested.

The uptake of harmful materials by inhalation is more difficult to avoid, because breathing is not optional. Liquids can become airborne as aerosol droplets and may carry radioactive compounds, pathogenic organisms and other harmful substances to the lungs. Aerosols form when a liquid surface is disrupted, for example through vortexing, pipetting, sparging, or sonication. A bubble forming and breaking on the end of a pipette can produce a spatter of airborne droplets carrying pathogenic bacteria and viruses, which is why biological containment hoods are used for culturing and handling such organisms. Solids can generate dust that is carried through the air and inhaled, especially if a reagent bottle is shaken before a sample is taken for weighing. Sodium dodecyl sulfate dust is particularly irritating to the lungs if inhaled, and acrylamide dust is toxic. In many cases, a dust mask or other respiratory filter can be worn to protect the laboratory worker from these types of airborne particulates. However, the best type of preventative measure is to minimize the amount of airborne contamination, and conduct problematic operations in a fume hood that can draw away the contaminated air.

Toxic vapors and fumes are more difficult to manage because they cannot be filtered out of the air by a simple dust mask. Operations that produce harmful fumes can be conducted in a fume hood, but a laboratory worker may also need to wear protective breathing gear such as a charcoal-filtered facemask that covers the nose and mouth. Fumes may have either a low density, and rise through an air column, or a high density and sink. A bottle of volatile liquid may fill with highly concentrated fumes that could be harmful if a researcher inserts his or her nose to smell. If a bottle or flask must be tested by smell, it is best to waft some of the air over the rim of the bottle to dilute the fumes before they are inhaled. Not all harmful chemicals smell bad; a researcher must anticipate when experiments need to be conducted in a hood, and when they can be conducted on a lab bench without special precautions.

Labeling and hazard information

Hazards in the laboratory need to be clearly labeled so that workers and visitors can avoid exposure and injury. For example, electrophoresis of a DNA sequencing gel may involve several thousand volts, and a laboratory operation such as this would rightly have a posted sign that reads 'Danger – High Voltage.' The researcher may be very well aware of the dangers of electrophoresis, but the laboratory visitor may not. Similarly, a laboratory bench that is being used for a procedure involving radioactive substances should have a posted sign or warning labels that read 'Caution – Radioactive.'

Reagent bottles should carry information that describes their chemical content, and their health hazards, for example: 'Poison', 'Corrosive', 'Irritant', 'Carcinogen', or 'Flammable'. Laboratories may be required by law to maintain records on substances being stored in a laboratory, including material safety data sheets (MSDS) that are provided by manufacturers and available in on-line databases. Chemical waste handling may also be regulated, and laboratories may be required to segregate, store, and label wastes in a particular way.

Researchers often develop complex solutions that involve many components, and give them names such as 'buffer A' or 'wash' solution. While these names may be meaningful to one person, they do not convey the bottle contents or hazards to others. For example, many 'wash' solutions used in DNA purification are made with ethanol and are flammable.

Labeling becomes problematic with biological or chemical samples that are being stored in small tubes or microtiter plates, because there is not enough space to write a full description of the contents. Nonetheless, a tube that contains a pathogenic or toxic substance must be marked and stored in a way that prevents harm to a researcher who may have forgotten its content, or to other researchers and visitors who were never apprised.

Accidents and spills

Accidents will happen, and when they do a laboratory worker may become mentally agitated and not think clearly. For example, if a laboratory fire starts then protection of laboratory workers must come before any other consideration. It would be a small comfort to have saved samples from a freezer in a burning laboratory if the worker subsequently dies of smoke inhalation. Small fires might be controlled with a fire extinguisher, but if there is a chance that toxic fumes are being generated then evacuation is the necessary course of action. Fire drills are important for training laboratory workers in how to evacuate in a timely fashion, and for identifying researchers who cannot be torn away from their experiments. These individuals are likely to ignore a real emergency, and may need to be pulled aside and retrained.

Laboratories often have to deal with the cleanup of minor spills, and workers need to have the materials and training to handle them. For example, cleaning up a minor spill of radioactive substance is a complicated procedure that requires training and guidance. A minor spill can be made much worse if the perimeter of the contaminated area is not controlled; if laboratory personnel walk through the area, contaminating their shoes and spreading the radioactivity widely. A large spill of a corrosive substance, for example breakage of a 5-l bottle of concentrated H_2SO_4, may require extensive cleaning and containment to prevent structural damage and leakage to the next building floor down. Spills of biological materials may require decontamination, in addition to routine cleaning.

As with any accident, a spill may not bring out the best behavior in laboratory workers, as they struggle with feelings of embarrassment and panic. A spill may be of such a composition or magnitude that it cannot be safely cleaned by laboratory workers, and this needs to be impressed on personnel during their training. Many laboratories are required to report spills of a certain magnitude to a safety officer, for record keeping and assistance with cleanup.

Safety equipment

Laboratories must have personal protective equipment, such as gloves, face and eye protection, and laboratory coats. In addition, some operations may require physical shielding, for example with vacuum flasks and other pressurized containers, or shielding to protect workers from ionizing radiation. Most laboratories are required to have safety showers and eyewash stations, to allow the rapid decontamination of an individual who has been exposed to a corrosive or toxic substance. A laboratory may have fire extinguishers, fire blankets, and respiratory devices to help deal with minor fire emergencies, and first aid kits to handle minor medical emergencies. The emergency exit doors are the most important piece of safety equipment in a laboratory, and they need to be kept free of obstructions and clutter.

Safety equipment must be kept in good supply, or good repair, so that it is available when needed. Eyewash stations and safety showers need to be tested regularly to make sure that they deliver water at a prescribed rate of flow. Monitoring equipment such as smoke detectors and radiation detectors must also be tested to make sure their batteries are charged, and that their electronics are functioning. Fire extinguishers must be kept charged, and promptly recharged if they are used.

Leadership and shared responsibility

Laboratory workers and researchers rely on each other for a safe working environment, but individuals must also show leadership in setting high standards of conduct. For example, if a junior researcher notices that a more senior member of a lab has unsafe behaviors, the matter of safety cannot be ignored because of 'rank'. Older researchers must demonstrate to younger scientists that they themselves wear protective gear and take appropriate precautions, and must not show off the bad habits that sometimes accumulate through a mixture of hubris and good luck.

A laboratory needs to develop a culture of serious work, so that dangerous practical jokes and horseplay are not tolerated. For example, a researcher who enjoys taking rides on laboratory carts, playing hockey in the hallway, or startling other researchers with exploding tubes, must be expelled from a laboratory if he or she cannot be dissuaded. The director or administrator of a lab is responsible for setting an expectation of responsible behavior, and laboratory workers must reinforce this expectation for their mutual safety.

2.6 Documentation

Research should be recorded in a way that is organized and easily understood, and this involves writing down the date, title, purpose, content, and procedures of an experiment, as well as results and a narrative discussion. Sometimes a researcher will depart a laboratory and leave a body of work that needs to be completed by others, or written into a manuscript. If the researcher's notebook is too difficult to read, the work may lie fallow or never be completed, which is a loss for both the laboratory that invested in the project and the departing researcher who was hoping to get a publication from it.

A notebook should be written in indelible ink, in the researcher's own hand, and writing mistakes should be crossed out instead of erased. Most laboratory notebooks have sewn-in pages to prevent tampering with the page order, and this is essential if the book becomes the subject of a legal or administrative inquiry. A researcher should bear in mind that if he or she is ever wrongfully accused of fabrication of data, a properly maintained notebook will be his or her best defense.

The date, title, and purpose of an experiment

Establishing the date of an experiment is particularly important if the work leads to a patentable invention, or if there is a need to establish a chronology of discoveries. Writing a date and numbering each page of a laboratory notebook helps to prove that the pages were written sequentially, and that the writer did not skip around in the book. Finding intervening blank pages in a notebook tends to imply the opposite; that the record may not be a chronology.

Each experiment that is discussed in a notebook should be titled with a brief description, to help the reader understand the flow of the writing and the boundaries between different experiments and procedures. For example, a notebook page might be titled 'April 1, 2015: Column purification of plasmid pImOK' with the purpose described underneath: 'Purpose: Preparation of plasmid for subcloning the UOK-2 gene.' The explanation of the purpose of the work helps the reader to follow the sequence of experiments, and the short-term or proximal goals. Recording longer-term goals such as 'Purpose: To find a cure for cancer' or 'Purpose: To finally get my Ph.D.' will probably not illuminate the writing. However, narrative reflections can make it easier to understand the thinking of the researcher at the time the page was written. For example: 'A different method of plasmid purification is being attempted because the previous approach did not provide sufficient yield and purity' can help the reader understand why a different approach is being attempted.

The materials and methods

The reagents and samples used in experiments come from many different sources, and maintaining an accurate record of these ingredients can be a significant help in the process of discovery. A reagent used in an experiment can be recorded by writing its full name, company name, catalog number, and a lot number. This information is helpful for two reasons: if the experiment succeeds, then it can be reproduced exactly, and if it fails then troubleshooting can be easily performed. The reagent name, company, and catalog number can help to identify errors in formulation.

For example, if the wrong chemical form of a reagent was selected this will be obvious if the notebook description is complete. The lot number may help to identify a specific bottle on a reagent shelf, and this is important in identifying reagents that have passed their expiration date or have become contaminated or decomposed. In a positive sense, the lot number increases the reproducibility of successful experiments, and speeds laboratory progress.

Methods should also be described with enough detail that they can be reproduced exactly. For example, the order of addition of reagents, the incubation conditions and times, the amount of mixing of samples, and other details of handling should be dutifully recorded. Again, if the experiment is successful this record can help the researcher reproduce the method exactly. If it is unsuccessful, the description of the protocol can be compared to other descriptions to see if some variance in method is responsible. Abbreviations and idiosyncratic writing should be kept to a minimum, so that others can interpret what is being written. An encrypted statement such as: 's.n. decanted to bluetop, q.s. buffer A to 10 ml, O.N. at R.T., then tx to −70' may be meaningful to one researcher, but not easily decoded by others.

In a similar way, tubes and other samples being stored in a freezer should be labeled unambiguously. Microfuge tubes don't have much writing space on them, but on the other hand there isn't much that can be done with a freezer that is full of racks of tubes that are all cryptically numbered 1 through 10! One way of solving this tube labeling problem is to give each tube a unique 'license plate' number, coded to a lab notebook page. For example, a tube marked 'BB107-5' could be traced to notebook page 107 of a researcher 'B.B.', and might be sample #5 on the page. If the contents of tube BB107-5 are used several months later as a sample in an experiment, leading perhaps to new tubes labeled BB155-1 through BB155-10, the composition of those new samples are readily traced back through the documentation provided in the notebook. This approach is a lot more tidy than trying to write a detailed description of the experiment on the side of the tube, which might not carry much information, or on the tube rack, which might be dropped and separated from the tubes.

The results

Experimental results may be a mixture of written observations and data. For example, a researcher might record a result such as: 'After overnight incubation at 37°C, the plate had 15 blue and 45 white colonies.' A researcher might also affix into the notebook the artifacts of analysis, such as photographs and scintillation counter printouts. In the latter case the taped-in materials should not cover any part of the writing, especially if the notebook is to be photocopied for safekeeping. Affixed materials will probably need to be annotated, for example pictures of gels need to have an indication of lane number, and scintillation counter printouts need to have the actual sample numbers written onto the page. These materials should also be dated so that they can be restored to their proper notebook page if they become unglued or separated from the page.

The critical issue in recording results is to be clear about the details of the data. If a printout from a spectrophotometer is included, for example, which line on the paper tape represents a measurement of the optical density of the sample? What was the wavelength setting? What was the dilution of the sample when it was tested? What was the source of the sample, and how does it relate to the earlier narrative in the book? It is easy for raw data to become uninterpretable if these types of facts are not recorded.

The discussion

Imagine trying to make sense of a notebook that just had the barest descriptions of the work, recipes and taped-in pictures of gels. The notebook might be adequate proof that work was accomplished, but what was the work and how should it be interpreted? Good laboratory notebooks include narrative descriptions that help the reader assemble the important facts. For example, a brief written comment such as: 'The co-migration of bands in lanes 2 and 4 suggests that the alleles are identical' can remind the researcher, years in the future, how an attached gel picture should be interpreted. A reflective comment such as: 'The smear is puzzling and unexplained – perhaps the DNA sample was sheared by excessive vortexing' can help with troubleshooting and suggest improvements in a method. A notebook can be made as formal or informal as a researcher wishes; however, informal personal interjections such as 'Yahoo!!!', 'Oops!', or 'Zounds! – why me?' may not wear well with the passage of time.

2.7 Special lab problems

Different laboratories have their own sets of problems specific to their methods, their organisms of study, and their wealth of funding. A few examples of laboratory situations are described below, to illustrate the different traditions and approaches that may become part of a lab culture.

The PCR lab

The polymerase chain reaction (PCR), described in Chapter 7, is an exquisitely sensitive method that can be used on very small samples of nucleic acid. Researchers in a laboratory conducting diagnostic PCR must be concerned by the accidental cross-contamination of samples and reagents. This concern may lead to an unusual amount of attention being given to sample handling techniques, and laboratory isolation procedures.

The problem with PCR contamination can be explained simply: a PCR might typically be initiated with 1000 molecules of a DNA template, but the yield from a successful reaction could be as much as 10^{10} to 10^{11} copies, which could be hundreds of nanograms. If those 10^{10} to 10^{11} copies are loaded on a gel and analyzed in the same laboratory workspace in which the next PCR is assembled, the pipetting devices, gel apparatus, benchtop, sink, and doorknobs could easily pick up femtograms to picograms of the DNA product. This is a minor amount by most measures, but an overwhelming amount of contaminant if incorporated in the next PCR. If a sporadic cross-contaminant emerges and becomes responsible for results, a researcher may reach experimental conclusions that are untrue. Criminals may go free while the innocent are incarcerated, patients may be given incorrect medical diagnoses, paternity may be incorrectly assigned, and published papers may need to be retracted.

What can be done to prevent this nightmare from being realized? A physical isolation of the PCR set-up from the PCR analysis is the most important factor. If the reactions are assembled in a clean area, using pipetting devices that are never used for other purposes, then the chances of cross-contamination are greatly reduced. Researchers involved in setting up PCR must think about their activities; if they have just paid a visit to a product-contaminated room, their lab coats, clothing, shoes, and hands may carry contamination into the clean area. Some laboratories relegate PCR set-up to biological containment hoods or clean rooms, and the workers wear disposable gowns, gloves, hair coverings, facemasks, and booties to protect the sample. Filter-plugged pipette tips are used to minimize aerosol contamination, and reaction buffers and components are aliquoted and jealously guarded. The reaction tubes are sealed in the clean area, and not opened until after the thermocyling is complete.

Careful handling and disposal of reaction products is a second important factor in minimizing cross-contamination. Reaction products may be loaded on a gel using a specific set of pipetting devices that are reserved for that one purpose, and stored separately. Spent pipette tips, gel buffers, and used gels can be disposed of in a way that minimizes contamination, or collected and autoclaved, or treated with a decontaminant to destroy the resident DNA. Dilute bleach might be used for this purpose; however, some safety protocols forbid the mixing of bleach and ethidium bromide stain.

The researcher must think about all forms of DNA trafficking between different parts of a lab. For example, if a gel containing PCR products is photographed on a UV transilluminator, then the transilluminator will be contaminated with DNA. If the next day, an innocent-looking plasmid-containing gel is photographed and brought back to the PCR clean area, then the residue from the previous gel may be introduced as a contaminant.

Negative controls are the third line of defense, but a single negative sample may not prove much. When solutions are contaminated at a low level, reaction artifacts may emerge sporadically. Multiple negative attempts to generate a product from a template-free sample may be the only convincing evidence that the solutions and instruments are free of contaminating template.

An interesting aspect of the whole problem is that the contamination is specific to the experiment being conducted. If Dr. Dexter and Dr. Sinistre worked next door to each other and were conducting PCR experiments with unrelated oligonucleotide primers, Dr. Dexter might use Dr. Sinistre's lab as a 'clean room' and Dr. Sinistre could use Dr. Dexter's lab as a 'clean room'. Aside from any unresolved problems of DNA trafficking between the rooms, Dr. Dexter's and Dr. Sinistre's problems with contamination would be completely independent.

A second general problem with the PCR lab is that small variations in PCR conditions can be compounded in each cycle, and have a profound effect on reaction success. Remaking a single reaction solution or performing the PCR method on a different thermocycler may be enough to change success to failure. The true causes of variation are discussed in more detail in Chapter 7, but suffice it to say that there may seem to be a very fine line between an experiment that works beautifully and one that yields a blank gel. The experimental basis of success and failure may not be easily discerned, and that can lead to general despair or a belief in superstitions among researchers in the PCR lab.

The RNA lab

Environmental RNase contamination is the bane of an RNA lab, and RNA researchers adopt patterns of behavior to help them avoid contaminating reagents and instruments with their own skin cells. Gloves are more often worn in the RNA lab to protect the sample from the researcher, rather than the other way around. An RNA laboratory may adopt additional practices and traditions to help maintain cleanliness. For example, a separate reagent shelf and lab bench may be maintained for RNase-free reagents, and these are only handled with gloves and sterile spatulas. Weighing paper taken from a box may be turned upside down as it is put on a pan balance, to expose the 'clean' side to the RNase-free reagent. DEPC-treated water may be reserved for RNA work, and disposable plasticware and baked glassware may be set aside as well. In a multipurpose research lab, the RNase-free areas are usually covered with disposable adsorbent paper and cordoned off with tape. Posted signs may advise the visitor not to touch anything in an RNase-free area (sometimes with threats of bodily harm explained fully as well).

Why are RNA labs so fastidious about their reagents and supplies? The reason is that RNase is not simple to inactivate. A lab surface contaminated with RNase cannot be decontaminated with a detergent. A solution contaminated with RNase cannot be made pure by autoclaving, and DEPC may not solve the problem either. A reagent that is used in a failed experiment may be presumed 'guilty of RNase contamination' by association, and summarily discarded. Furthermore, one sloppy scientist in an RNA laboratory can ruin everyone else's work, and this leads to an interesting laboratory dynamic. RNase-free reagents and instruments can be effectively shared between RNA researchers, but only if everyone in the lab is trustworthy. A breakdown of trust can lead to hoarding of materials, and lack of sharing.

The tissue culture lab

Mammalian cell tissue culture is complicated and expensive, and unless the researcher is only growing a 'lab weed' such as HeLa cells, there may be intermittent problems. Nothing stops laboratory progress quite like the catastrophic death of every living culture in the incubator. Ongoing work and irreplaceable cultures may be lost overnight, and reestablishing cultures from frozen archived samples may take weeks of work before the lab is back on its feet.

Successful tissue culture requires growth medium and supplemental solutions that are consistently high in quality. In laboratories in which culture medium is made from individual chemical reagents, or a pre-mixed powder, the purity of the water used to formulate the solution may be a critical component. Where glassware is being used, a change in brand of washing detergent may cause experiments to fail just by the residue that is left after rinsing. Many laboratories avoid this issue entirely by purchasing medium supplied in a liquid form, and using only disposable tissue-culture-grade plasticware. If serum is added to the growth medium, the cells may grow better in some types or specific lots of serum than others. Laboratories that conduct cell culture may set aside and jealously guard certain bottles of serum that are successful with fastidious cell lines, and use a more commonly available serum for the tolerant cells.

Cell handling technique is also a critical issue for tissue culture labs, particularly if reagents are being shared. For example, pipettes that are used in tissue culture should never be used repeatedly, because this can cross-contaminate cultures. If a researcher plans to add 2 ml of a trypsin solution to each of five different adherent cell lines, then filling a single pipette with 10 ml of trypsin and dispensing 2 ml, sequentially, to each plate is a poor technique. Even if the researcher plans to dispense the solution with the pipette held well above the plates, and never touches the pipette tip to the plate surface, the falling solution spatters when it hits the plate surface and creates aerosol droplets that may rebound and contaminate the tip. When the pipette is moved to the next plate, some of the culture from the previous plate may be introduced with the added solution. If the first culture has a faster growth rate than the second, it will form a mixed culture and the faster-growing culture will eventually displace the slower culture.

Worse still, if the one pipette is assumed to be sterile and is returned to a communal bottle of trypsin for replenishment, then every other researcher's cultures may be cross-contaminated as well. Bad technique with pipettes may spread mycoplasma, and other bacteria and fungi between cultures. Contaminants of this kind can wipe out a culture, and spread throughout other cultures in an affected incubator if they are not stopped. A single sloppy laboratory worker in a tissue culture lab can set other researchers back in their work by months, and this is another example of a situation in which sharing of reagents and incubator space is based on mutual trust.

The indigent lab

Whether due to lack of funding, or the parsimonious nature of the princi-pal investigator, some laboratories operate on a shoestring budget. They find inventive ways to wash disposable glassware, save and re-use electrophoresis buffers, and construct experimental apparatuses out of throwaway bits of plastic and metal. Everything is made from scratch, and reagent kits are only purchased as a last resort. When kits are finally purchased, they may be picked apart and studied, with an eye toward reverse engineering them in the future.

Can good research be done in indigent labs? Absolutely! Provided that money is being spent where it will do some good, there is no point in wasting money where it won't help. Researchers in an indigent lab must become experts in making their own solutions, and in troubleshooting common techniques. Having not purchased a reagent kit, the indigent researcher has no corporate technical support operator on the telephone line to provide troubleshooting advice and replacement components.

Researchers in the indigent lab may feel deprived of the simple pleasures to be found in wealthier labs: pipette tips that arrive loaded in racks instead of bags, and gels that are pre-made and ready to load. Experimental work may be slow, because of the distractions that poverty brings. A worker from an indigent lab cannot easily pass by a dumpster without taking a quick peek over the rim, just to see if someone in another lab has thrown out a salvageable piece of equipment. Quite aside from the thrill of dumpster diving, there is also pride to be found in self reliance, in knowing exactly how an experiment was made to succeed, and in a research program that provides a high return on its investment.

The wealthy lab

Some laboratories have plenty of funding, and that leads to a different style of science. There may be little wear and tear on the pan balance, because solutions are purchased pre-made, instead of being made on site. Nothing is washed and re-used, so a dishwasher doesn't need to be hired to sort pipettes and put away glassware. The water supply to the laboratory is of only incidental concern, since laboratory water is purchased from a commercial supplier in bottles.

Commercial enterprises will provide almost any type of material and technical outsourcing, for a price. Research can be conducted through the principal investigator's mailbox in many cases, with custom-made libraries, DNA clones and antibody reagents ordered from companies that provide rapid turnaround time and a guarantee of quality. In these wealthy laboratories, researchers conduct the key experiments that bring together materials and reagents from suppliers, and they are not usually fettered by the reliability of these components.

Can good research be done in wealthy labs? Absolutely! In fact, it may be easier to make rapid progress in a wealthy lab than an indigent lab, in the same sense that it is faster to drive on a paved superhighway than a dirt road. However, the superhighway only permits travel in limited directions, and may be less flexible for exploring new terrain. Wealthy labs may have a smaller measure of self-reliance, because they adapt to the pace and style of their research. If a laboratory can only make progress with reagent kits, then the kits circumscribe what can be accomplished.

Further reading

Chemical materials

O'Neil, M.J. (ed.) (2001) *The Merck Index: An Encyclopedia of Chemicals, Drugs, & Biologicals*, 13th Edn. Merck & Co. Whitehouse Station, NJ.

Laboratory protocols

Sambrook J, Russell D (2001) *Molecular Cloning: A Laboratory Manual*, 3rd Edn. Cold Spring Harbor Laboratory Press, Cold Spring Harbor, NY.

Laboratory Safety

Furr AK (2000) *CRC Handbook of Laboratory Safety*, 5th Edn. CRC Press, Boca Raton, FL.

Solutions to problems

These are the solutions to problems posed in Section 2.4.

(i)   The volume of the gel will be approximately $8 \times 10 \times 0.6$ cm = 48 cm^3 = 48 ml. The height of each well will be approximately 5 mm (because the gel is 6 mm thick but the well will be 1 mm above the plate), so the maximal volume of the well will be $2 \times 6 \times 5$ mm = 60 mm^3 = 60 µl. In practice it is better to allow for some room at the top of the well so the samples do not spill into neighboring wells.

(ii) The migration distance of the dye is proportional to both voltage and time, under normal gel conditions, so the quotient (migration distance)/(volts)(hours) is reasonably constant. From 9:30 AM to 10:00 AM the quotient was (1.5 cm)/(100 V)(0.5 h), or we could say the dye ran at a rate of 0.03 cm per Vh. We want the dye to run an additional 6 cm in 7 h, so the expression (6 cm)/(z V)(7 h) can be set equal to 0.03 cm Vh^{-1}, and we can solve for z as the unknown voltage setting: the solution is z = 28 V. Another way of looking at this is that we want the dye to go four times the distance in 14 times the time, so the voltage should be 4/14 of its previous value.

(iii) We first graph the logarithm of the standard DNA fragment size as a function of distance of migration, either by using semi-logarithmic paper or by taking the logarithms manually. For example, we could plot the following (x,y) ordered pairs on regular graph paper, where the y-axis is $\log_{10}$(DNA size): (9.0, 3.00); (6.0, 3.30); (4.4, 3.48); (3.3, 3.60); and (2.7, 3.70). The unknown DNA has a migration of 6.5 cm, and interpolating between the first two points graphically or mathematically we find the corresponding value on the y-axis is 3.25. This number is the logarithm of the apparent DNA size, so the solution is $10^{3.25} \approx 1800$ bp. However, it is important to bear in mind the underlying assumption that the unknown DNA is linear.

(iv) If we have equal fluorescence staining among the five fragments, that means that the fragments are present in the same gram amounts. The molar amount of a DNA fragment is the gram amount divided by the molecular weight, and molecular weight is proportional to DNA length. Therefore the 5000-, 4000-, 3000-, 2000-, and 1000-bp fragments are present in molar ratios of 1 to 1.25 to 1.67 to 2.5 to 5, respectively.

Solutions, buffers, stocks and cocktails

3

3.1 Solutions, suspensions, and slurries

Why is it that some mixtures are called solutions and others are not? When we dissolve a salt completely in water, for example, it makes a mixture that is homogeneous and we say it is a solution. It is homogeneous because if we take samples of the solution from different parts of the flask, it will be the same everywhere. Suppose that we lower the temperature of the solution and that some of the salt precipitates and forms crystals that drop to the bottom of the flask. Is it still a solution? We might call the liquid supernatant a saturated solution, because taken alone it would be homogeneous. However, the liquid and solid parts together could not be called a solution, even if the flask were swirled so that the solids were suspended.

The structure of water

Liquid water has an interesting structure; there exists both disorder of the water molecules and intermittent hydrogen bonding between them. The disorder is favored as a matter of entropy, and this explains why many nonpolar 'oily' surfaces are not soluble. A protein surface has amino acid side chains that are a mixture of polar and nonpolar surfaces, and whether the protein is soluble or insoluble depends on energetics. If the nonpolar surfaces of a protein are too extensive, admitting the protein to solution may entail a decrease in the entropy of the water. The organized water molecules become something like a 'fence' that surrounds and excludes the insoluble protein from solution. Insoluble proteins aggregate or drop out of solution because they can be pushed together, and less fencing is required if they are corralled into a group and surrounded with a single fence.

The hydrogen bond

Understanding the behavior of water molecules requires some thinking about the relative electronegativities of atoms. Oxygen is much more electronegative than hydrogen, meaning that in a covalent oxygen-hydrogen bond the oxygen nucleus will tend not to share the electron distribution evenly with the hydrogen nucleus. The oxygen will take more than half of the electron cloud, so it bears a partial negative charge. The hydrogen, being left with less than half of the electron cloud, bears a partial positive charge. This slight charge separation is important, as will be discussed later, and the covalent bond is said to be polar. We can contrast this to a nonpolar covalent bond between carbon and hydrogen, where the electron density is more evenly distributed because the two atoms have similar electronegativities.

We may imagine electron clouds around atoms as being like a chocolate frosting on a dessert. Suppose we want to make up a batch of cupcakes with a single candied cherry on the top of each. We pour a batch of melted chocolate over the cherry and cupcake, and the frosting bonds them together. We'll imagine that the cake represents either an oxygen or a carbon nucleus, the candied cherry represents a hydrogen nucleus, and the chocolate frosting holding them together represents the shared electron cloud. In the case of the carbon nuclei cupcakes, the chocolate makes a thick covering that evenly coats the cherry and the cupcake. The cherry is firmly held in place, and this is analogous to a nonpolar covalent bond between a carbon and hydrogen. However, in the case of the oxygen nuclei cupcakes, the chocolate frosting drains away so that the cherry is only thinly covered, and the cupcake receives the lion's share of the frosting. The chocolate frosting represents a distribution of negative charge, and a 'thinly frosted' hydrogen nucleus would carry a slight net positive charge, while the 'thickly frosted' oxygen nucleus would carry a slight net negative charge. This uneven distribution is what is called a polar covalent bond.

Just as we may imagine that a candied cherry might be so thinly covered with frosting that it cannot easily stick to its own cupcake, a covalent bond may become so polarized that the hydrogen nucleus also departs with a full positive charge. This is acid dissociation, and we can appreciate why acid dissociation of a carbon-hydrogen bond is not usually significant; the nonpolar covalent bond keeps the hydrogen nucleus firmly in place. The slight positive charge on the hydrogen in a polar covalent bond is attractive to the slight negative charges that may be found on the polar surfaces of other molecules, particularly the oxygen and nitrogen atoms that are electronegative. A hydrogen bond involves three nuclei in roughly a straight line, the two on the outsides being electronegative ones such as oxygen or nitrogen, and the one in the middle being a hydrogen nucleus. The hydrogen nucleus belongs to one of the electronegative atoms, being engaged as it were in a polar covalent bond, but it is nonetheless electrostatically attracted to the other electronegative atom in a hydrogen bond by virtue of their opposite partial charges.

Slurries and suspensions, emulsions and colloids

We've discussed the characteristics of a solution, the most important being that it is homogeneous. We can also have solids that are suspended in a liquid but are not in solution. This mixture might be called a suspension, or slurry if it were very thick. Often these suspensions or slurry mixtures can be separated into liquid and solid components by centrifugation.

The use of the language is a bit nuanced here. When is a mixture a solution and when is it a suspension? We might mix some lymphocytes into a buffer and this would properly be called a suspension of cells and not a solution of cells. The reason is that, if left alone, the cells would gradually settle to the bottom of the tube and samples taken from different parts of the tube would differ. On the other hand if we suspended something much smaller, such as ribosomes, they would not settle to the bottom of the tube at $1 \times$ gravity, and the mixture would be homogeneous. We could take a sample from one part of the tube and be confident that it would be the same as any other sample taken from a different part. Nonetheless, we might be reluctant to call this mixture a solution of ribosomes because the mixture could be separated into solid and liquid parts by ultracentrifugation. A protein or rRNA extracted from the ribosome would be a small enough particle that it could not be readily sedimented by ultracentrifugation, so we usually refer to macromolecules or particles at this scale as being 'in solution' rather than 'in suspension'.

Some very small particles dispersed in a mixture might properly be called colloids, which is a word that helps to bridge the gap between solution and suspension. Colloid particles are in the range of 1 to 1000 nm in diameter, and they are too small to settle out of solution. A colloid of two liquids is called an emulsion, and an example is a shaken mixture of phenol and water. This milky liquid may settle out into organic and aqueous phases gradually, or rapidly through centrifugation, but immediately after mixing it is reasonably homogeneous. Protein can act as an emulsifying agent for phenol and water, meaning that it helps to stabilize the emulsified state by inhibiting aggregation of the different parts of the colloid.

3.2 Solubility of macromolecules

Protein solubility

Why does water admit large molecules such as DNA or proteins into solution? Many of the oxygen and nitrogen atoms on these molecules can form hydrogen bonds with the water, either as donors or acceptors. The dipole moment of the water molecule also allows water to orient to charges on the surfaces of macromolecules. In a sense, the surfaces of a soluble macromolecule are wettable and create a low surface tension with water, and this allows the macromolecule to be admitted into solution. Some proteins are not readily soluble in water, and this may be a function of the number of nonpolar amino acid side chains exposed on the surface.

Salting in, salting out, dielectrics, pH and temperature effects

Charged amino acid side chains on a protein can lead to either attraction or repulsion between two molecules, and this causes several interesting effects. Under low salt conditions, the charges on a protein are not well shielded by the solution, and some protein molecules may attract each other, aggregate, and drop out of solution. In this case, increasing the ionic strength of the solution may increase the solubility of a protein, a method called salting in. On the other hand, adding extra salt to a solution can change the properties of a solvent to the extent that a protein is no longer soluble, a method called salting out. Different anions and cations vary in their ability to assist with salting in or salting out, but ammonium sulfate is a classic example of a salt that is used to decrease the solubility of proteins in solution.

Solution properties may also be changed with strong hydrogen bonding solutes, such as polyethylene glycol, or organic solutes such as acetone that reduce the dielectric constant of the solvent. With these substances present in high concentration, the properties of the water are changed and the water no longer admits the macromolecule or particle into solution. The polar and nonpolar surfaces on a protein determine its solubility, but changes in the solvent amplify the effects of these surfaces.

Electrical charges on amino acid side chains are pH dependent, and the pH at which a protein is least soluble is usually the same pH at which the protein has a net neutral charge. This specific pH is called the isoelectric point, or pI, and is different for each protein species. When the pH of a solution equals the pI of a protein, there are as many positively charged amino acid side chains as negatively charged amino acid side chains, and less repulsion between proteins. A protein may aggregate more easily under conditions in which the solution pH equals the pI of the protein.

Many soluble proteins have nonpolar amino acid side chains that are buried in the interior of the protein, and not exposed to the solvent. These proteins may become insoluble if they are heated to the point of denaturation. A simple culinary example of this is when a chef makes a chicken soup, placing spices, vegetables and chicken into a pot of water and bringing it to the point of boiling. As the temperature rises, some of the protein in the soup becomes unfolded or denatured, and collects as an unappetizing scum at the surface of the pot. This might seem a bit counterintuitive; wouldn't we expect that increasing the temperature of the water would increase the solubility of solutes? The unfolding of the protein exposes nonpolar amino acid side chains to the water that were formerly concealed in the interior of the protein, so it isn't really the same solute. Molecular biologists face the same problem as the aforementioned chef, because most enzymes used in the laboratory become denatured at high temperatures. For example, a DNA polymerase from a mesophilic organism such as *E. coli* could not survive the high-temperature (94°C–100°C) phase of a polymerase chain reaction without denaturation. While there might not be enough enzyme present to make a visible scum on the surface of the tube, the high temperature can still unfold the protein and make the enzyme nonfunctional.

Nucleic acid solubility

DNA and RNA are soluble in most aqueous solutions, and this is due in large part to their negatively charged phosphodiester backbone and their capabilities for hydrogen bonding with the solvent. However, nucleic acids are not entirely composed of polar surfaces. The ring structures in the nucleotide bases have nonpolar surfaces that tend to detract from the solubility of DNA or RNA, just as nonpolar amino acid side chains make proteins less soluble.

At this point we are only considering how a double-stranded DNA goes into solution as a double-stranded molecule. The discussion has not yet turned to how a double-stranded DNA may be divided into two single strands, which is the denaturation of DNA, but many of the physical principles related to electrostatic repulsion are the same.

Cationic shielding

Nucleic acids are polymers that bear negative charges uniformly along their lengths, and in solution these negative charges must be surrounded by equal numbers of cationic charges, such as Li^+, Na^+, K^+, and Mg^{+2}. The effect of surrounding the negative charges with a cloud of positive charges reduces the strength of electrostatic repulsion between molecules, and is called shielding. As the ionic strength of the solution increases, the shielding of the negative charges along the backbone is improved and the nucleic acids are more likely to aggregate and drop out of solution. For example, DNA hybridization is often performed under high ionic salt conditions because the shielding allows probe and target molecules to approach each other without repulsion. A related effect of this shielding is seen in the conformation of a single nucleic acid polymer; increases in ionic strength cause the polymer to adopt a more compact conformation in solution, because there is less intra-polymer repulsion.

If the cations form a shield around the phosphate backbone, how far can they stray from the nucleic acid? That depends in part upon the characteristics of the solvent, and in particular its dielectric constant. By mixing a high dielectric solvent such as water with a lower dielectric solvent such as ethanol or isopropanol, the ionic environment around the nucleic acid can be changed in a way that decreases the solubility of the nucleic acid. This effect has a practical application, as it relates to the methods of alcohol precipitation of DNA and RNA (see Section 3.5).

DNA and RNA are called nucleic acids for a reason! At a pH of about 5 to 6, the phosphate backbone of a nucleic acid may become acidified, or paired with protons, and this neutralizes the negative charges on the backbone. This neutralization of the charge decreases the solubility of the nucleic acid in aqueous solutions, and as will be discussed below it also has the effect of increasing its partitioning into the organic solvent phenol. Bringing a nucleic acid into solution is therefore a balance of energetic effects: the hydrogen bonding between the nucleic acid and the solvent, the electrostatic repulsion along the phosphodiester backbone, the ionic shielding of those charges, the dielectric characteristics of the solvent, and the change in entropy of the solvent molecules.

Solubility in organic solvents

Phenol extraction

A phenol extraction is a common laboratory technique for removing proteins from an aqueous solution. Proteins in an aqueous solution are folded so that their nonpolar surfaces are hidden, but in phenol they can unfold and still be soluble. Forming an emulsion of the phenol and water by vigorous mixing will tend to partition the nucleic acids into the aqueous phase and the proteins into the organic phase. Chloroform and isoamyl alcohol are often added to the phenol to help break the emulsion so that the aqueous and organic phases can be cleanly separated. The addition of these de-emulsifiers does change the partitioning of some substances between the aqueous and organic phases; for example, agarose in a water solution will partition into phenol, but not into a 1:1 mixture of phenol and chloroform.

The phenol and water mixture is interesting, because while these substances separate into two phases when they are not emulsified, phenol is somewhat soluble in water as well. A gram of phenol will dissolve in about 15 ml of water, which explains why phenol extractions are usually followed immediately by a chloroform extraction to lure the dissolved phenol out of the aqueous phase. Pure phenol is a crystalline solid at room temperature, and the bottles of liquid phenol that are usually kept in laboratory refrigerators are actually water-saturated solutions of phenol. The denser, bottom phase is phenol containing dissolved water, while the less dense, upper phase is water containing dissolved phenol. When taking a sample of phenol from such a two-phase system, the researcher sticks a clean glass pipette through the top phase and withdraws liquid from the bottom phase. It is common in phenol extractions to use phenol saturated with a buffer, such as Tris at pH 7.5, so that nucleic acids remain negatively charged and partition into the aqueous phase during an extraction.

Miscible alcohols

Some organic solvents, such as ethanol and isopropanol, are perfectly miscible with water. That is, they can be mixed in any proportions and they form a continuous liquid phase. The hydroxyl groups in these substances become hydrogen-bonded with the water molecules in solution, and the nonpolar surfaces of the alcohol molecules are not so extensive as to prevent them from being admitted into solution. This allows the use of alcohol-water mixtures to solve certain types of problems of reagent insolubility. For example, the antibiotic tetracycline cannot be dissolved in water to a high concentration but can be dissolved in an alcohol such as ethanol. Once dissolved in ethanol, tetracycline can be added in small amounts to a purely aqueous solution and will remain dissolved.

3.3 Solute concentration

The idea of the concentration of a solute is a simple one: we have a certain amount of a solute that is dissolved into a homogeneous solution, and together they occupy some volume. The amount of solute in a unit volume of this solution is its concentration. If there are crystals of undissolved solute at the bottom of a bottle, or even in suspension, they are not part of the solution and are not counted towards the total amount of solute. The presence of undissolved crystals under a liquid phase may be an indication that the solution is saturated. For example, an ammonium sulfate solution that has undissolved crystals at the bottom of the bottle, at equilibrium, is saturated.

Why is solute concentration so important in laboratory work? One reason is that the concentration of a substance in water can change the behavior of the solvent. If we want to precipitate a nucleic acid, for example, we have to add enough alcohol to bring the concentration to a level that the nucleic acids will clump together, but not so much that the salt precipitates along with it. The concentration of salts also changes the way other solutes behave, for example through electrostatic shielding.

The meaning of concentration

One way we can think about the concentration of a solute is that it is related to how frequently the solute molecule or atom will bump into other things in the mixture. Imagine a reaction in which two molecules, A and B, must collide in order to make products C and D.

$$A + B \rightarrow C + D$$

If A and B are in solution at very low concentrations, they will collide infrequently. That is why the rates of bimolecular reactions depend on concentration. Getting an experiment to work often depends on setting things up in a way that the reactants and enzymes have some chance of finding each other.

Perhaps we can develop a better concept of concentration by using our imagination; what would 100 mM NaCl look it like in solution, if we could actually see the ions? Suppose we had a tiny graduated cylinder with a radius of 3 nm, and a height of 7 nm. On the scale of a DNA molecule, that cylinder would be approximately three times the radius of a DNA double helix and would have a length spanning two turns of the double helix, or 20 bp. We will ignore the howls of outrage from physical chemists and imagine that this cylinder represents the scale of a neighborhood of a segment of DNA. The cylinder would have a volume of about 2×10^{-22} l, and would contain, based on a concentration of 100 mM, about 2×10^{-23} mol of NaCl. That's about a dozen sodium ions and a dozen chloride ions in the 'neighborhood', which if they were bees at a picnic might be considered a fair swarm!

In the laboratory setting, 100 mM NaCl is enough to provide partial shielding of the electrostatic charges on the phosphate backbone, and on the enzyme. On the scale of what is possible with NaCl, it is not very much, and some methods use substantially higher concentrations of NaCl. During the hybridization of single-stranded DNA molecules in Southern blotting, for example, the DNA is typically incubated in salt concentrations that are about 16 times higher. Once again the salt is serving a purpose of shielding and reducing electrostatic repulsion, although in this case it is to encourage base-pairing.

The components of laboratory recipes

When we first get involved in molecular laboratory work, the recipes for combining solutes into a solution may seem mysterious and arbitrary. For example, a recipe for a buffer used in restriction endonuclease reactions might read as follows:
• 100 mM NaCl

- 10 mM Tris-Cl, pH 7.5
- 6 mM $MgCl_2$

How are the concentrations for each solute established, and why are they so different? We've already come to understand how NaCl serves a role of shielding, but let's think about the next ingredient, the Tris buffer. Since the recipe says the molar concentration of the Tris is one tenth that of NaCl, we know there will be 10 times fewer particles of Tris per unit volume than of Na or Cl ions. That's one of the beautiful things about recipes that are given in molar amounts; if the recipe had stated the same concentration requirements in mass, calling for 5.84 g l^{-1} NaCl and 1.21 g l^{-1} Tris, we would have to consult the formula weights of each substance to figure out that this recipe represents the same 10:1 numerical relationship in particle concentrations.

The 10 mM Tris in the recipe provides a small amount of buffering capacity, and the selected pH of 7.5 has an effect on the activity of the enzyme. The DNA samples used in the restriction endonuclease reaction may vary in pH, and having buffering capacity in the reaction salts corrects and standardizes the pH to a level that will be optimal for the specific enzyme being used. Buffering is a matter that will be discussed in more detail later (see Section 3.7), but to give an extreme example, if 1 millimole of a strong acid were added to a liter of this buffer the pH would fall only half a pH unit, to about 7.0. That's pretty good, considering that without any buffer in solution the 1 mM strong acid would cause the pH to drop to 3!

Isn't it odd that the $MgCl_2$ concentration in the recipe is given as 6 mM? Why 6 mM and not 5 mM? The $MgCl_2$ concentration seems to have been stated with a bit more exactness than the NaCl and Tris concentrations, and indeed, the Mg ion concentration is very important in restriction enzyme activity. The divalent cation is involved both in the enzyme catalysis and as a counterion for the DNA, and slight changes in concentration can have significant effects on the reaction rate. In the polymerase chain reaction discussed in Chapter 7, the Mg ion concentration is particularly critical because slight changes in efficiency are compounded over the many cycles that make up the reaction.

When concentration does not equal activity

To this point we've been discussing ideal solutions in which the solute particles take up no volume. A more representative way of thinking about solutions is to talk about the activities of the solutes, rather than their concentrations. What's the difference? Let's remember that what we're really interested in is how often the particles collide with each other. As a concentration of a solute increases, the solute may begin to take up a significant volume of the solution and that requires a more sophisticated model. The difference between concentration and activity of a solute is usually not important, but there are a few interesting exceptions.

Polyethylene glycol

It is common to find protocols that call for the addition of polyethylene glycol (PEG) to a DNA ligation reaction. Ligation is a reaction in which two DNA ends are brought together and made covalently connected through their phosphodiester backbones (see Chapter 4). PEG is a polymer with a structure $H\text{-}(CH_2\text{-}CH_2\text{-}O)_n\text{-}H$, and PEG can be obtained with a variety of molecular weights and corresponding values of n. For the type of PEG added to a ligation reaction the molecular weight of the PEG is approximately 4000 to 8000. What this long polymer does in solution is to participate in hydrogen bonding with the water, and to tie up the solution volume. Other solutes have less volume in which they can effectively move, and so their activity is higher than their concentration. In a DNA ligation reaction, the DNA ends can collide with each other more frequently than their concentration might suggest, because their activities have been increased by the PEG. The amount of DNA in the solution hasn't changed, but its behavior has been changed by the addition of another polymer that limits its freedom.

3.4 Solution pH

The pH of a solution is a measure of the concentration of H^+ ions, or more accurately the activity of H_3O^+ ions. For the purposes of discussion we will use the first, informal explanation of pH. If the concentration of H^+ ions in solution is 10^{-7} M, then the $\log_{10}$ of the concentration is the exponent, or -7. To find the place of this solution on the pH scale we just switch the sign by multiplying by -1, so the solution has a pH of 7. If the concentration of H^+ ions in solution were 10^{-8} M, or 10 times lower than 10^{-7} M, then the pH of the solution would be one more than 7, or 8. We write a formula for this as $pH = -\log_{10}[H^+]$. Why do we switch the sign when we're calculating a pH? There's no reason for it, except for the convenience of not having to place a '$-$' sign in front of every value. Switching the sign does cause us to have to remember that as pH values go up, the $[H^+]$ is going down, and *vice versa*.

The meaning of pH

The pH value of a solution tells us something about the solution environment and how molecules in it will behave. Some molecules have a place where a H^+ could bind, and we'll call them bases. Others have a H^+ that could be lost, and we'll call them acids. HCl is a strong acid, because the chloride ion loses its H^+ readily in solution. Some acids tend to lose their H^+ even at a low pH, while others will hang on to their H^+ ions until the pH is quite high. We've already discussed several important examples of this effect in macromolecules. The phosphates in the backbone of nucleic acids will tend to lose their H^+ ions above pH 5 to 6, and so nucleic acids are polyanions at neutral pH. Proteins also vary in their net charge, depending on the pH of the solution, and enzymes have specific pH optima because the binding of H^+ ions changes their chemical behavior. The side chain of aspartic acid, for example, tends to lose its H^+ ion when the pH rises above approximately 4. The side chain of arginine hangs on to its H^+ ion to a higher pH, not losing its H^+ ion until the pH rises above approximately 12.5. Using an analogy might help us understand why these chemical groups behave so differently.

Let us imagine a city in which people often wear hats, and that the hats represent H^+ ions. People in this city who are wearing hats are like acids, and those with uncovered heads are like bases. Now let's imagine a windy city where the air is full of hats. As people walk around this city, a gust of wind might suddenly strip a hat off of their heads, uncovering them briefly, or blow a hat so that it lands onto their heads. Because the hats only come in one size, some peoples' heads fit the hats better than others. That is, some people may be wearing hats most of the time, because the hats fit perfectly, and others will very rarely find themselves wearing a hat. It is also true that the fraction of time a person in the city will be wearing a hat depends on how many hats are blowing through the air; once a hat is stripped away by a wind gust another hat may not land on the person's head very soon if the hats are few in number. We could imagine a world full of such cities, some with a lot of free hats blowing in the wind, which we will call low pHat or acidic cities, and others with very few hats blowing in the wind, which we will call high pHat or basic cities. There would be some people who never have a hat stick to their heads, even in the cities with the most hats blowing in the wind (the lowest pHat cities), and we would call these individuals 'strong acids'.

This provides us with an analogy for the pH of a solution: The pHat of a city represents an environmental condition of hats, just as the pH of a solution represents an environmental condition of H^+. A significant immigration of 'strong acid' people into a city might cause the pHat of a city to drop, and that might cause other people to discover that they're wearing hats more frequently. Similarly, an immigration of hatless but well-shaped heads into a city might cause the pHat to rise, as more of the windborne hats are absorbed.

To make sure that the analogy is not lost on the reader, we could think of the arginine as being a well-shaped head, in comparison to aspartic acid. The 'arginine head' would wear hats in many cities in which the 'aspartic acid head' would find itself usually hatless.

In laboratory experiments, it is often critical to set the pH of a solution appropriately so that the components of a reaction have the expected biochemical properties. In the earlier example of a recipe for a restriction endonuclease buffer, the asked-for pH of 7.5 might be optimal for the restriction enzyme. At a pH that is higher or lower than 7.5, the number of acidic groups in the protein carrying H^+ ions might be respectively too low or too high for optimal function. Buffers help to stabilize the pH of a solution, and so help to bring reproducibility to experiments involving macromolecules. This is a matter that is discussed later (see Section 3.7).

3.5 Precipitation methods

Nucleic acid precipitation

A common technique in the molecular biology lab is to precipitate a macromolecular solute from a solution, in order that it might be concentrated and made more pure. For example, we may take 100 µl of a solution containing DNA or RNA, add a sufficient amount of NaCl stock solution to bring the concentration of NaCl to 200 mM, then mix in 250 µl of ethanol. These two additions will cause the DNA or RNA to clump together and be cast out of the liquid phase. Upon centrifugation, the DNA or RNA may be isolated at the bottom of a centrifuge tube and separated from the supernatant ethanolic liquid. The DNA or RNA precipitate might then be re-suspended and brought into a solution in a different volume and solute composition.

What made the nucleic acid clump together and drop out of solution? DNA and RNA are soluble in a water solution, but if the characteristics of the solution are changed by the addition of new solutes then the nucleic acid may no longer be welcomed in solution. In this case the two additions were a salt, NaCl, and an alcohol, ethanol. The salt shields the nucleic acid and reduces the electrostatic repulsion along the phosphate backbone, and the alcohol causes the solution to be less easily polarized (have a lower dielectric constant), which reduces further the separation between the charges. With less repulsion between molecules, the DNA clumps together and leaves the solution.

The choices of salts and alcohols that can be used to shield the phosphate backbone are varied, and they are used at different concentrations. NaCl at 100–200 mM has already been mentioned, but NaOAc (sodium acetate, pH 5.2) at 300–500 mM, NH_4OAc (ammonium acetate) at 2.0–2.5 M, and LiCl 750–800 mM are also common. Ethanol and isopropanol are used as alcohols, although at different concentrations. In the case of ethanol, a researcher would look at the volume of the aqueous solution containing nucleic acids and salts and add two and one-half times that volume in ethanol. So, starting with 0.5 ml of sample, for example, 1.25 ml of ethanol (that is, 2.5×0.5 ml) would be added. The total volume of the sample plus ethanol would be a bit less than the sum of $0.5 + 1.25 = 1.75$ ml, because the ethanol and water become interleaved in solution. Isopropanol is more effective than ethanol in casting nucleic acids out of solution, and typically only 0.5 to 0.7 volumes are added, depending on the details of the protocol. For a 0.5-ml sample, that would mean that only 0.25 to 0.35 ml of isopropanol would be added.

The wide variety of methods of alcohol precipitation might cause disconcertion, particularly on the occasions when we don't achieve precipitation of a visible pellet of nucleic acid after centrifugation. The worried researchers may wonder: do we not see a pellet because the DNA was lost at an earlier step, or did the precipitation method fail in some way?

A 'failsafe' method

Accordingly, here is a failsafe method that will precipitate DNA or RNA. For each 100-µl sample of nucleic acid, add 4 µl of 5 M NaCl, 1 µl of 20 mg ml^{-1} glycogen, and 265 µl of ethanol. That is, the solution is made to be approximately 200 mM NaCl, and 20 µg of glycogen carrier is added, then two and one-half volumes of ethanol are added. The solution is mixed to homogeneity and then subjected to centrifugation ($20\,000 \times g$)

for 10 min. If the centrifugation is conducted in a fixed angle rotor, the researcher must be attentive to the orientation of the tube in the rotor so that the likely site of deposition of a pellet is known. After centrifugation, the ethanolic solution should be carefully removed using a pipette, and disturbance of the likely site of a pellet should be avoided. The ethanolic liquid should be saved in a clean tube, rather than immediately discarded, in case the pellet was accidentally dislodged and removed from the tube along with the liquid, or in case additional efforts to recover a pellet are required. Careful inspection of the tube may reveal signs of a pellet of nucleic acid. Centrifugation in a fixed angle rotor may produce a pellet that is spread out over one side of the tube and difficult to see. However, holding the tube up to the light and turning it slowly, perhaps viewing the tube bottom through a magnifying glass, may reveal an irregular feature on the wall of the tube that is the nucleic acid pellet. The glycogen is added to help produce a visible pellet, particularly when samples of less than 1 µg of nucleic acid are being precipitated. Highly purified glycogen is commercially available as a carrier for precipitation and does not interfere with subsequent use of the nucleic acid. Linear acrylamide is also available for this purpose. The efficiency of precipitation of small DNA oligonucleotides can be improved by the addition of $MgCl_2$ to a concentration of 10 mM, prior to the addition of ethanol. However, if a sample contains more than 4 to 5 mM of EDTA, the EDTA may come out of solution in the ethanol precipitation as a dense oily liquid that prevents the pelleted DNA from fixing to the bottom of the tube. In this case, the pellet may be accidentally discarded along with the supernatant. The only solution is to dilute or dialyze the sample to the point where the residual EDTA is soluble in the ethanol solution.

If a nucleic acid pellet is recovered by ethanol precipitation, it may be necessary to wash the pellet with 100 µl of 70% (v/v) ethanol, to remove traces of NaCl. The decision on whether to wash a pellet is based on the planned use of the sample. For example, if the nucleic acid is to be resuspended and loaded on an agarose gel for electrophoresis, then a residual amount of sodium and chloride ions would not be expected to interfere. On the other hand, if the next planned step involves use of an enzyme that is inhibited by these ions, then washing the pellet to remove this residue may be important. Washing a nucleic acid pellet involves addition, and then removal, of the 70% ethanol. The pellet is not dissolved, but salt residue diffuses into the 70% ethanol and is removed. A short period of centrifugation may be required if the pellet is dislodged during addition of the 70% ethanol, or if the sample is purposefully mixed. Some researchers use 70% ethanol that is ice-cold; however, this may not be necessary for every application.

The choice of a salt and an alcohol

Ethanol and sodium chloride work well to precipitate nucleic acids. However, they are not always the method of choice. Isopropanol is often used in place of ethanol and is used very successfully in precipitating DNA out of solutions that have a very high ionic strength, without causing salt to be co-precipitated. For example, a crude plasmid DNA prepared by alkaline lysis of bacteria may have a very high ionic strength, as a consequence of adding large quantities of sodium hydroxide and potassium acetate. Addition of 0.7 to 0.8 volumes of isopropanol, which is to say addition of 70 to 80 μl of isopropanol for every 100 μl of lysate, is commonly recommended. No additional salts are needed to shield the nucleic acid and cause precipitation, in that case. Isopropanol is more effective in precipitating nucleic acids than ethanol, in the sense that only one-quarter to one-third as much volume needs to be added. Furthermore, if the concentration of DNA is sufficiently high then isopropanol can induce precipitation even from a low ionic strength solution.

Ethanol and sodium acetate are often used together to precipitate nucleic acids, and there are several advantages to using this combination. For one thing, when a nucleic acid is going to be used in conjunction with an enzyme that is sensitive to residual sodium chloride, then precipitation using sodium acetate may remove that potential source of inhibition. Sodium acetate is also highly soluble in a 70% ethanol solution, and so its use may result in less salt carryover. The typical method involves the addition of sodium acetate at pH 5.2 to a concentration of 0.3 M, perhaps by addition of 0.1 volumes of a 3 M sodium acetate, pH 5.2 stock, followed by addition of 2.5 volumes of ethanol. It is possible that the sodium acetate, in addition to its ionic shielding of the negative charges on the phosphate backbone, also eliminates some of the negative charges on the phosphates by acidification. Where protons are added to the phosphate backbone, as a consequence of the pH 5.2 condition, the negative charges will be neutralized.

Ammonium acetate may be used instead of sodium chloride or sodium acetate, when a researcher wishes to prevent the precipitation of very small nucleic acids and nucleotide substrates. For example, a researcher conducting a radiolabeling reaction with [^{32}P]-dCTP might precipitate DNA that is radiolabeled by adding ammonium acetate to a concentration of 2 M, perhaps by addition of 0.5 volumes of a 4 M solution, then adding 2.5 volumes of ethanol. The products of the labeling reaction will be efficiently precipitated, but only a small percentage of the unincorporated [^{32}P]-dCTP will be co-precipitated. The pellet may be washed with 70% ethanol to remove residual traces of ammonium acetate.

When a nucleic acid is precipitated using a sodium or ammonium salt, those cations become the counterions for the phosphate backbone of the nucleic acid. The excess salts may be washed away with 70% ethanol, but the counterions are not separated from the nucleic acid and are carried into solution when the nucleic acid is re-suspended. Many protocols call for precipitation of nucleic acids using lithium salts, for example adding LiCl to a concentration of 0.8 M and then adding 2.5 volumes of ethanol, and in that case lithium becomes the counterion for the precipitated nucleic acid. Some consideration needs to be given to the planned use of the nucleic acid, and to whether any of these options for counterions may inhibit a forthcoming enzymatic reaction. Higher concentrations of lithium salts, for example 2 M LiCl, are sometimes used to precipitate nucleic acids directly, without addition of an alcohol. This method may be particularly useful for precipitation of RNA.

Protein precipitation

While we have focused on alcohol precipitation of nucleic acids, there are analogous methods for precipitation of proteins that make use of organic solvents. Soluble proteins can be made insoluble by the addition of 1 volume of acetone or 4 volumes of ethanol, but proteins are also susceptible to denaturation in these organic solvents. The problem is that once a protein is denatured, it may not easily renature when transferred to an aqueous solvent. Temperature is a factor in denaturation, and so a protein sample would typically be precipitated on ice, with a pre-chilled solvent.

The effect of pH

The pH and ionic shielding of the sample affect precipitation because the charge distribution on the surface of a protein affects its ability to aggregate. Proteins are generally least soluble at a pH at which they have no net charge, a pH that is termed the isoelectric point or pI of a protein (see Section 3.2). A protein can be precipitated using a smaller amount of organic solvent, and with less accompanying denaturation, if the pH is adjusted to this optimal level. It is important to note that a heterogeneous sample of proteins will have a wide variety of pI values, so there is no single pH that will be optimal for all proteins. Proteins differ in pI because they have different compositions of charged amino acids. Ribosomal proteins, for example, are often very basic and have a high pI, whereas nonhistone proteins are often acidic and have a low pI. If the pI of a specific protein is known, the precipitation method can be adjusted to favor the precipitation of that protein, and, perhaps, disfavor the precipitation of other unwanted proteins in a sample.

The effect of salts

Proteins have different degrees of solubility in salt solutions, as was discussed earlier, and ammonium sulfate is particularly effective in shielding the electrostatic charges on protein surfaces and encouraging aggregation. Proteins have different solubilities in ammonium sulfate, and so sequential precipitation with increasing concentrations of the salt can be used as a simple fractionation of a sample. A saturated ammonium sulfate solution can be added slowly to a protein sample, with thorough mixing, until a desired percentage of saturation is achieved. In a heterogeneous protein sample this may lead to some proteins remaining soluble and others becoming insoluble. A researcher can then separate the soluble and insoluble proteins by centrifugation, and continue to work with either fraction. The insoluble proteins can be brought back into solution by the addition of a well-buffered aqueous solvent, then dialyzed to remove traces of the ammonium sulfate salt. The soluble fraction can be treated with additional ammonium sulfate to generate another 'cut' from the mixture, or can be dialyzed to remove the ammonium sulfate. It is only the ammonium sulfate salt that is causing the precipitation of the proteins; no organic solvents are involved.

Driving large particles out of solution

Large particles in suspension, such as bacteriophage, can sometimes be made to aggregate and precipitate by the addition of PEG and sodium chloride. The effects of PEG on increasing the activity of DNA ends have already been discussed, and precipitation using PEG is based on a similar effect. A typical precipitation method may call for the addition of PEG-8000 to a concentration of 3–10% (w/v) and NaCl to a concentration of 0.4–1.0 M. The sodium chloride shields electrostatic charges on the coat proteins of the bacteriophage, allowing a closer approach of bacteriophage to each other and encouraging aggregation. The exact amounts of PEG-8000 and sodium chloride needed to cause aggregation and precipitation will vary, depending on the bacteriophage species.

3.6 Preparing a solution

When we prepare solutions for use in the laboratory, we want to be certain that they are made in a way that their composition is known and that they can be remade in the future in exactly the same way. The underlying principle is that we want to be certain of the concentration of each solute, meaning the amount of solute per unit of volume. The amount of solute could be expressed in terms of moles, or mass, or as a percentage of weight or volume. The unit volume would typically be liters, milliliters, or microliters. For example, we might prepare a solution of mouse genomic DNA that is $1 \, \text{mg ml}^{-1}$, a solution of NaCl that is 5 M, or a solution of sodium dodecyl sulfate (SDS) that is 10% (w/v). In all three examples we could say exactly how much solute would be present in 10 ml of the solution; that is, 10 mg DNA, 0.05 mol NaCl, and 1 g SDS, respectively.

Making solutions with known concentrations

Solutes take up space in a solution, and this makes solution preparation somewhat awkward. We cannot start with a liter of water, dissolve 292.2 grams or 5 moles of NaCl into that liter, and have a final volume that is still 1 liter. The volume expands as the solute is dissolved, and so a solution prepared in this way would have 5 moles of NaCl in a volume that was greater than 1 liter, and consequently a concentration that was less than 5 M. To make a NaCl solution that is exactly 5 M, we must therefore dissolve 5 moles of NaCl into a volume of water that is less than 1 liter, then use water to increase the volume of the solution to exactly 1 liter when the salt is completely dissolved. A well-mixed, homogeneous solution of 5 mol NaCl, in a total volume of 1 l, has a concentration of 5 M.

Accuracy and precision

When using a balance or other measuring tool in the laboratory, we must be attentive to the accuracy and the precision of measurement. Accuracy means that, in a number of trials, the instrument will tend to yield an average measurement that is close to the true value. Precision means that there is little variation between independent measurements. Of course, an instrument in need of calibration could be highly precise but not very accurate. For example, suppose a researcher sets a pipetting device to deliver 1.00 ml and finds in five sequential trials that the device actually delivers the following volumes: 0.805 ml, 0.803 ml, 0.805 ml, 0.804 ml, and 0.804 ml. The pipetting device has an error in accuracy of nearly 20%, but is highly precise because there is little variation. On the other hand, suppose a researcher sets up a very good analytical balance on a table that is subject to vibration and in five sequential trials a 50.0-mg mass yields these measurements: 44.3 mg, 55.0 mg, 53.8 mg, 46.2 mg, and 51.1 mg. Taken together, the average of these measurements is about 50.0 mg, meaning that the balance is accurate. However, the variation between measurements is large which means that it is not very precise.

Weighing out the solute

There are several important points to attend to during solution making, starting with the technique for weighing out a solute on a balance or spring scale. The balance must be in proper repair, on a level surface, and in a part of the laboratory that is reasonably free of strong vibrations and air currents. The selection of a balance must also be appropriate for the application; for example, 10 mg of a solute cannot be accurately weighed on a digital scale that has a readout in 0.01 g units. A sheet of weighing paper or a plastic weighing boat is placed on the pan of the balance, and the balance is either zeroed with the article in place or the tare weight determined for later subtraction.

Some researchers hold a bottle of granulated reagent horizontally and close to the surface of the weighing boat, and either slowly rotate the bottle or tap it to transfer reagent to the boat. Provided that it does not lead to a sudden avalanche of crystals from the bottle, or release of dust, this method has an advantage in that no weighing spatula needs to be used. Plastic or stainless steel weighing spatulas are usually necessary when a small amount of material needs to be transferred for weighing, but they must be clean and dry before use, and they carry some risk of contaminating a reagent bottle. In either case, if the transfer is not controlled and some reagent misses the weighing boat, spilling onto the balance pan, the scale will need to be cleaned and the weighing process re-commenced.

Solid reagents differ in their forms, with some being free-flowing crystals and others being powders or pellets. Some reagents are hygroscopic, meaning that they absorb water from the air, and these must be stored tightly capped or under desiccation, so that they do not gain water weight. A reagent bottle may have a mixture of free-flowing granules and 'rocks' that are pressed conglomerates of granules. The bottle can often be made more uniform by shaking it vigorously, which has the effect that the free granules may mill the rocks into smaller pieces that are easier to dispense. However, with some reagents this may be inadvisable because shaking the bottle may create dust that escapes when the bottle is opened. Some researchers will attack rocks in the bottle by stabbing them with a spatula, but this has inherent risks of launching some of the reagent into the air, or contaminating the rim of the bottle with the researcher's hand. Weighing spatulas are not designed to be used as hand chisels, or screwdrivers for that matter, and a researcher always runs some risk of personal injury when a piece of equipment is misused.

If too much reagent is added to a weighing boat, then a spatula may need to be used to remove some of it and reduce the weight. Whether that reagent is returned to the reagent bottle or disposed of is a matter of laboratory policy. Certainly an inexpensive reagent might be disposed of without concern, since the cost of the reagent is greatly exceeded by the risk of contaminating the reagent bottle by transferring some back. If the reagent is expensive and the weighing boat is otherwise clean, then there might be a reasonable case made for saving the excess. Little argument can be made for saving a reagent that is spilled outside of the confines of a clean weighing boat, unless the reagent is so expensive that it has value even when contaminated in this way. All spills need to be cleaned up, so that they do not damage the equipment or contaminate the next sample.

On some analytical balances, attempting to remove excess weight from a pan balance by shoveling reagent with a spatula may damage the knife-edge of the balance. In these cases it may be necessary to put the balance pan in a safe mode, or turn off the scale, before removing the excess weight. Alternatively, when preparing solutions that are based on milligram amounts of solute, for example a 50 mg ml^{-1} solution of ampicillin, a researcher might decide that any volume between 0.50 and 1.0 ml would be acceptable. In that case, it is not necessary to add and subtract reagent until an exact weight is measured. The researcher can simply tap ampicillin onto the weighing paper until the measure exceeds 25 mg, then calculate the volume of the solution from the weight obtained. If the final weight is determined to be 28.3 mg, for example, then the researcher prepares the solution so that it has a final volume of 28.3/50 ml, or 566 µl. This not only saves time in weighing out small amounts of a chemical, but also saves some wear and tear on the analytical balance.

Transfer of solute to liquid

The next step in preparing a solution is to transfer the reagent from the weighing boat or sheet of weighing paper to a graduated cylinder or flask. A weighing boat can be squeezed at its ends to form a more narrow pouring lip, and a sheet of weighing paper can be similarly folded to direct the pouring of reagent into a vessel. It is important that this pouring be done carefully so that no reagent is spilled outside the vessel. If a powdery reagent adheres to the plastic weighing boat or weighing paper, it may be chased into the vessel with a bit of water. Many researchers prefer to place a powder funnel on top of the vessel to collect the reagent, rather than rely on the folded weighing boat or paper to direct the pouring. With many granulated reagents, the material can be transferred to a dry vessel and water added next. Some reagents, particularly those that are in the form of a fine powder, may be more easily solubilized if they are added slowly to a vessel that already contains water and has a stirring bar to provide continual mixing. In these cases, adding water to a mound of dry powder may cause the formation of large lumps that dissolve very slowly, whereas allowing a dry powder to gradually adsorb water from below may result in more uniform dissolution. In other cases, a solute may be wetted with a solvent until a slurry forms, and a clean glass rod can be used to break lumps into smaller pieces that can be more easily dissolved as the volume is increased.

Bringing a solute into solution

Mild heat can be applied to increase the rate of dissolution of a reagent, but it is usually not necessary to do so. Heat can damage some reagents, and in particular with solutions that contain proteins or nucleic acids the application of heat can lead to denaturation of the materials. Using a stirring bar can be helpful in maintaining agitation, particularly if the granules of reagent are heavy and quickly drop to the bottom of a vessel. As an alternative, a sheet of ParaFilm® can be stretched over the surface of a graduated cylinder and held in place with a firm hand while the cylinder is inverted to mix the contents.

Some materials vary in solubility, depending on the pH of the solution, and in that case attention may need to be given to the pH of the solution while the reagent is being dissolved. When the pH of a solution must be adjusted by the addition of a liquid acid or alkali, that addition has to take place prior to the solution being brought to its final volume. It is important to note that some buffer stocks may not perform well with a pH meter because the activity coefficient is significant in the concentrated solution. In that case the true pH of the buffer on dilution may be different than the measured pH of the stock, and it may be necessary to test a dilution of the stock. It is generally better to prepare buffers based on calculated additions of reagents, as will be discussed later, than to rely on a pH meter to provide accurate results.

There are situations where dissolution of a reagent requires patience, above all else. Large macromolecules may need to rehydrate gradually, particularly if they have been stored in a dried, desiccated form. For example, the dissolving of high-molecular-weight genomic DNA into solution can take days!

Bringing a solution to its final volume

The last step in preparing a solution is bringing the solution to its final volume. If a stirring bar was used in the vessel to maintain agitation, then it must be removed so that it does not displace solution volume during this measurement. There are two important points to be made concerning reading volume measurements in a graduated cylinder. First, the solution volume is read from the bottom of the meniscus, and second, the reading must be taken with the meniscus at eye level to prevent parallax errors. A pipette may be used to control the addition of the last few drops of solvent, so that the total volume does not exceed the planned volume.

Solutions based on molarity

When we prepare a solution that has a concentration based on molarity, we need to have a way of converting from units of mass to units of moles, or from units of volume to units of moles in the case of a liquid solute. The formula weight is a conversion factor used to calculate the number of moles from the mass. For example, the formula weight of NaCl is 58.44, meaning that 58.44 g of NaCl is 1 mol of NaCl. To use this information to make a solution, we need first to decide what volume of solution will be needed, and what will be the solute molarity. The total moles of solute needed to prepare the solution is then simply these two numbers multiplied by each other:

volume (liters) × molarity (moles liter^{-1}) = total solute (moles)

For example, if we decide to make 250 ml of a 1.50 M solution, then the total solute needed is:

0.250 (liter) × 1.50 (moles liter^{-1}) = 0.375 (moles)

Note that in setting up this calculation, it is important to use consistent units for volume. We had decided to make 250 ml of solution, but we used 0.250 l as the factor in multiplication because the molarity was written in units of moles per liter, not moles per milliliter. Once we know the number of moles of solute that are needed to make a solution, we can determine the number of grams that are needed by multiplying by the formula weight:

solute (moles) × formula weight (grams mole^{-1}) = solute (grams)

The number of grams needed to make the NaCl solution example described above is:

0.375 (moles) × 58.44 (grams mole^{-1}) = 21.9 (grams)

So, if we were to weigh out 21.9 g of NaCl, dissolve it in water and bring the total volume to 250 ml, the solution would be 1.50 M NaCl.

The formula weight

It is important to look at the formula weight of a reagent, as it is printed on the reagent bottle, because some reagents are sold in different forms having different formula weights. For example, the formula weight of $MgCl_2$ is 95.21; however, it is common for laboratories to purchase a hexahydrate form of the reagent ($MgCl_2 \cdot 6H_2O$), in which the $MgCl_2$ is co-crystallized with six water molecules. The formula weight of $MgCl_2 \cdot 6H_2O$ is 203.3, with the extra 108.1 g mole^{-1} coming from the water molecules. If we are following a recipe that calls for a certain number of grams of anhydrous $MgCl_2$, and we substitute the same number of grams of $MgCl_2 \cdot 6H_2O$, then we will have a solution with less than half of the intended amount of $MgCl_2$. How could we overcome this type of problem? Recipes that are written with gram amounts can be converted to molar amounts by dividing by the formula weights. For example, in the case of anhydrous $MgCl_2$:

$$9.52 \text{ g MgCl}_2 \div 95.21 \text{ g mol}^{-1} = 0.1 \text{ mol MgCl}_2$$

Once the recipe instruction is written in units of moles, the correct number of grams of a substitute form, for example $MgCl_2 \cdot 6H_2O$, can be calculated by the usual method:

$$0.1 \text{ mol MgCl}_2 \cdot 6H_2O \times 203.3 \text{ g mol}^{-1} = 20.3 \text{ g MgCl}_2 \cdot 6H_2O$$

Does it matter that we are weighing out mostly water when we use the hexahydrate form of $MgCl_2$? No, because 9.52 g of anhydrous $MgCl_2$ and 20.3 g $MgCl_2 \cdot 6H_2O$ both have exactly 0.1 mol of $MgCl_2$ in them. The water in the reagent crystal simply mixes together with the rest of the water that goes into making up the solution. The final volume is the same, and it doesn't matter that some of the water came from the solid reagent.

Molarity and normality

With acids or bases it is common to express concentrations in either molarity or normality, but there is an important difference. Normality (N) is the molarity (M) times the number of equivalents of H^+ or OH^- generated during dissociation in solution. For example, sulfuric acid has the formula H_2SO_4 and dissociates in solution to yield two protons (or hydronium ions) and one sulfate ion. A 1 M solution of H_2SO_4 would therefore be 2 N H_2SO_4. On the other hand, a 1 M solution of hydrochloric acid, HCl, would be 1 N HCl because upon dissociation HCl yields only one equivalent of protons. There is a parallel situation with bases, where equivalents of OH^- are used to determine normality from molarity. For example a 0.1 M solution of NaOH is 0.1 N NaOH, but a 0.1 M solution of $Ba(OH)_2$ is 0.2 N $Ba(OH)_2$ because it yields two equivalents of OH^- upon dissociation.

Solutions based on mass per volume

When solutes have formula weights in the thousands or millions, it is common to prepare these solutions based on a gram measure. For example, protein and nucleic acid solutions are often prepared in concentration units of mg ml^{-1} or µg ml^{-1}. This does not preclude us from determining the molar concentrations of the solutions, if we know the molecular weights of the macromolecules, but many biochemical methods are based on gram measures so it is usually a convenience to express concentrations in this way. For example, if we have a 10 µg ml^{-1} solution of DNA then we know that 1 µl of the solution will contain 0.01 µg, or 10 ng. This amount would be detectable with the stain ethidium bromide, which has a limit of detection of about 1 ng. It wouldn't matter how large the DNA was, because the stain would detect 10 ng of 100 bp fragments just as easily as 10 ng of 10 kbp fragments. The same cannot be said for DNA concentrations reported in molar amounts. A sample of 10 fmol of a 10-kbp DNA has 100 times more mass of DNA than a sample of 10 fmol of a 100-bp DNA. The 10 fmol of 10-kbp DNA would be easily detected with ethidium bromide, but the 10 fmol of 100-bp DNA would be difficult to detect.

Bulk solutions of DNA or protein are sometimes prepared from solutes weighed out on a pan balance; however, usually this does not give a satisfactory level of accuracy. The concentration of DNA or protein in a solution can be determined by a chemical or physical assay, such as a spectrophotometry or fluorometry, rather than by a gravimetric analysis of a solute.

Percent weight/volume

One way of recording concentrations of solutions is the percent weight/volume – %(w/v) – method, which is just a special case of gram concentrations. Saying that a solution has a solute concentration of 1% (w/v) means that the solute concentration is 1 g per 100 ml total volume. For example, a solution that is 0.900% (w/v) NaCl, also referred to as normal saline, has 0.900 g NaCl per 100 ml solution volume, or 9.00 g NaCl per l. We can use the formula weight to determine that the molar concentration of NaCl in this normal saline solution is 0.154 M. As another example, we sometimes prepare a solution of 10% (w/v) ammonium persulfate as a catalyst for polyacrylamide gel casting. This means 10 g per 100 ml, but we certainly wouldn't prepare such a large volume. We might weigh out 50 mg of ammonium persulfate, an amount sufficient to prepare 500 µl of a 10% (w/v) solution since 50 mg/500 µl is the same concentration as 10 g/100 ml.

Liquid solutes

Up to this point we have discussed solutes that are solids before being dissolved in solution, but we also make solutions from solutes that are liquids, for example acetic acid, ethanol, and glycerol. In these cases the liquid solutes are usually measured by volume, and we start with some knowledge of their density or concentration. For example, pure (glacial) acetic acid is 17.5 M, so if we want to prepare 100 ml of a 1 M solution of acetic acid we must measure a volume of acetic acid that is equivalent to 0.1 mol. The volume required is the total moles of solute divided by the molarity:

solute (moles) ÷ concentration of solute (moles liter^{-1}) = volume (liters)

So in the case of the acetic acid, if we want only 0.1 mol then we divide that number by the molarity:

0.1 mol ÷ 17.5 mol l^{-1} = 0.00571 l

We could measure 5.71 ml of glacial acetic acid, add it to water and bring the final volume to 100 ml, and we would have 0.1 mol of acetic acid in 0.1 l, or a 1 M solution.

Percent volume/volume

The concentrations of liquid solutes are sometimes expressed by a percent volume/volume – %(v/v) – method, in which both solute and total solution are measured volumetrically. A 1% (v/v) solution has 1 ml of the solute dissolved into a total volume of 100 ml. For example, a 10% (v/v) glycerol solution could be prepared by taking 10 ml of glycerol and adding water, with thorough mixing, to make a homogeneous solution with a total volume of 100 ml. As another example, a 70% ethanol solution would be prepared by taking 70 ml of 100% ethanol and mixing it with water to bring the total volume to 100 ml. It is important to note that in this last example, 30 ml of water would not be a sufficient amount to bring the total volume to 100 ml because the volumes of ethanol and water are not conserved when they are mixed together. That is, 70 ml of ethanol and 30 ml of water would make a solution that was more concentrated than a 70% (v/v) ethanol solution.

There is a third way of expressing concentrations as a percent, and that is by weight. A solution that is 1% weight/weight, or 1% (w/w) has 1 gram of solute in every 100 grams total mass of solution. This has some advantages for the analytical chemist, since mass is conserved even when volume is not. However, it would be difficult to manage a molecular biology experiment using this method.

Problems

The solutions to these problems are provided at the end of the chapter.

(i)   The formula weight of NaCl is 58.44. How many grams of NaCl would you weigh out if you wanted to make 100 ml of a 1 M solution?
(ii) The formula weight of $MgCl_2$ hexahydrate is 203.3. How many grams of $MgCl_2 \cdot 6H_2O$ would you weigh out if you wanted to make 50 ml of a 0.5 M solution?
(iii) If you have weighed out 19 mg of ampicillin on an analytical balance and want to make a $50\ \mu g\ \mu l^{-1}$ solution, what should be the final volume of the solution?
(iv) How would you prepare 50 ml of a 10% (w/v) solution of sodium dodecylsulfate?
(v)   Acetic acid has a formula weight of 60, and pure (glacial) acetic acid is 17.5 M. What is the density of acetic acid?

3.7 Buffer considerations

Why do we conduct most experiments in buffers? We want the pH to be stable, so as to provide conditions that are reproducible. Imagine that we set up a series of restriction digestions using different DNA samples, but without a buffer. Slight variations in pH in the samples might cause wildly different pH values in the reaction, and this might cause some reactions to succeed while others fail. With a buffer present in the reaction mixture, these differences are settled and the reaction can proceed under standardized conditions.

How do pH buffers work?

We can think of pH buffering compounds in solutions as being analogous to the spongy potting soil we use to keep houseplants healthy. When we water a houseplant, the potting soil soaks up the extra water so that the pot is not flooded. Then, over a period of days as evaporation claims water from the plant, the potting soil releases its water gradually so that the roots do not dry out. The total amount of water in the pot may change dramatically between waterings; however, the roots experience a nice even level of humidity. The pH buffer also acts as a stabilizer, but rather than H_2O it is H^+ ions that are absorbed and released. A buffered solution can withstand the addition of small amounts of strong acids or bases and not change significantly in pH.

The pK_A

The absorptive power of a buffer is related to its behavior in acid dissociation, the reaction in which an acid loses a H^+ ion to the solvent and takes on the form of a base. This behavior can be characterized by a single number, the equilibrium acid dissociation constant K_A. If K_A is high, the buffering compound readily dissociates from its H^+, and if K_A is low, the buffering compound does not easily lose its H^+. The constant K_A is characteristically different for every buffering compound, which leads to the fact that we use different buffers to achieve stability in different parts of the pH range. Sodium acetate might be used to buffer a solution at pH 5, for example, while Tris-Cl might be used to buffer a solution at pH 8.

We can define the pK_A of a buffer as $pK_A = -\log_{10}K_A$, which should remind us of our definition of pH as the negative logarithm of the H^+ concentration. This important number, pK_A, tells us the pH range in which a buffer will do its best work in stabilization, so it has a lot to do with helping us decide which type of buffer to use. A buffer is effective when the solution pH is within approximately one pH unit of the buffer's pK_A. For example, sodium acetate has a pK_A of approximately 4.7 and works well as a buffer between pH 3.7 and pH 5.7. Tris has a pK_A of approximately 8.2 and works well as a buffer between pH 7.2 and pH 9.2. The very best stabilization occurs when the solution pH is exactly equal to the pK_A of the buffer, and above or below that pH the stabilization becomes less impressive.

Let's consider a couple of examples to illustrate the point. If we have 100 ml of a 100 mM Tris-Cl buffer solution at pH 8.2, which is a pH that is equal to the pK_A of the buffer, and we add one drop of concentrated HCl, the pH would fall only to about 8.1. That's pretty good, considering that if the drop contains 40 µl of 12.4 M HCl then we have added about 0.5 mmol of acid to 10 mmol of the buffering compound. The solution would have similar resistance to pH change upon addition of a strong base, such as NaOH. On the other hand, if we started with the same amount of Tris-Cl buffer but set the original solution at pH 7.2, adding a drop of concentrated HCl would cause the pH to fall to about 6.8. Why did the Tris buffer not perform as well at pH 7.2? It is because the pK_A of Tris is 8.2, and pH 7.2 is one pH unit below that pK_A. If the solution pH had originally been set to 9.2, then adding a drop of concentrated HCl would cause the pH to fall to about 9.0. The buffer is a bit more resistant to this one drop of HCl when the pH is one unit above the pK_A, because the pH is moving towards the point of greatest stability, the buffer's pK_A, instead of away from it. Of course, adding a drop of concentrated HCl is a severe test, and for most experimental applications a Tris buffer at pH 7.2 or pH 9.2 would perform satisfactorily.

It is important to bear in mind the workable range of a buffer, because stepping too far outside of that range can lead to an unstable pH. Suppose, in the aforementioned example, we had substituted 100 ml of a 100 mM sodium acetate buffer for the Tris buffer. If we started with a pH of 8.2 we would be well outside the buffering range of sodium acetate, which has a pK_A of 4.7, and adding a drop of concentrated HCl to such a solution would make the pH drop to 6.0. To be fair, at a pH of 4.7 the sodium acetate solution would be an excellent buffer and the Tris-Cl would be an unsatisfactory buffer.

The molar concentration

A second factor to consider when working with buffers is the molar concentration of the buffering compound. A dilute buffer will have less absorptive capacity for H^+ ions than a concentrated buffer, per unit of volume. For example, DNA samples are commonly dissolved in a solution containing 10 mM Tris-Cl, pH 7.5. That is a sufficient amount of buffering capacity to guard against wild fluctuations of pH, but not so strong as to overwhelm the pH of a reaction solution to which the DNA sample might be added next. Reactions using restriction enzymes, polymerases, or other DNA-modifying enzymes are usually run in buffers ranging from 10 to 100 mM in concentration. This slightly higher absorptive capacity may be helpful in bringing uniformity to experiments involving a collection of different samples, and in establishing a stable pH around the optimum for each enzyme. When a much higher absorptive capacity is needed, for example when neutralizing an agarose gel that has been soaked in 0.5 M NaOH, a buffer might be used at a concentration in the 100 mM to 1 M range. These statements about absorptive capacity assume that we have the same unit of volume, because it is the molar amount of a buffering compound that is responsible for absorptive capacity and not the concentration, *per se*. For example, 10 ml of a 100 mM Tris-Cl buffer at pH 7.5 would have roughly the same absorptive capacity as 100 ml of a 10 mM Tris-Cl buffer, at the same pH.

When the pH and pK_A are one and the same

The reason that buffers are most stable when the pH is equal to the pK_A is because this is the pH at which exactly half of the buffering compound has absorbed a H^+ ion. When the pH is less than the pK_A then more than half of the buffering compound has absorbed a H^+ ion, and when the pH is greater than the pK_A then less than half of the compound has absorbed a H^+ ion. The pH is an environmental condition, reflecting the concentration of H^+ ions, and the pK_A is representative of a particular affinity for H^+ ions.

Returning to our farfetched example of a city with hats blowing in the wind (Section 3.4), we remember that the hats were analogous to H^+ ions and that the people in the city had heads that didn't always fit the hats perfectly. Some people wore hats most of the time, because their heads fit the hats well, and others found themselves hatless most of the time. Furthermore, some cities had lots of hats blowing in the air and others had fewer hats. We might now imagine that for each person, there would be a characteristic 'pK$_A$ city' in which he would find that he wore a hat exactly half the time. This specific pK$_A$ city would reflect an individual's affinity for hats, and would depend on the shape of his head. If the person traveled from his pK$_A$ city to a city in which there were ten times more hats blowing in the wind, then he would find himself hatless exactly one-eleventh of the time. That is, the ratio of time spent wearing hats to the time spend hatless would be ten to one. Similarly, if he traveled from his pK$_A$ city to a city in which there were ten times fewer hats blowing in the wind, then the ratio of time spent wearing hats to the time spent hatless would be one to ten.

The Henderson–Hasselbalch formula

We can think similarly about buffering compounds, because each buffering compound has a characteristic pK$_A$ that represents its affinity for H$^+$ ions. The buffering compounds are not wearing hats, of course; they are 'wearing' H$^+$ ions that fit their 'heads' more or less well. When the pH is exactly pK$_A$ − 1, then the ratio of H$^+$-wearing (acidic form) to H$^+$-less (basic form) compound is 10:1, and when the pH is exactly pK$_A$ + 1 then the ratio is 1:10. The exact relationship between pH and pK$_A$ is described by the Henderson–Hasselbalch formula:

$$pH = pK_A - \log_{10}R$$

R is the ratio between the amounts of acid form and base form of the compound. Where pH = pK$_A$ then it must be true that $\log_{10}R = 0$, and so R = 1. Similarly, when pH = pK$_A$ − 1 then it must be true that $\log_{10}R = 1$ and R = 10, and when pH = pK$_A$ + 1 then $\log_{10}R = -1$ and R = 0.1.

The important message of the Henderson–Hasselbalch formula is that the ratio of acid form to base form reflects the pH, and conversely, the pH reflects the ratio. When we add a drop of a pH indicator to a solution during a titration, the color of the solution reflects the ratio between the acid and base forms of the pH indicator compound. The indicator is added in such small amounts that the pH is not significantly changed. However, if we make a solution with a substantial concentration of a buffering compound, and in doing so decide to set a specific ratio of acid to base form of the buffer, then we take control of the pH.

Setting the molar ratio directly

This may be an epiphany for researchers who feel discomfort with the whole issue of pH, or those who view the pH meter as a taskmaster instead of as a servant. We can make a buffered solution at a desired pH by establishing the correct molar ratio between acid and base forms of the buffering compound. For example, we can make a sodium acetate buffer with a pH of 4.7 simply by adding together acetic acid and sodium acetate in equimolar amounts. Why would the pH of this solution necessarily be 4.7? The pH will be pK$_A$ − $\log_{10}R$, and when acetic acid and sodium acetate are present in equimolar amounts their molar ratio R will be 1, and $\log_{10}R$ will be 0. Under those conditions, the pH of the solution equals the pK_A for sodium acetate. Similarly, we could make a solution of Tris-Cl buffer with a pH of 8.2 simply by adding together Tris-HCl (an acid form) and Tris base in equimolar amounts. When their molar ratio is 1, the pH then equals the pK_A for Tris, which is 8.2. A pH 7.2 Tris-Cl buffer could be made by mixing together Tris-HCl and Tris base in a 10:1 molar ratio, and a pH 9.2 Tris-Cl buffer could be made by mixing together Tris-HCl and Tris buffer in a 1:10 molar ratio.

If we decide on a desired pH for a buffer, we can solve for the molar ratio R from the Henderson–Hasselbalch formula:

$$R = 10^{-(pH-pKA)}$$

For example, if we wish to make a sodium acetate buffer with a pH of 5.2, then

$$R = 10^{-(5.2-4.7)} = 10^{-0.5} = 0.316$$

That is, if acetic acid and sodium acetate are present in a 0.316:1 molar ratio, then the pH of the solution will be 5.2. If we wish to make a sodium acetate buffer with a pH of 4.5, then

$$R = 10^{-(4.5-4.7)} = 100.2 = 1.58$$

That is, if acetic acid and sodium acetate are present in a 1.58:1 molar ratio, then the pH of the solution will be 4.5. We could make a Tris-Cl buffer at pH 8.0 by adding Tris-HCl and Tris base in the same 1.58:1 ratio. Why do we get the same ratio with these two different situations? A pH of 8.0 is 0.2 units below the pK_A of Tris, just as a pH of 4.5 was 0.2 units below the pK_A of sodium acetate. It is the arithmetic difference between the pH and the pK_A that gives us the molar ratio R, and so the solution will be the same regardless of the buffer being used. *Table 3.1* shows the relationship between this difference, $pH - pK_A$, and R. The table also shows the molar fractions of acid and base form that would yield these ratios, and the resulting pH values for the two examples of Tris and sodium acetate buffers.

In this method of buffer preparation, we add the acid form and base form to a solution separately, and thereby take direct control over both the numerator and denominator of the fraction R. It will probably seem intuitive that adding the acid form in excess over the base form brings the pH down, because we are adding an acid that will partially dissociate and contribute H^+ ions to the solution. Similarly, adding the base form in excess over the acid form brings the pH of the solution up above the pK_A.

Setting the molar ratio indirectly

There is an alternative approach to making buffers that is useful in many circumstances, and that is to make a solution with only the base form and then convert a portion of it into the acid form by addition of a strong acid. For example, if we combined 1.0 mol of Tris base and 0.5 mol of HCl, we would find that 0.5 mol of the Tris base absorbs a H^+ from the dissociated HCl and becomes the acid form of Tris, with a counterion of

Table 3.1 Solution pH and molar ratio R

pH - pK	R	Molar fractions		pH examples	
		Acid fraction	Base fraction	Tris buffer	Acetate buffer
−1.50	31.6	0.969	0.031	6.7	3.2
−1.40	25.1	0.962	0.038	6.8	3.3
−1.30	20.0	0.952	0.048	6.9	3.4
−1.20	15.8	0.941	0.059	7.0	3.5
−1.10	12.6	0.926	0.074	7.1	3.6
−1.00	10.0	0.909	0.091	7.2	3.7
−0.90	7.94	0.888	0.112	7.3	3.8
−0.80	6.31	0.863	0.137	7.4	3.9
−0.70	5.01	0.834	0.166	7.5	4.0
−0.60	3.98	0.799	0.201	7.6	4.1
−0.50	3.16	0.760	0.240	7.7	4.2
−0.40	2.51	0.715	0.285	7.8	4.3
−0.30	2.00	0.666	0.334	7.9	4.4
−0.20	1.58	0.613	0.387	8.0	4.5
−0.10	1.26	0.557	0.443	8.1	4.6
0.00	1.00	0.500	0.500	8.2	4.7
0.10	0.794	0.443	0.557	8.3	4.8
0.20	0.631	0.387	0.613	8.4	4.9
0.30	0.501	0.334	0.666	8.5	5.0
0.40	0.398	0.285	0.715	8.6	5.1
0.50	0.316	0.240	0.760	8.7	5.2
0.60	0.251	0.201	0.799	8.8	5.3
0.70	0.200	0.166	0.834	8.9	5.4
0.80	0.158	0.137	0.863	9.0	5.5
0.90	0.126	0.112	0.888	9.1	5.6
1.00	0.100	0.091	0.909	9.2	5.7
1.10	0.0794	0.074	0.926	9.3	5.8
1.20	0.0631	0.059	0.941	9.4	5.9
1.30	0.0501	0.048	0.952	9.5	6.0
1.40	0.0398	0.038	0.962	9.6	6.1
1.50	0.0316	0.031	0.969	9.7	6.2

chloride. The amount of Tris base that is left over is the starting amount minus the converted amount, or $1.0 - 0.5 = 0.5$ mol. The ratio between the acid form and base form of Tris would then be 0.5/0.5 or 1, and the pH would be the same as the pK_A, or 8.2. We could prepare a Tris buffer in which acetate was the counterion by substituting acetic acid for the HCl.

This is an indirect approach to setting the ratio between the acid form and the base form of a buffer because the solution is initially prepared with a ratio R = 0, where the Henderson–Hasselbalch formula is undefined, and then R is subsequently increased to the desired level. For example, suppose we want to prepare a Tris-Cl buffer at pH 7.5. The ratio R can be established from the solution of the Henderson–Hasselbalch equation, where $7.5 = 8.2 - \log_{10}R$, so R = 5.01 (see *Table 3.1*). A liter of 1 M Tris-Cl buffer at pH 7.5 might be prepared by placing 1 mol of Tris base into a flask, adding approximately 800 ml of water to the flask to bring the Tris base into solution, and then adding enough concentrated HCl so that the molar ratio between Tris-HCl and Tris base is 5.01:1. The solution volume could be brought to a total of 1 l and the preparation would be complete.

That still leaves the question of how much HCl is enough to give the desired pH. We can solve this question algebraically. If the number of moles of HCl needed to achieve a pH of 7.5 is denoted x, and we know the ratio R = 5.01, then we know that:

$$R = x/(1-x) = 5.01$$

The fraction on the left side of the equation represents the molar ratio of the acid and base forms of Tris. The amount of acid form of Tris is equal to x mol, and the amount of Tris base left over is represented in the denominator by $1 - x$ mol. Solving the equation for x, we find that x = 5.01/6.01 = 0.834 mol. That is, if we add 0.834 mol of concentrated HCl, an amount equal to 67.3 ml of 12.4 M HCl, then we will convert 5.01 out of every 6.01 Tris base molecules to the Tris acid form, leaving 1.00 out of every 6.01 Tris base molecules 'unconverted'. The ratio R is then 5.01:1, and the pH will be 7.5.

Table 3.1 includes a pair of columns indicating the fraction of a buffering compound in acid and base forms, at different pH values above and below the pK_A. The column entry in the table under 'acid fraction' can be multiplied by the total number of moles of buffer being placed in solution, as a way of establishing the number of moles of strong acid (for example HCl) that need to be added. As a general way of solving these types of problems, we can write $x = tR/(1 + R)$, where t is the total moles of buffer in solution and x is the moles of HCl. Preparing 250 ml of the aforementioned 1 M Tris-Cl buffer with a pH of 7.5 would only require us to use x = (0.25)(5.01)/(1 + 5.01) = 0.208 mol of HCl.

While adding HCl is a useful method for making Tris buffers, it is not an appropriate method for converting many other bases. For example if we took 1 mol of sodium acetate and added 0.5 mol of HCl, the pH of the solution would be equal to the pK_A of sodium acetate, or 4.7, but the solution would also be 0.5 M NaCl, due to the donated chloride from the HCl and the sodium ions from the sodium acetate. Assuming that the researcher doesn't want this NaCl in the buffer, the method would lead to an incorrect formulation.

Calculating the components that go into buffer solutions and measuring them by weights and volumes gives reproducible and accurate results. pH meters do not provide uniformly accurate results, and the manner in which they are typically used during buffer preparation does not always provide valid pH readings. As discussed previously (see Section 3.3), highly concentrated solutes can have significant activity coefficients that interfere with pH measurement, and some pH electrodes do not function properly with Tris solutions. A pH meter can be used to check a dilute solution, but it is not good form to place a pH electrode into a concentrated buffer solution as it is being adjusted in pH, adding HCl drop by drop until the pH meter reading matches the desired pH. When that buffer solution is diluted for use in an experiment, the true pH may differ greatly from the measured value.

Another difficulty with adjusting the pH of a buffer 'on the fly', while relying on a pH meter for accuracy, is that a researcher may overshoot the desired pH by adding too much acid. This is partly due to the fact that pH changes registered by the meter are not proportional to the amount of acid added; for example, it takes 4.27 ml of 12.4 M HCl to bring the pH of a liter of 1 M Tris-Cl from pH 8.0 down by one-tenth of a unit to pH 7.9, but only 0.937 ml of the same HCl to bring the pH from 7.0 down to 6.9.

What is to be done with a buffer solution if the researcher added too much HCl and overshot the desired pH? Can it be saved? No, because unless the buffer components are reformulated into a larger preparation it is not possible to take back the extra H^+ ions in any sensible way. If the researcher adds NaOH to raise the pH, then the solution may have the correct pH, finally, but an incorrect formulation due to the extra NaCl that results from the addition of both HCl and NaOH.

Watching a researcher stand in front of a pH meter and overshoot the pH with a strong acid, and then try to rescue the solution by adding a strong base, is a little like watching a person trying to even out a wobbly table by sawing off the end of one of its legs. After making a first cut, the table is wobbly in a different direction, and a second leg needs to be shortened. As the mistakes mount the table gets shorter and shorter, and finally has to be used with a telephone book stuck under one of its legs. So it is with pH meters and buffer solutions that are made improperly. There is not much you can do with either of them.

Making a buffer solution with the right pH and concentration

There are two issues that have to be considered when making a buffer solution: the concentration of the buffer, and the ratio of acid and base forms. When the buffer compound is added entirely in one form, as in the method of mixing Tris base and HCl, the concentration is established first and the ratio R is established later.

For example, if we want to make 100 ml of a 900 mM Tris-acetate buffer, pH 8.3, we would first calculate the amount of Tris base (FW 121.1 g mol^{-1}) that is needed, and then add an appropriate amount of glacial (17.5 M) acetic acid. The amount of Tris base to add is calculated as follows:

$(0.100 \text{ l}) \times (0.900 \text{ mol l}^{-1}) = 0.0900 \text{ mol}$

$(0.0900 \text{ mol}) \times (121.1 \text{ g mol}^{-1}) = 10.9 \text{ g Tris base}$

This Tris base could be put into a solution of approximately 80 ml, in preparation for the addition of acetic acid. How much acetic acid is needed to bring the pH down to 8.3? We determine that a Tris buffer at pH 8.3 would have a molar ratio R of 0.794. That is, 0.794 out of every 1.794 mol of Tris will be in the acid form, representing an acid fraction of $R/(R + 1)$ = 0.443 (see *Table 3.1*). In this specific example, we multiply the total moles by the acid fraction as follows: $(0.0900 \text{ total mol}) \times (0.443 \text{ mol of}$ acid form/total mol$) = 0.0399 \text{ mol}$. This represents the acetic acid that needs to be added to the 0.0900 mol of Tris base to generate 0.0399 mol of Tris acid, and leave an amount of unconverted Tris base equal to 0.0501 mol. Glacial acetic acid is 17.5 M, so the volume of acetic acid needed can be calculated as: $(0.0399\ \text{mol})/(17.5\ \text{mol l}^{-1}) = 0.00228$ l, or 2.28 ml. With the Tris base and acetic acid combined, and brought to a total volume of 100 ml, the total concentration of Tris acid and Tris base combined is 0.900 M. Furthermore, the pH will be 8.3 because $\log_{10}(0.0399/0.0501) = -0.10$, so by the Henderson–Hasselbalch formula the pH of the solution will be: $\text{pH} = 8.2 - ^-(0.10) = 8.3$.

This may seem like a complicated calculation, so let us break it down into its two separate steps. First, all of the Tris buffer is added in the form of Tris base. That makes things easy because the question of concentration is solved before we turn our minds to the pH question. Second, we convert some of the Tris base to Tris acid. A critical point to understand is that the total amount of Tris, the sum of acid and base forms, remains unchanged as some of the base form is converted to acid form.

Getting the correct buffer concentration and pH is a more different calculation when using the direct method of setting pH, because the acid and base forms are added separately and the total concentration and pH of the solution are solved simultaneously. How would we prepare 100 ml of a 900 mM Tris-Cl solution at pH 8.3 by the direct method, using Tris base and Tris-HCl? We can approach the problem algebraically by writing out two equations:

total moles = acid form + base form

molar ratio R = acid form/base form

We know what we want for total moles of Tris in the first equation; it is simply the desired molarity (M) times the desired volume (l), or 0.0900 mol, the same as in the last example. We also know that if the Tris buffer is to be pH 8.3, our molar ratio R should be 0.794, the same solution to the Henderson–Hasselbalch equation as in the last example. This gives us our two linear relationships between the quantities of acid form and base form:

0.0900 = acid form + base form

0.794 = acid form/base form

We can solve these two linear equations simultaneously by a variety of methods, but one way is to make a substitution for one of the terms in the equation. For example, the first equation can be rearranged to read:

acid form = 0.0900 − base form

The right side can be substituted into the second equation:

0.794 = (0.0900 − base form)/base form

This new combined equation has only one unknown quantity, and can be rearranged algebraically and solved:

base form = (0.0900)/(0.794 + 1) = 0.0501 mol

Now that we have solved the equation for one of the two unknowns, we can substitute that result into one of the first equations:

0.0900 = acid form + 0.0501

Solving this equation for the acid form, we find that acid form = 0.0399 mol.

Now that we have our two molar amounts, we can calculate the number of grams to weigh on the scale by multiplying each by its formula weight. The formula weight of Tris base is 121.1 and the formula weight of Tris-HCl is 157.6, so the two amounts that we need to make our solution are:

$$(0.0399 \text{ mol Tris-HCl}) \times (157.6 \text{ g mol}^{-1} \text{ Tris-HCl}) = 6.29 \text{ g Tris-HCl}$$

$$(0.0501 \text{ mol Tris base}) \times (121.1 \text{ g mol}^{-1} \text{ Tris base}) = 6.07 \text{ g Tris base}$$

These two amounts can be combined in a solution, brought to a total volume of 100 ml, and the total concentration of Tris, the sum of both acid and base forms, will be 0.900 M. Furthermore, the ratio between acid and base forms of Tris will provide a pH of 8.3.

The details of a sodium acetate buffer solution might similarly be calculated. Suppose we want to prepare 250 ml of a 3 M sodium acetete buffer, at pH 5.2. The total amount of acetate needed is $(0.250 \text{ l}) \times (3 \text{ M}) = 0.750$ mol, so the moles of sodium acetate plus the moles of acetic acid must equal 0.750.

$$0.750 = \text{acid form} + \text{base form}$$

That is our first equation. The molar ratio R between the acetic acid and sodium acetate needs to be 0.316, as determined from the Henderson–Hasselbalch equation:

$$pH = pKa - \log_{10}R$$

$$5.2 = 4.7 - \log_{10}R$$

$$R = (\text{acid form})/(\text{base form}) = 0.316$$

That is our second equation. We can solve our linear equations by the same method used in the previous example. We will need 0.180 mol of acetic acid and 0.570 mol of sodium acetate. To prepare this solution, we could weigh out 46.75 g of sodium acetate, the FW being 82.03 g mol^{-1}, and add it to about 180 ml of water. Then we could add 10.3 ml of 17.5 M (glacial) acetic acid, and bring the volume of the solution up to 250 ml. Once again, the total concentration of acetate would be 3 M and the pH would have the expected value of 5.2.

Buffers with multiple pK$_A$ values

Some acids carry multiple protons and hold them with different affinities. For example, carbonic acid (H_2CO_3) has two protons, phosphoric acid (H_3PO_4) has three, and EDTA in its most acidified form has four. Each of the acid dissociations associated with these compounds will have a characteristic equilibrium constant, and a characteristic pK$_A$. For example, besides phosphate in the form of phosphoric acid, we can purchase sodium phosphate in the monosodium (NaH_2PO_4), disodium (Na_2HPO_4), or trisodium (Na_3PO_4) forms. These represent three acid dissociations, and in each the acid form is shown below on the left, and the base form on the right:

phosphoric acid (H_3PO_4) $\leftrightarrow$ H^+ + monosodium phosphate (NaH_2PO_4)

monosodium phosphate (NaH_2PO_4) $\leftrightarrow$ H^+ + disodium phosphate (Na_2HPO_4)

disodium phosphate (Na_2HPO_4) $\leftrightarrow$ H^+ + trisodium phosphate (Na_3PO_4)

The respective pKa values for these acid dissociations in an ideal solution are 2.1, 7.2, and 12.3, and they are often denoted pK_A1, pK_A2, and pK_A3. Mixing equimolar amounts of monosodium phosphate and disodium phosphate would ideally yield a pH of 7.2, the same as pK_A2. However, the activity coefficient for phosphate is significant, and so the pH is strongly affected by concentration. An equimolar mixture of the two will yield an experimental pH of 6.88 at a concentration of 25 mM, at 20°C.

Accounting for temperature and concentration effects

This particular example reveals two problems with buffers. First, the true condition of a solution, including its pH, reflects the activities of the solutes and not their concentrations, *per se*. Some ions have significant activity coefficients that require the use of corrected pK_A values that are appropriate for a nonideal solution. Phosphate buffers are notably difficult in this regard. Furthermore, the activities of ions depend on what other solutes are in solution, a matter that was discussed for the example of polyethylene glycol (see Section 3.3). Second, many buffering compounds have pK_A values that are temperature-dependent. We have assumed that the pK_A for Tris base is 8.2, and so it is at approximately 22°C where a pH buffer might be constructed. However, at 5°C the pK_A is approximately 8.66 and at 37°C the pK_A is approximately 7.78. That means that a Tris buffer that is made to have a pH of 8.2 at room temperature will have a higher pH on ice, and a lower pH at body temperature. The pK_A of Tris base drops by approximately 0.027 pH units for each °C increase over this range.

With the corrected value of pK_A2 for phosphate, we can prepare 100 ml of a 0.250 M sodium phosphate buffer, pH 7.2, as follows: the total amount of phosphate in solution, the sum of the monosodium (acid) and disodium (base) forms, will be the product of the volume and molarity, or 0.0250 mol. That provides us with our first equation:

monosodium + disodium = 0.0250

The molar ratio R between monosodium and disodium forms should be 0.479, assuming that we resolve to use the phosphate buffer only when it is diluted 10-fold, and at 20°C, so that pK_A2 has a predictable value of 6.88. That gives us our second equation:

monosodium/disodium = 0.479

We can solve these two equations by substitution, and find that:

disodium = 0.0250/1.479 = 0.0169 mol

monosodium = 0.0250 − 0.0169 = 0.00809 mol

Supposing that we have on hand the heptahydrate form of disodium phosphate ($Na_2HPO_4 \cdot 7H_2O$, FW 268.1), and the anhydrous form of monosodium phosphate (NaH_2PO_4, FW 120.0), then the recipe would be:

- 4.53 g disodium phosphate (heptahydrate)
- 0.971 g monosodium phosphate (anhydrous)

These weights of the base and acid forms would be brought into solution in a volume of 100 ml and would have a total concentration of 0.250 M. When diluted to a concentration of 0.0250 M, the pH will be 7.2.

Mixed buffer systems

Some buffer solutions, particularly those used as running buffers in gel electrophoresis, are based on a mixture of different buffering compounds. This approach provides greater pH stability than a single buffer system. Typically, the mixture involves a cationic and an anionic acid with similar pK_A values, so that the buffering effects are overlapping. For example, Tris is neutral as a base and positively charged in its acid form. Boric acid (H_3BO_3) has a pK_A of 9.2, and is neutral in its acid form and negatively charged in its base form. The buffer TBE is used in the electrophoresis of nucleic acids, and includes Tris base and boric acid in equimolar (90 mM) amounts. The pH of this mixed buffer is approximately 8.3. Glycine is another example of a buffer that is overall neutral (zwitterionic, to be precise) when its amine is in its acid form and its carboxylic acid is in its base form. The acid dissociation of the amine occurs with a pK_A of approximately 9.7. Tris-glycine buffer is used in the electrophoresis of proteins, and a mixture of 25 mM Tris base and 192 mM glycine has a pH of just over 8.5.

Problems

The solutions to these problems are provided at the end of the chapter.

(vi)   The formula weights of Tris base and Tris-HCl are 121.1 and 157.6, respectively, and the pKa of Tris is 8.2. If you combine 10 g of Tris base and 10 g of Tris-HCl in a 200 ml solution, what is the molar ratio of the acid form and base form of Tris? What is the pH of the solution?

(vii) The formula weight of Tris base is 121.1, and the pKa of Tris is 8.2. If you add 12.11 g of Tris base to a solution, then add enough HCl to the solution so that the pH is 7.9, what is the molar ratio of the acid form and base form of Tris? How much HCl did you add?

(viii) Suppose you decide to prepare 100 ml of a 1 M solution of potassium acetate, with a pH of 5.5. The formula weight of potassium acetate is 98.14, and it has a pKa of 4.7. You plan to make the 1 M acetate solution as a mixture of potassium acetate and 17.5 M (glacial) acetic acid. How would you proceed to make this solution?

(ix)   Your next-door neighbor in the lab doesn't know how to make buffers! To make a 1 M solution of potassium acetate at pH 5.5, he weighs out 9.814 g of potassium acetate (0.1 mol), brings it into solution in 90 ml of water, inserts a pH probe and adds acetic acid until the pH meter reads 5.5. Then, he removes the pH probe and brings the final volume of the solution to 100 ml with water. The pH is 5.5, but what is the molarity of acetate in the solution?

(x) If you took 100 µl of 50 mM Tris-Cl, pH 8.2, and added 1 µl of 0.1 M NaOH, what would be the change in pH of the solution?

3.8 Stock solutions

When we make solutions in the laboratory for immediate or future use, we typically make them to as high a concentration as practicable, and then dilute them as needed for an experiment. We call these concentrated solutions our stock solutions because they represent a 'stock' of something that is being saved for later. For example, suppose we are using the following conditions for a restriction enzyme digestion:

- 33 mM Tris-acetate, pH 7.9
- 10 mM magnesium acetate
- 66 mM potassium acetate
- 0.5 mM dithiothreitol

Would we want to make that solution just as it is written? Probably not, because a restriction enzyme digestion also needs to have a DNA sample and a restriction enzyme. If we added those components to the solution later, then the concentrations of the individual components would drop; the 33 mM Tris-acetate might become 30 mM, the 10 mM magnesium acetate might become 9 mM, and so on for each component. It just wouldn't be quite right, nor would it be reproducible if the volumes of our additional components varied.

If we have a version of this buffer that is 10 times more concentrated then we can mix it into a larger volume and leave space for the DNA and restriction enzyme. Such a $10 \times$ stock solution would have the following components:

- 330 mM Tris-acetate, pH 7.9
- 100 mM magnesium acetate
- 660 mM potassium acetate
- 5 mM dithiothreitol

Let's check how this works with an example. If we were to take 10 µl of this $10 \times$ stock solution and add it to a reaction that is being conducted in a total volume of 100 µl, then the 10 µl would carry with it 3.3 µmol of Tris-acetate, 1.0 µmol of magnesium acetate, 6.6 µmol of potassium acetate, and 0.05 µmol of dithiothreitol. When the reaction is brought to a total volume of 100 µl, each of the solution components would have their new intended concentrations: for the Tris-acetate, 3.3 µmol/100 µl is 33 mM; for the magnesium acetate, 1.0 µmol/100 µl is 10 mM; for the potassium acetate, 6.6 µmol/100 µl is 66 mM; and for the dithiothreitol, 0.05 µmol/100 µl is 5 mM. Furthermore, the calculation would work just as well if we diluted 4 µl into 40 µl, 5 µl into 50 µl, or 100 µl into 1000 µl.

However, it would be wrong to take 4 µl of a $10 \times$ stock and add 40 µl, of water, because then the final volume would be 44 µl instead of 40 µl. The concentrations of each component in the solution would be less than their intended $1 \times$ concentrations. The right way to use a $10 \times$ stock solution is to consider the final intended volume, and then use an amount of the $10 \times$ stock that is equal to one-tenth of that volume.

Stock solutions can be made to any arbitrary concentration, but we typically see numbers like $2 \times$, $5 \times$, $10 \times$, $20 \times$, and $50 \times$ on our laboratory stock solutions, not $3 \times$, $7 \times$, or $21 \times$. It is a convenience in mental arithmetic to use numbers with twos and fives in them. For example, we can easily divide a number by 50 in our heads simply by doubling the number and then moving the decimal point by two places to divide by 100. This is just an inherent part of our base 10 number system, and the fact that we have 'decimal' places. It probably has a lot to do with the fact that we have two hands and five fingers on each hand. If we ever receive a sabbatical scientist from a planet around Alpha Centauri, and she has three fingers on each of her seven arms, then she may explain to us that $3 \times$, $7 \times$, and $21 \times$ stock solutions are much easier to use in the laboratory, because 'to divide by 147 you just triple the number and move the unicosimal point by two places to divide by 441!'

Regardless of how a stock solution is formulated, having a concentrated stock of components provides a measure of consistency and reproducibility between different experiments. If a solution is constructed from scratch each time an experiment is conducted, then not only does that take additional time but there are potential problems with errors in formulation and variability of reagents. Stock solutions make it easy to set up an experiment under nearly identical conditions on two different days.

Mixing and unmixing

Some types of stock solutions may be stored frozen, and therein lies a problem because the act of freezing a solution causes some un-mixing of the components. When water freezes into a crystalline structure, it initially excludes solutes and they increase in concentration around the developing ice. The solutes may eventually be frozen as inclusions in the ice, but when the solution is thawed they may be pulled to the bottom of the bottle by gravity because they have a higher density than the surrounding solution. A frozen stock may also have some pure water ice crystals that have developed around the top of the bottle or tube, as the water in the solution gradually sublimes and re-freezes. This water is an important part of the total volume and must be mixed back into the solution when it is thawed. As a general rule, solutions that have been frozen and thawed need to be re-mixed to homogeneity before a sample is taken from them. A pipette that is inserted into the middle of a thawed but unmixed stock may withdraw a solution that is too dilute. There is a related problem with solutions that are not completely thawed before a sample is taken, because even if they are mixed thoroughly, the liquid portion may have more solute in it than the remaining water ice. Sometimes researchers are in a hurry to set up an experiment and may be tempted to take a liquid sample from a tube that is only partially thawed. While this practice may appear to save time, it can cause the sample taken to have too much solute and may ruin the experiment. Furthermore, the removal of a more concentrated fraction from the partially thawed stock reduces the concentration of what is left, and this may ruin future experiments as well.

Microbial growth

Concentrated stock solutions are often more resistant to microbial growth than diluted solutions, which is an added benefit because molds secrete digestive enzymes such as RNases that may interfere with an experiment. A solution that has mold in it, usually detected by the appearance of a puffy white colony resting at the bottom, should be decontaminated and discarded. Bacteria may be visible by an increase in turbidity in a solution, or by sedimented cells on the bottom or sides of a bottle. Bacterial growth is most commonly seen in stock bottles of bacteriological medium that have not been properly sterilized, or have not been handled under sterile conditions. In either case the contents of contaminated bottles should be decontaminated before they are discarded.

Using stock solutions in recipes

Suppose that we are conducting an experiment to determine the optimal concentration of Mg ion for a DNA polymerase. We might plan to set up a series of 10 reaction tubes, each containing a 50 µl reaction volume with 0.05 ng µl^{-1} single-stranded DNA template, 0.05 µM oligonucleotide primer, 0.2 units µl^{-1} of DNA polymerase, 50 µM dATP, 50 µM dTTP, 50 µM dGTP, 5 µM dCTP, 0.25 µCi µl^{-1} α-[^{32}P]-dCTP (3,000 Ci mmol^{-1}), 50 mM Tris-Cl, pH 7.5, 100 mM NaCl, and a variable concentration of MgCl$_2$ ranging from 3 to 12 mM in 1 mM steps. Including the water we would add to bring these reactions to volume, each reaction would have 12 components, and we would need to pipette solutions 120 times to assemble the 10 tubes from independent stocks. Even if we had use of a robotic liquid handler to prevent hand fatigue, we would still have problems with the poor precision of pipetting such small volumes. How could we set up this experiment most effectively, realizing the greatest benefit from stock solutions?

We would want to have highly concentrated stocks of individual components; for example, we would want a 1 M Tris-Cl, pH 7.5 stock, a 5 M NaCl stock, and a 1 M MgCl$_2$ stock. We would want individual 100 µM stocks of dATP, dGTP, dTTP, and dCTP, each buffered with 10 mM Tris-Cl, pH 7.5, to maintain a stable neutral pH during freezing and thawing of the nucleotide stock. When diluted 2000-fold into the final reaction, this small amount of buffer would be a negligible contribution and could be ignored.

Master mixes and cocktails

On the day of the experiment, we might assemble these stocks into 'master mixes' or 'cocktails', as dedicated stocks are sometimes called. The first mixture would include all of the nucleotides and salts, and the DNA polymerase enzyme.

- 2 × dNTP/enzyme mix
- 100 mM Tris-Cl, pH 7.5
- 200 mM NaCl
- 6 mM MgCl$_2$

- 100 µM dATP
- 100 µM dTTP
- 100 µM dGTP
- 10 µM dCTP
- 0.5 µCi µl^{-1} α-[^{32}P]-dCTP (3000 Ci mmol^{-1})
- 0.4 units µl^{-1} DNA polymerase

This master mix includes the enzyme, so it cannot include salts that are so concentrated as to disable the enzyme. It also includes the expensive parts of the experiment, the enzyme and the radiolabeled nucleotide, and so would only be prepared in an amount that was needed for that one day.

The second mixture would include all components that need to be denatured in a boiling water bath for 5 min before the reaction is assembled on ice.

- 10 × DNA master mix
- 0.50 ng µl^{-1} single-stranded DNA template
- 0.50 µM oligonucleotide primer

The DNA template and oligonucleotide primer would be prepared in a 10 mM Tris-Cl, pH 7.5 and 0.5 mM EDTA solution, and once again, we would tend to ignore the small contribution those make to the final reaction. The third stock solution made that day would be 50 mM $MgCl_2$, diluted from the 1 M $MgCl_2$ stock, and used to adjust the total concentration of $MgCl_2$ in each tube. The reactions in the 10 tubes might then be constructed from the master mixes, added in the following order:

- Water: 20 µl, 19 µl, 18 µl, ..., 12 µl, and 11 µl in the 10 tubes, respectively
- 50 mM $MgCl_2$: 0 µl, 1 µl, 2 µl, ..., 8 µl, and 9 µl in the 10 tubes, respectively
- 2 × dNTP/enzyme mix: 25 µl
- 10 × DNA master mix: 5 µl

Each reaction would be assembled from four components. The first two components, the water and $MgCl_2$, amount to a volume of 20 µl and add to the 3 mM $MgCl_2$ that is contributed by the 2 × dNTP/enzyme mix. When all four components are assembled, the $MgCl_2$ concentration will vary between 3.0 mM and 12 mM.

It is important when planning stock solutions and master mixes to consider the limitations of the pipetting equipment. It would not have been prudent to add 0, 0.2, 0.4, ..., 1.6, and 1.8 µl of a 0.25 M $MgCl_2$ stock to the 10 tubes, respectively, because most pipetters cannot achieve that level of accuracy and precision. The most critical component of the experiment, the concentration of the $MgCl_2$, would be irreproducible. In fact, if the laboratory pipettes in question cannot accurately deliver 1 µl then it would be better to generate a series of ten 2.5 × stocks of $MgCl_2$ and water, prepared at a larger volume, so that 20 µl of each respective 2.5 × stock would deliver the correct amount of $MgCl_2$ to the reaction tube. $MgCl_2$ is not expensive, so there is no harm in wasting it to build the reaction conditions correctly.

Order of addition

As the 10 reactions were assembled in our example, the water and $MgCl_2$ were combined first and second, then the $2 \times$ dNTP/enzyme master mix was added third. Would it have mattered if we had added the $10 \times$ DNA master mix third, instead of the $2 \times$ dNTP/enzyme master mix? Yes, because the DNA would have been exposed to concentrations of $MgCl_2$ varying between 0 and 18 mM, during the time it took to add the last component. That temporarily high $MgCl_2$ concentration could easily precipitate the DNA in many of the tubes, preventing it from participating in the reaction. It would have been even more serious if the 50 mM $MgCl_2$ and $10 \times$ DNA master mixes were added to each other, because the $MgCl_2$ concentration could have varied between 0 and 32.1 mM in the different tubes.

Would it have mattered if we had added the $10 \times$ DNA master mix, $2 \times$ dNTP/enzyme master mix, and water first, then added the 50 mM $MgCl_2$ fourth? Yes, because the reaction could commence with only those first three components, and the DNA polymerase might have made considerable progress before the $MgCl_2$ concentration was finally set. Since the entire point of the experiment was to see how $MgCl_2$ affects the activity of the enzyme, the results would be inconclusive if the first 3 min of the reaction were conducted at $MgCl_2$ concentrations between 3 and 5 mM, and then, after the components of the reaction were finally completely assembled and mixed, the last 27 min were conducted at $MgCl_2$ concentrations between 3 and 12 mM.

The order of addition of components is an important part of the description of a solution, and it is a good idea to write recipes for complex solutions so that the order of recipe listings is also the recommended order of component addition. It is also a good idea to follow this advice consistently when building complex solutions, so that component incompatibilities are avoided and experiments are reproducible.

Experiment-to-experiment consistency

If we were planning on doing many similar labeling experiments with α-[32P]-dCTP on different days, we might want to simplify the construction of the $2 \times$ dNTP/enzyme master mix. This could be accomplished by combining the nonradioactive nucleotide substrates into a common $50 \times$ stock consisting of 2.5 mM dATP, 2.5 mM dTTP, 2.5 mM dGTP, and 0.25 mM dCTP. Would we want to add the α-[32P]-dCTP to this dNTP stock as well? Probably not, because the α-[32P]-dCTP has a radioactive half life of just over 2 weeks and we would want to have it be a reasonably fresh component in future experiments. Adding it to the stock would just spoil the nonradioactive dNTP components. We might also want to assemble the Tris-Cl, NaCl, and $MgCl_2$ into a $10 \times$ stock consisting of 500 mM Tris-Cl, pH 7.5, 1 M NaCl , and 30 mM $MgCl_2$. With these intermediate stocks prepared, 25 µl of a $2 \times$ dNTP/enzyme master mix could be assembled on the day of the experiment by combining an appropriate amount of water, 1 µl of the $50 \times$ dNTP stock, 5 µl of the $10 \times$ Tris-Cl/NaCl/$MgCl_2$ stock, the α-[32P]-dCTP, and the DNA polymerase.

Obtaining consistent results in the laboratory depends in part on having stock solutions that are formulated correctly, and minimizing errors in assembly of these stocks into more complex mixtures. Master mixes help to reduce the total amount of repetitive pipetting, but more importantly, they can help to minimize differences between reaction tubes when consistency is important.

Further reading

Camoes MF, Lito MJG, Isabel M, Ferra A, Covington AK (1997) Consistency of pH standard values with the corresponding thermodynamic acid dissociation constants. *Pure Appl. Chem.* **69**(6): 1325–1333.

Chang R (2005) *Physical Chemistry for the Biosciences.* University Science Books, Sausalito, CA.

Deutscher M, Abelson J, Simon M (1990) *Guide to Protein Purification.* Academic Press, New York.

Sigma (9/96) *Sigma Technical Bulletin, No. 106B.* Sigma Chemical Co., St. Louis. http://www.sigmaaldrich.com/sigma/bulletin/106bbul.pdf.

Stephanson FH (2003) *Calculations for Molecular Biology and Biotechnology: A Guide to Mathematics in the Laboratory.* Academic Press. New York.

Solutions to problems

Solutions to problems posed in Section 3.6.

(i) The first thing to do is to figure out how many moles of NaCl are needed. We want 0.1 l of a 1 mol l^{-1} solution, so the number of moles of NaCl in that solution will be the product: $(0.1 \text{ l}) \times (1 \text{ mol } l^{-1}) = 0.1$ mol NaCl. Now we use the formula weight to figure out how many grams that is: $(0.1 \text{ mol}) \times (58.44 \text{ g mol}^{-1}) = 5.844 \text{ g}$.

(ii) This is a similar problem to the first one, and we solve it the same way. The number of moles we need in solution is $(0.05 \text{ l}) \times (0.5 \text{ mol } l^{-1}) = 0.025$ mol. The formula weight of $MgCl_2$ hexahydrate is 203.3, so the number of grams we need to weigh out is $(0.025 \text{ mol}) \times (203.3 \text{ g mol}^{-1}) = 5.08 \text{ g}$.

(iii) 50 µg µl^{-1} is the same concentration as 50 mg ml^{-1}. If we have 19 mg of ampicillin, we figure that there is some final solution volume of x ml such that 19 mg/x ml = 50 mg ml^{-1}. We then calculate $x = 19/50 = 0.38$ ml.

(iv) When we write 10% (w/v), we mean 10 g per 100 ml solution. Since we want 50 ml final volume, the amount of sodium dodecylsulfate we need to weigh out is: $(50 \text{ ml}) \times (10 \text{ g}/100 \text{ ml}) = 5 \text{ g}$. We bring that 5 g SDS into solution and adjust the total volume to 50 ml, and that makes a 10% (w/v) solution.

(v) If we have 1 ml of glacial acetic acid, or 1/1000 of a liter, the number of moles of acetic acid that contains is 1/1000 of 17.5 mol, or 0.0175 mol. Since the formula weight of acetic acid is 60, that 1 ml of acetic acid will have a mass of 0.0175 mol $\times$ 60 g mol^{-1} = 1.05 g. The density of acetic acid is therefore 1.05.

Solutions to problems posed in Section 3.7

(vi) To figure out the molar ratio of the Tris-HCl (acid form) to Tris base, we need to convert the gram measures into molar measures, using the specific formula weights.

$(10 \text{ g Tris-HCl})/(157.6 \text{ g mol}^{-1}) = 0.0634 \text{ mol Tris-HCl}$

$(10 \text{ g Tris base})/(121.1 \text{ g mol}^{-1}) = 0.0826 \text{ mol Tris base}$

The molar ratio of the acid form and base form of Tris is therefore 0.0634/0.0826, or 0.767. We can use this information to determine the pH of the solution through use of the Henderson–Hasselbalch equation.

$pH = pKa - \log_{10}R$, where $R = $ (acid form/base form)

In this case:

$pH = 8.2 - \log_{10}0.767 = 8.2 - (-0.11) = 8.31$

(vii) When the pH of the solution reaches 7.9, the molar ratio can be determined by the Henderson–Hasselbalch equation:

$pH = pKa - \log_{10}R$, where $R = $ (acid form/base form)

therefore, $7.9 = 8.2 - \log_{10}R$

and solving for the molar ratio, $R = 10^{-(7.9-8.2)} = 10^{-(-0.3)} = 1.99$

To solve the second part of the question, we convert the measure of 12.11 g of Tris base to moles, using the formula weight:

$(12.11 \text{ g Tris base})/(121.1 \text{ g mol}^{-1}) = 0.1 \text{ mol Tris base}.$

If we add x mol of HCl (where $x < 0.1$), then the amount of Tris base remaining is $(0.1 \text{ mol} - x)$, and the amount of Tris acid generated from Tris base is x. The molar ratio of acid form and base form of Tris is therefore $x/(0.1 - x)$. Going back to our solution of the molar ratio R, we now know that:

$x/(0.1-x) = 1.99$

therefore, $x = 0.199/2.99 = 0.0665 \text{ mol HCl}$
We can check our work by inserting the numbers into the Henderson–Hasselbalch equation: $7.9 = 8.2 - \log_{10}(0.0665/ (0.1- 0.0665))$.

(viii) We need to establish two equations to solve this type of problem. First, we figure that the total amount of acetate in solution will be the product of volume and molarity, and that the total acetate will be the sum of the potassium acetate and the acetic acid:

$(0.1 \text{ l}) \times (1 \text{ mol l}^{-1}) = 0.1 \text{ mol acetate} = $ potassium acetate + acetic acid

Second, we determine the molar ratio R between the acetic acid and potassium acetate forms at pH 5.5 by the Henderson-Hasselbalch equation:

$5.5 = 4.7 - \log_{10}R$

$R = $ acetic acid/potassium acetate $= 10^{-(5.5-4.7)} = 0.158$

We could rewrite the first equation as:

potassium acetate $= 0.1 \text{ mol} - $ acetic acid

Substituting this into the second equation, we have:

acetic acid/(0.1 mol − acetic acid) = 0.158

acetic acid = 0.0158/1.158 = 0.0136 mol

Therefore, the potassium acetate needed is (0.1 − 0.0136), or 0.0864 mol. To prepare the solution, we would weigh out an amount of potassium acetate equal to 0.0864 mol, or (0.0864 mol)(98.14 g mol^{-1}) = 8.48 g. We would bring this into a solution of approximately 90 ml with water, then add an amount of acetic acid equal to 0 0136 mol, or (0.0136 mol)/(17.5 mol l^{-1}) = 7.82 × 10^{-4} l = 782 μl. Then we would bring the solution to a final homogeneous volume of 100 ml.

(ix) Let's figure out how much acetic acid must have been added to get the pH of 5.5. The ratio of acid form to base form at pH 5.5 was determined to be 0.158 in the previous problem, and so we know that to get that pH our neighbor added 0.158 mol of acetic acid for every mole of potassium acetate. Since he started with 0.1 mol of potassium acetate, he must have added 0.0158 mol of acetic acid. The total moles of acetate is the sum, 0.1 + 0.0158, and the volume is 0.1 l, so the molarity of the neighbor's acetate solution is 1.158 M.

(x) We need to figure out how many moles of acid and base form of Tris we are dealing with in our 100 μl of 50 mM Tris-Cl, pH 8.2. The total moles, the sum of the acid and base forms, is the product of the volume and molarity:

$$(1 \times 10^{-4} \text{ l}) \times (5 \times 10^{-2} \text{ mol l}^{-1}) = 5 \times 10^{-6} \text{ mol}$$

The molar ratio of the acid and base forms is determined from the Henderson–Hasselbalch equation:

pH = pKa − log$_{10}$R, where R = (acid form/base form)

However, since pH = pKa in this example we know that R = 1. There are as many moles of acid form as there are base form in solution, at the buffering point, so we have 2.5 × 10^{-6} mol of Tris-HCl (acid form) and the same amount of Tris base.

Now, if we contemplate adding some NaOH, we only need to figure out how the ratio R will change in response. 1 μl of 0.1 M NaOH is:

$$(1 \times 10^{-6} \text{ l}) \times (0.1 \text{ mol l}^{-1}) = 1 \times 10^{-7} \text{ mol NaOH}$$

When we add that to the Tris buffer, we will convert a portion of the Tris-HCl (acid form) to Tris base, and the molar ratio R will decrease slightly:

R = (2.5 × 10^{-6} − 1 × 10^{-7} mol)/(2.5 × 10^{-6} + 1 × 10^{-7} mol) = 0.923

The new pH can be calculated from the Henderson–Hasselbalch equation:

pKa − log$_{10}$(0.923) or 8.2 + 0.03 = 8.23

Note that the pH didn't change much! This is an example of how buffers can help keep your solutions at a stable pH.

Cloning vectors

4

4.1 Plasmids

Plasmids are DNA molecules that are separated from, and replicate independently of the chromosome of a cell. In the recombinant DNA laboratory, plasmids *can* be used for the replication and maintenance of foreign DNA in a cell. Plasmids may be isolated from a cell as well, and can be worked on *in vitro* using DNA modifying enzymes such as restriction endonucleases (discussed in Chapter 5) and DNA polymerases (discussed in Chapter 6). Most plasmids that we use in the laboratory are designed so that they will replicate in *E. coli*. However, some will also replicate in a different prokaryotic or eukaryotic host cell.

Nearly all of the commonly used plasmids are double-stranded, circular DNA molecules that are much smaller than the chromosomes in the host cell; a typical plasmid used in *E. coli* might be between 3 and 10 kbp, approximately 0.1% to 0.3% of the size of the *E. coli* genome. Plasmids can represent a significant portion of the DNA of a cell, if they are replicated to a high copy number per cell. For example, if a 3-kbp plasmid is replicated in *E. coli* to the extent that it is present at 100 copies per cell, then about 10% of the DNA in the cell is plasmid DNA.

Supercoiling

Plasmid DNA extracted from a cell is typically supercoiled, and a few words about DNA structure are required to explain what supercoiling means. In a DNA molecule, the two single strands are twisted around each other in a double helix. Our usual way of thinking about DNA is that there are 10 base-pair steps per turn of the double helix, as in B-DNA, but DNA can be a bit under-wound or over-wound as well. Let's imagine that we could make ourselves small enough to work with DNA with our hands. If each of us walked to the different ends of a double-stranded DNA molecule and grabbed the single-stranded ends with our hands, we could look down the DNA molecule. Even though we would be facing each other, our perspectives would be the same; the two strands would appear to be twisting around each other in a clockwise or right-handed screw.

Now suppose that I hang on to my two ends tightly and don't let them turn, while you turn your ends either to the right or the left, like the steering wheel of a car. If you turn your ends to the left, the DNA will become a bit under-wound, meaning that the average number of base pairs per turn increases. If you turn your ends to the right, the same direction as the natural turn of the helix, then the DNA helix will become a bit over-wound, meaning that the number of base pairs per turn decreases. In either case, your turning of the helix will cause some tension in the molecule that I will feel at my end. If I lose my grip and let go with both hands, the two strands will whip around each other as they relieve this tension, and the molecule will return to its equilibrium state of about 10 base-pair steps per turn. In fact, I don't need to let go of both strands to see this tension relieved; if I accidentally let go of just one of the two strands, the other strand will whip around the one that I am still holding fixed.

However, if I don't let go, and if you keep turning the helix to the right or left, then before long you and I will see that the double helix begins to writhe around itself like a rubber band that is being wound up. The double helix is coiling itself into a higher-order helix that is called a super-coil. If you had turned your end to the right to over-wind the DNA, then we call it a positive supercoil, and if you had turned your end to the left to under-wind the DNA, then we call it a negative supercoil. In either case, you and I would find ourselves being pulled closer together as the double helix formed these supercoils, because the supercoiling pulls the DNA into a more compact structure.

If we walked towards each other, we could join the two ends of the DNA molecule that we were holding, and this linking of the molecule into a covalently closed circle would preserve the supercoils if we let go. The way we could do this is if you and I each held a free 5' DNA end in our right hands and a free 3' DNA end in our left, we could face each other and join your right with my left, and my left with your right. Once again, we would have to make sure to join both strands, because if we only joined one of the two strands then the other one would be free to whip around and relieve the tension of supercoiling. The covalently closed circular molecule would have positive or negative supercoils in it, depending on how you turned your end before we joined the two ends together, or it could be relaxed and have no supercoils if we just joined our ends together without any extra turning.

Plasmid DNA extraction

When we want to do some work on a plasmid, for example digesting it with a restriction enzyme or making a labeled probe, we must first grow the plasmid in a host cell, extract it, and purify it. *Escherichia coli* is the usual laboratory host for plasmids, so this discussion will be focused on methods using these cells. The yield of plasmid depends on the type of plasmid being used, and on the culture conditions. Some plasmids, for example those with a ColE1 origin of replication, replicate to a high copy number in cells. Other plasmids, for example those used as bacterial artificial chromosomes (BACs), replicate to a low copy number in cells. Plasmid-bearing cells may provide a higher yield of plasmid per cell if they are grown in a richer medium (for example 'terrific broth'), and if they are vigorously shaken (250–300 r.p.m.) to provide more aeration during growth.

Alkaline lysis methods

The most widely used method of extracting plasmid DNA from *E. coli* involves gently opening suspended cells with an alkaline solution, then neutralizing the released contents using a high concentration of potassium acetate (1). Provided that the lysed cells are not handled too roughly, the bacterial chromosome and insoluble debris from the cell are separated from the plasmid by centrifugation. In brief, the steps in this method are:

i.   Centrifugation of bacterial cells. A bacterial cell culture is grown overnight in a culture medium containing a selective antibiotic, then placed in a centrifuge tube and subjected to centrifugation at approximately $2000 \times g$ for 10 min. The bacterial pellet is saved, and the spent culture medium is disinfected and discarded. The pellet may be frozen for future extraction, or used fresh.

ii.   Re-suspension of bacterial cells. The pellet of bacterial cells is resuspended in a simple buffered solution of 25 mM Tris-Cl, pH 8.0, 10 mM EDTA. Re-suspending the cells before lysis helps to separate the cells, so that the released genomic DNA does not turn the bacteria into a gummy lump of cells, and a typical scale for the re-suspension would be 1 ml for each 50 ml of cultured cells. Many researchers add lysozyme to this solution, to a final concentration of 0.2 to 1 mg ml^{-1}, and incubate the re-suspended cells on ice to allow the enzyme to work. Lysozyme weakens the cell wall of the *E. coli*; however, it is not always necessary for complete lysis and can usually be omitted. It is common to add RNase A to a concentration of 1–100 μg ml^{-1}, because RNase degrades RNA that is released upon cell lysis. If the plasmid DNA is going to be purified by CsCl gradient centrifugation, RNase A might be omitted because the RNA is more easily separated from the DNA in the CsCl gradient tube if the RNA remains largely intact. Furthermore, in the special case of laboratories that study RNA (as discussed in Chapter 2), the use of RNase for plasmid preparations may be frowned upon because it can contaminate other sensitive experiments.

iii. Alkaline lysis. One to two volumes of an alkaline solution, usually 0.2 M NaOH and 1% sodium dodecyl sulfate (SDS), is added and gently mixed together with the re-suspended *E. coli* cells. For example, if the cells had been re-suspended in 1 ml in the previous step, then 1–2 ml of the alkaline solution would be added at this point. The two-fold range in suggested volume of alkaline solution reflects differences in published protocols; if the researcher is using a kit provided by a commercial supplier, then the recommendations accompanying that kit should be followed. The high pH and ionic detergent causes the cells to burst open and release their cytoplasmic contents, and the solution becomes clear and viscous. It is important not to vortex or mix the solution vigorously after lysis, because that will fragment the chromosomal DNA into small pieces that might co-purify with the plasmid. In some methods, it is recommended that the lysate be incubated for 5 min at room temperature prior to neutralization.

iv.   Neutralization. One volume of an acidic solution, usually 3 M KOAc (potassium acetate), pH 4.8, is added and gently mixed into the lysed cell solution. That is, if the bacteria had been originally re-suspended in 1 ml, and lysed by the addition of an additional 1–2 ml of alkaline solution, then 1 ml of the potassium acetate solution would be added in this step. The potassium acetate causes precipitation of the bacterial capsule debris, denatured proteins, and most of the chromosomal DNA and SDS, producing a mass that is white and gooey, or flocculent. In some methods, it is recommended that the sample be stored on ice for 10 min. The precipitated debris is separated from the clear plasmid-containing supernatant by centrifugation at $10\,000 \times g$ for 10 min. At times, it can be a challenge to effect a good separation because the gooey precipitate has a density that is very similar to the supernatant and can sometimes float on the surface of the liquid. If the volume of the extraction is large, for example 20 to 100 ml, some researchers will filter the solution through sterile gauze or layers of cheesecloth to separate most of the gooey precipitate from the liquid portion prior to centrifugation. Otherwise, the researcher must simply do his or her best to transfer the clear supernatant to a clean tube, and subject the solution to further centrifugation if there is carryover of cell debris. The clear supernatant contains the plasmid DNA, contaminating RNA, and a lot of salts that are left over from the extraction.

v. Purification. The plasmid DNA in a crude supernatant isn't sufficiently pure to be used for most applications, except perhaps for determining the relative sizes of supercoiled DNA molecules on a gel. The DNA has to be purified, and there is a variety of ways in which this can be done, as described below. At the simplest level, some researchers add 0.7 volumes of isopropanol to precipitate the nucleic acids and separate them from the salts, which remain in the alcohol solution. For example, if the lysate solution volume is 3 ml, then 2.1 ml of isopropanol would be added. The nucleic acid pellet can be collected by centrifugation ($10\,000 \times g$ for 15–30 min), and re-suspended in a smaller and more manageable volume of a solution, such as 10 mM Tris-Cl, pH 8, and 1 mM EDTA. However, an alcohol-precipitated plasmid DNA will very likely be contaminated with RNA and other potential inhibitors of enzymes, and so a more sophisticated method of purifying the DNA is usually required.

Affinity binding matrices

Affinity binding matrices are available that selectively bind nucleic acids, based on their chemical and electrostatic properties. Affinity binding methods do not enrich a sample for supercoiled plasmid DNA, but they are simple and rapidly conducted. There are many DNA-binding matrices that can be used to 'grab' the DNA out of solution, and hold it to a solid support so it can be washed free of contaminants. Often we cannot tell exactly what these solid supports are, the formulations being proprietary information that a company guards as a trade secret. Some affinity matrices are based on silica-impregnated porous gels and membranes; a simple version of this material consists simply of powdered glass. Anion-exchange resins may also be silica-based, but they are modified to carry a cationic surface and operate on a different principle. While the silica glass-like matrices bind DNA in high salt and release DNA in a low ionic strength solution, the anion-exchange resins bind DNA under variable salt concentrations and pH conditions, depending on their composition, and release the DNA in a solution with a higher salt or pH. For the purposes of clarity we will refer to the silica glass-like matrices as being a 'silica-based' affinity matrix, and the anion exchange resins as 'anion exchange resins' (whether or not they are built on a silica foundation).

When plasmid DNA is bound to silica glass-like matrices, it is either added to the top of a silica-based purification column or mixed in solution with a suspension of silica-based particles. The DNA binds to the column bed or suspended particles under the high salt conditions that are present in the crude supernatant, and the proteins and other soluble contaminants do not bind (see *Figure 4.1A, B*). A researcher can vigorously wash the solid support with a 'wash buffer', usually a mixture of ethanol, salt, and buffering compound that does not dislodge the DNA, and the contaminants such as soluble proteins are flushed away (see *Figure 4.1C, D*). The binding and washing of silica-bound DNA in salt and ethanol might remind us of the similar conditions that are used to shield the phosphodiester backbone and bring DNA molecules together during alcohol precipitation, a technique described in Chapter 3. The DNA is eluted from the silica by addition of water or a low ionic strength buffer (see *Figure 4.1E*). Elution of DNA from silica-based affinity matrices is more successful if there is no residue of wash buffer around the solid support, because salt contamination will prevent separation of the DNA from the support. Release of DNA from silica matrices may require several sequential washes of the silica with water or low ionic strength buffer, and incubation at a higher temperature to help dislodge the bound DNA.

Diethylaminoethyl (DEAE)-derivatized solid supports are anion exchange resins, and these require different binding and elution conditions than the aforementioned silica surfaces. Anion exchange resins have a positively charged surface to which the negatively charged DNA binds. This interaction is favored in low concentration salt solutions, because the phosphodiester backbone and DEAE groups are electrostatically attracted to each other when they are unshielded (see *Figure 4.1F, G*). The process of washing such a resin involves raising the salt concentration to a point that plasmid DNA is retained but other contaminants are released (see *Figure 4.1H, I*). For example, RNA is released from a DEAE surface at a lower salt concentration than DNA, so the wash buffer may contain an intermediate salt level that differentially washes away RNA and leaves the DNA bound for later elution. Controlling the pH of the wash solutions is also a critical element in DNA binding, because if the pH is too low (near pH 5) the phosphodiester backbone may be uncharged and not attracted to the DEAE, and if it is too high (above pH 9) then the DEAE may be uncharged and not attracted to the DNA. Elution of the DNA is partly based on increasing the salt concentration of the wash solution to shield the electrostatic attraction between charged groups, and increasing the pH of the solution towards the pKa of the DEAE groups to make them uncharged (see *Figure 4.1J*). DNA eluted from an anion exchange resin is alcohol precipitated and re-suspended in a lower ionic strength buffer, before being put to experimental use.

Plasmid DNA purification is only one application of affinity binding matrices in the molecular lab. For example, a fragment of DNA isolated on a gel might be purified by binding to an affinity matrix or column,

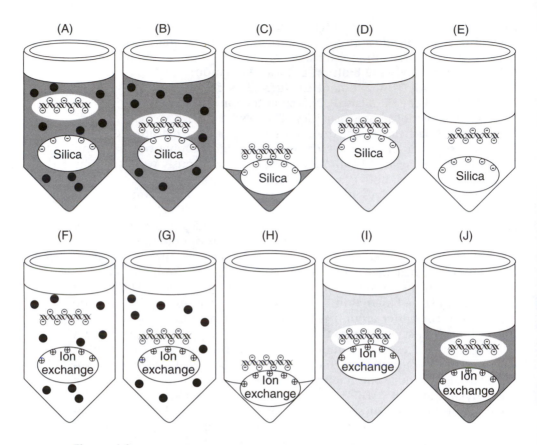

Figure 4.1

Purification of DNA on affinity matrices. (A–E) Binding and release of DNA from an unmodified silica glass-like matrix. The DNA binds in a high-ionic-strength solution and is released in a low-ionic-strength solution. (A) DNA and contaminating proteins (black circles), are mixed with a silica matrix, in a high-ionic-strength solution (gray). (B) In the high-ionic-strength solution, the DNA and silica bind. (C) The silica is separated from the liquid supernatant and unbound contaminants by centrifugation or filtration. (D) A buffered solution consisting of diluted ethanol and salt is used to wash away unbound contaminants. (E) The silica is separated from the wash solution, and DNA is released from the matrix by addition of purified water or a low-ionic-strength buffer. (F–J) Binding and release of DNA from an anionic exchange resin. The DNA binds in a low-ionic-strength solution and is released in a higher-ionic-strength solution with an elevated pH. (F) DNA and contaminating proteins (black circles), are mixed with an anionic exchange resin, in a low-ionic-strength solution (white). (G) In the low-ionic-strength solution, the DNA and anionic exchange resin bind. (H) The anionic exchange resin is separated from the liquid supernatant and unbound contaminants by centrifugation or filtration. (I) A buffered salt solution is used to wash away unbound contaminants. (J) The anionic exchange resin is separated from the wash solution, and DNA is released from the matrix by addition of a high-ionic-strength solution with elevated pH.

separated from contaminants by a controlled set of washes, and then eluted under a different ionic and pH condition. Genomic DNA or cellular RNA might also be differentially purified after being bound to an affinity matrix. The kits and materials that are available for these applications are not generally interchangeable, and some caution in their use is required. For example, a 'wash buffer' from one kit might behave as an 'elution buffer', were it to be substituted in a different kit. Another problem is that some affinity matrices will bind small DNA fragments very tightly, and a researcher might inadvertently lose a 150-nt fragment of DNA isolated from a gel because it is not released from the matrix during the elution step.

Affinity matrices have different capacities for binding DNA, and may not perform well if they are overloaded with material. With plasmid DNA purification, for example, a researcher should consider whether the plasmid was replicated to a high or a low copy number in the host cells. Plasmid DNA extracted from a 50-ml culture of bacterial cells might overload one type of DNA purification column if the plasmid was replicated to a high copy number, but not if the plasmid was replicated to a low copy number.

Cesium chloride density gradient centrifugation

Plasmid DNA can be purified from a sample by CsCl density gradient, a method that takes some time to complete but gives excellent results (2). In this approach, solid CsCl and ethidium bromide are added to a solution of DNA so that the final concentration of ethidium is 250 μg ml^{-1} and the final solution density is 1.55 g ml^{-1}. The amount of CsCl needed to bring the sample to this density is approximately 1 g solid CsCl for every ml of DNA solution. If the DNA sample is a crude lysate, the high salt in the mixture may result in precipitation of proteins, and these solids can be removed by centrifugation (10 000 × g for 10 min) before the ultracentrifuge tubes are loaded (see Figure 4.2A). Ethidium bromide is mutagenic, so researchers need to wear protective gloves and take other safety precautions when working with and disposing of the materials used in this procedure.

The ethidium bromide binds to the DNA in the sample, and so much binds that the density of the ethidium/DNA complex is significantly reduced compared to DNA without ethidium bound. As was explained in Chapter 2, ethidium intercalates between the nucleotide bases and winds the DNA double helix more tightly. Negatively supercoiled DNA extracted from a cell becomes first an open circle, then an over-wound positively supercoiled DNA as more and more ethidium binds. At that point no more ethidium can bind to the positively supercoiled DNA because the molecule is under strain. A nicked circular or linear DNA can continue to absorb ethidium beyond the point that a covalently closed circular DNA would stop, because there can be rotation around one or two free DNA ends to relieve the strain. At the high levels of ethidium bromide added, supercoiled DNA can be distinguished from linear and nicked circular DNA by having a higher density. The density of the supercoiled DNA is higher than that of the linear and nicked circular DNA, because less of the ethidium was absorbed into it.

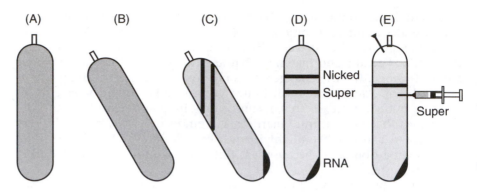

Figure 4.2

Purification of supercoiled DNA by CsCl gradient centrifugation, using a fixed-angle rotor and sealed tubes. (A) A tube is filled with a crude DNA solution containing CsCl and ethidium bromide. The tube is heat-sealed. (B) The tube is loaded into a fixed-angle rotor, which holds the tube at an angle. The center of the rotor is to the left. (C) After 20 hours of centrifugation, the CsCl forms a density gradient, and the DNA-ethidium complexes migrate to form bands of matching density. The diagram shows vertical bands because the tube is still under centrifugation. (D) Following centrifugation and removal of the tubes from the rotor, the density gradient re-orients to the Earth's gravity. The band containing linear and nicked circular DNA is marked 'Nicked', and the higher density band containing supercoiled DNA is marked 'Super'. The RNA has a greater density than the CsCl at the bottom of the gradient, and it forms a pellet at the bottom of the tube. (E) The supercoiled DNA band is extracted using a syringe needle, inserted through the soft tube side. A second needle at the top of the tube allows air to enter the tube as the solution is withdrawn.

A CsCl density gradient can be formed in an ultracentrifuge tube by ultracentrifugation for approximately 20 to 50 h, and the DNA will migrate to the part of the tube in which its density matches the CsCl density (see *Figure 4.2B, C*). The rotor speed and time of the centrifugation depends on the geometry of the rotor. Samples in CsCl density gradients approach equilibrium, and the DNA bands become sharper as centrifugation time is extended. This is unlike a sedimentation gradient, for example, where samples will eventually sediment to the bottom of the tube (see *Figure 4.2D*). After centrifugation, two red bands may be visible in the upper half of the tube. If they are not visible under regular lighting, they may be visualized by exposure to ultraviolet light, in which case the bands fluoresce pink. A UV-opaque face shield or UV-protective goggles are important pieces of safety equipment if ultraviolet light sources are used.

The lower, denser band is the supercoiled plasmid DNA, and the upper, less-dense band is a mixture of linear and nicked circular plasmid DNA, and contaminating linear chromosomal DNA fragments. The position of the bands in the gradient says nothing about their relative sizes. In fact, any two covalently closed circular DNA molecules, say a 5-kbp and a 25-kbp circle, would migrate to the same position in the gradient. They would populate the lower band, because they have the same density after being exposed to ethidium bromide. Similarly, the upper band on the gradient is a mixture of all different sizes of linear and nicked DNA molecules.

In CsCl density gradients, contaminating protein has a low density and will rise to the top of the gradient during its formation. On the other hand, RNA is denser than the CsCl at the bottom, and may form a pellet at the bottom of a tube that has been placed in a swinging bucket rotor, or along the sides of a tube that has been placed in a vertical or fixed angle rotor. The ethidium bromide stains the precipitated RNA a deep red. Many researchers omit RNase from a plasmid preparation so that the contaminating RNA can be efficiently precipitated and discarded during CsCl gradient centrifugation.

If too much CsCl has been added to the tubes, or if the centrifuge rotor has been subjected to too high a g-force, solid CsCl may precipitate along the bottom or sides of a tube. This precipitate is easily distinguished from RNA because it is not stained with ethidium, but it should be avoided by lowering the rate of centrifugation if it is detected repeatedly. The formation of solid CsCl may lead to unequal radial distribution of mass between tubes in a rotor, and could cause a catastrophic rotor failure during a centrifuge run.

The supercoiled DNA may be collected from a CsCl tube by fixing the tube in a clamp and using a hypodermic syringe and needle to pierce the tube just below the lower band to withdraw the DNA in that band (see *Figure 4.2E*). A second needle may be used to pierce the top of the tube to allow entry of air during this process, if the tube is otherwise sealed. Only the DNA in the lower band should be collected; the DNA in the upper band should be left behind because it contains contaminating chromosomal DNA. Care must be exercised when performing this procedure. It is bad enough that syringe needles are involved in the method, but the danger of needle sticks is heightened by the fact that the sample carries high concentrations of mutagenic ethidium bromide. A tray or beaker should be placed underneath the tube while the DNA band is being collected, to catch any spilled ethidium from the pierced tube. All ethidium-contaminated solutions should be appropriately decontaminated or collected as hazardous waste, as should sharps and other ethidium-contaminated solid wastes that are accumulated during the procedure. Contaminated surfaces in a lab should be decontaminated and wiped clean.

The ethidium can be removed from the plasmid DNA sample by repetitive extraction with 2-butanol (sec-butanol), and the CsCl removed by dialysis against a storage buffer such as 10 mM Tris-Cl, pH 8, 0.5 M EDTA. The sample can be further purified and concentrated by ethanol precipitation, and the plasmid re-suspended in a suitable storage buffer.

DNA handling and storage

We often need to handle DNA solutions in small volumes, and the fear of losing a sample on the side or cap of a tube may cause some consternation to a new researcher. Fortunately, if a sample is being maintained in a tightly capped microcentrifuge tube there is an easy way to recover all of the tiny little droplets of liquid; the tube can simply be placed in a centrifuge and subjected to a short spin of about 10 s. The centrifugal force collects the droplets at the bottom of the tube, and a sample can be withdrawn. So, a small sample volume can be vigorously shaken or mixed with a vortexer, and the volume will not be lost.

There are a few good ways of storing DNA in the laboratory, but there is always some tradeoff between maintaining the stability of the DNA and being able to withdraw a sample of it at a moment's notice. If we want to store a sample of purified DNA for a century, and won't need to use any of it in the meantime, storing the DNA as a dry pellet in a desiccator at room temperature is probably the best solution. If the DNA is of high molecular weight, such as genomic DNA, it may take many days to go back into solution when it is rehydrated. However, the advantage of desiccation is that the DNA is not susceptible to attack by DNase enzymes, nor does the sample need to be maintained in a working freezer or refrigerator. DNA may also be stored dry after binding to a derivitized solid support, and there are commercially available systems for archiving collections of DNA in this way, in a small space.

Most molecular laboratories store plasmid DNA in a frozen solution, at $-20°C$ or $-70°C$. The DNA is stored in a buffer containing 10 to 25 mM Tris-Cl, pH 7.5 to 8.5, and 0.5 to 2 mM EDTA. This mixture of Tris-Cl and EDTA is often abbreviated as 'TE' buffer. The low temperature reduces the activities of DNase enzymes, and the EDTA chelates residual magnesium ions that are needed for DNase activity. Furthermore, the EDTA chelates heavy metal ions that might damage the DNA nonenzymatically. The Tris buffer helps to maintain a stable pH in the solution, which is particularly important when solutions are stored frozen.

DNA may be stored in nuclease-free water, but it may gradually degrade over time in this solution. Evidence of degradation may include smearing of the DNA bands on a gel due to exonuclease contamination and DNA shortening, but that is a rather insensitive test for DNA damage. If a DNA molecule has single-stranded cohesive ends, for example after digestion with a restriction enzyme, these ends may be susceptible to contaminating exonucleases. Removal of a single nucleotide from the ends can make cohesive ends incompatible during ligation procedures, and this type of 'nibbling' damage would happen long before a size change could be detected on a gel.

Endonuclease-caused nicks are usually not detectable, unless the DNA is supercoiled and can be converted to a relaxed circle by breakage of one of the phosphodiester backbones. However, if the nicking of a linear or circular fragment is so extensive that two nicks happen to lie within a few base pairs of each other on opposite strands, the DNA may be divided at that point, leaving short overhanging ends. Heavily nicked DNA would be expected to more easily denature at high temperatures, because the number of hydrogen bonds holding together the strands of DNA between nicks is also reduced. Nicking of relaxed circles and linear DNA does not change their electrophoretic mobility as double-stranded molecules, because the hydrogen bonds keep the strands together. However, if the DNA is analyzed on a denaturing gel, for example an alkaline agarose gel or a polyacrylamide gel containing urea, the single strands will be shorter than the length of the double-stranded DNA and may appear as a broad smear of different sizes. This is a common problem when a researcher uses a double-stranded DNA size standard on a denaturing gel. If a sample has become nicked over time, during storage in the freezer perhaps, it will not perform well on a denaturing gel though it still works perfectly on a nondenaturing gel.

High-molecular-weight DNA, for example genomic DNA, can be stored in the same way, but is often stored instead at 4°C. The reason for this may be that freezing tends to 'un-mix' samples, a matter that is discussed in Chapter 3. High-molecular-weight DNA does not go into solution quickly, and freezing and thawing may cause a sample to partially precipitate. If a thawed genomic DNA contains a translucent lump of solid DNA in the tube, it should be allowed to reenter solution before a sample is taken. This dissolution takes time and patience, and can only be accelerated at the expense of the integrity of the sample. For example, if we vigorously agitate or vortex the sample, we will shear the DNA in solution so that its average lengths are reduced to approximately 10 to 50 kbp. Shearing is not a problem when handling small plasmids, but with high-molecular-weight DNA that has to be intact for library construction or Southern blot analysis, it should be avoided. When transferring high-molecular-weight DNA solutions, researchers will often use pipettes or pipette tips with wide bores, to minimize the hydrodynamic stress as the solution enters and exits the tip. High-molecular-weight DNA should not be mixed into a solution by swishing it into and out of a pipette, because this will cause shear. Instead, a solution might be gently stirred or rocked to mix components. A regular plastic pipette tip can be trimmed with a single-edge razor blade to make it a wide bore tip, or if this is a constant need in a laboratory, special wide bore tips can be obtained from commercial suppliers.

With some applications, for example pulsed-field gel electrophoresis, the DNA shearing associated with even gentle handling and pipetting of solutions would be unacceptable. In these cases, DNA may be stored encased in blocks of agarose, at 4°C, in a buffer containing a high concentration of EDTA. The blocks can be handled and transferred between tubes using a sterile spatula, and equilibrated in different buffers by gradual diffusion through the agarose matrix.

DNA can be stored while precipitated with ethanol; in other words, a researcher need not rush a sample to centrifugation during an ethanol precipitation procedure, or immediately re-suspend the pellet in an aqueous buffer. However, a mixture of ethanol, buffer, salts, and DNA is not homogeneous, so a sample of the ethanolic liquid will not contain a proportional amount of the precipitated DNA. For a representative sample, the ethanol precipitation method must be brought to completion: DNA must be collected by centrifugation and re-suspended in an aqueous solution such as TE.

Plasmid origins of replication

If plasmids are to be stably maintained in a population of replicating cells, they must themselves be copied so they can be fairly distributed to the daughter cells. Plasmids are not integrated into the chromosome DNA, so they cannot benefit from replication forks that pass through and copy the host cell DNA. DNA replication is initiated at specific sequences, called origins of replication, and these replication origins may be regulated to control the number of times a piece of DNA can be copied during a cell division cycle. For example, if an origin of replication is tightly regulated and only ever activated once during each cell division cycle, the plasmid will probably be present at a low copy number in the cell. If an origin of replication is less tightly regulated, it may occasionally be activated several times in a cell division cycle and a plasmid carrying that origin may be replicated to a high copy number in a cell.

Many recombinant plasmids used in *E. coli* have origins of replication derived from pMB1 or ColE1, and replicate to a high copy number per cell. These plasmids are suitable for day-to-day cloning work, with sizes that are in the range of 3 to 15 kbp. Larger plasmids such as cosmids, bacterial artificial chromosomes (BACs) and bacteriophage P1 artificial chromosomes (PACs) typically have origins of replication that allow them to be maintained as low copy number plasmids in an *E. coli* cell. BACs are based on the F factor origin of replication, and PACs are based on the bacteriophage P1 lysogenic origin of replication. The amount of plasmid DNA in a cell is related to both copy number and plasmid size; for example, a cell containing 100 copies of a 5-kbp plasmid and a cell containing two copies of a 250-kbp BAC both contain the same number of grams of plasmid. Cosmids, PACs, and BACs are discussed in Section 4.3.

Origins of replication are not universal; just because they are active in one host species does not mean they will necessarily work in another. The activation of origins of replication require *trans*-acting factors such as helicases, which are host-cell specific. A plasmid with a narrow host range may only function in a single species or genus, while a plasmid with a broad host range may function in several different species or genera. The host range of an origin of replication may be extended if the auxiliary factors can be transferred along with the *cis*-acting origin sequences. For example, the SV40 origin of DNA replication is activated in many mammalian cells provided that SV40 T antigen is also expressed. Some plasmids are designed to work in several different hosts, and these are called shuttle vectors. They contain independent origins of replication for each host in which they are capable of being maintained. If a plasmid is transferred to a host in which it cannot replicate, it may persist briefly, but will gradually be diluted out in the growing population of cells.

Many bacterial plasmid vectors contain an origin of replication from ColE1 or pUC, and also a second type of origin of replication that is derived from a filamentous bacteriophage such as f1. This f1 origin of replication yields a single-stranded DNA product when activated by infection of the cells with a helper virus, such as a strain of M13 filamentous bacteriophage. For this to work, the host cells must carry an F′ episome that permits expression of pili, making them male. The cell infected with the helper bacteriophage is able to express the *trans*-acting factors that activate the f1 origin, and the coat proteins that are necessary for bacteriophage assembly. A single-stranded copy of the plasmid is packaged into a bacteriophage capsid and secreted into the growth medium, where it can be collected and the DNA extracted.

The exact strand that is produced depends on the orientation of the f1 origin sequence, so some plasmids such as pBlueScript® (Stratagene®) are available in '+' and '−' versions, denoting these two orientations. The f1 origin of replication can be used to generate single strands of a plasmid for site-directed mutagenesis, DNA sequencing, or probe preparation. The '+' version of the vector can be used to generate single-stranded DNA carrying the sense strand of the plasmid's inherent β-galactosidase gene, and the '−' version used to generate single-stranded DNA carrying the antisense strand of the gene. Origins of replication from f1 are present in plasmids that are called phagemids, meaning a plasmid that can be copied and packaged into a filamentous bacteriophage capsid. Phagemid activation is discussed in more detail in Section 4.3.

The yield of plasmid from a cell culture depends on many factors, including the specific host strain, growth conditions, and nature of the origin of replication. For example, if a 5-kbp plasmid is replicated to a level of 500 copies per cell, and each ml of culture contains 10^9 cells, then the yield of plasmid may be 2 to 3 µg per ml of culture. A 50-kbp cosmid replicated to a level of 10 copies per cell may only yield 0.5 µg per ml of culture. The richness of the growth medium, rate of agitation and aeration, and timing of plasmid extraction are also significant factors in determining yield.

Researchers sometimes add the antibiotic chloramphenicol to bacterial cultures during logarithmic growth to inhibit protein synthesis, limiting host chromosome replication and resulting in amplification of the number of copies of a plasmid per cell. This method is not suitable in every case, because some plasmids confer chloramphenicol resistance and some hosts do not have an appropriate genetic makeup. Furthermore, many high-copy-number plasmids already give a high yield of plasmid without chloramphenicol amplification.

Can a cell carry more than one type of plasmid? Yes, if the plasmids are in different compatibility groups. However, if a single bacterial plasmid vector were used to prepare two different recombinant clones, then those two clones could not be maintained in a single cell. Perhaps that is fortunate for cloning work, because there is some assurance that a single colony growing on a plate of transformed cells will have been founded upon only one plasmid candidate. *Agrobacterium tumifaciens* is sometimes manipulated so that it carries two different plasmids, a so-called binary vector system, where one plasmid carries sequences that can be mobilized and transferred to plant cells and the other 'disarmed' Ti plasmid provides *trans*-acting factors that permit mobilization of the first plasmid.

Selectable marker genes

Most plasmids that we use in the laboratory carry genes that provide an advantage to a host cell when the cell is grown under selective conditions. That is, a cell that is not carrying the plasmid and selectable marker gene will either stop growing or be killed in the selective medium. A common example of this is a plasmid-borne β-lactamase (*bla*) gene that, when expressed, confers ampicillin resistance on bacterial cells. A plasmid in *Saccharomyces cerevisiae* might be maintained by complementation of an auxotrophic mutation in the host cell; for example, a *S. cerevisiae* strain that is *URA3−* and unable to grow in media lacking uracil is rescued by a plasmid carrying a complementary *URA3* gene. A plasmid maintained in a plant cell might provide a selective advantage through expression of the bialophos resistance (*bar*) gene, conferring resistance to the herbicide Basta®.

Not every antibiotic is suitable for use as a selectable marker in a given host cell. For example, kanamycin cannot be used to maintain a plasmid in mammalian cells because the kanamycin does not kill those cells; on the other hand, a related antibiotic G418 (Geneticin®) is effective in killing most mammalian cells, and is widely used as a selective agent.

Resistance to an antibiotic can be based on different principles. In the case of ampicillin resistance, an expressed β-lactamase degrades the drug in the medium, while in the case of Zeocin™ resistance a protein expressed from the *ble* gene binds to the drug and prevents it from intercalating into the host DNA. An important practical difference between these is that when ampicillin is destroyed in the vicinity of a resistant cell, other nonresistant cells in the area may share in the benefit. A colony of ampicillin resistant cells may be surrounded by satellite colonies of cells that are plasmid-free. On the other hand, the *ble* gene provides Zeocin™ resistance only to the cell that expresses it. In a similar way, expression of a plasmid-borne *URA3* gene in one yeast cell will provide no benefit to a neighboring *URA3−* cell that lacks the plasmid.

A resistance gene must be embedded in regulatory sequences that are appropriate to the host cell, usually conferring constitutive (continual) transcription and translation. Like origins of replication, transcriptional and translational control sequences can be host-specific, and a promoter that is active in one species may be nonfunctional in another. For example, a neomycin phosphotransferase gene (*npt*, neo^R, kan^R) will confer kanamycin resistance and neomycin resistance to bacterial cells, and G418 resistance to mammalian cells. The *npt* gene product is functional in either case, but the regulatory sequences surrounding the gene have to be adjusted to the context of the host cell.

Cloning sites

A cloning site is a place to insert a piece of DNA in a plasmid vector, for example a unique restriction endonuclease site that allows a plasmid to be opened in one place for the purposes of receiving an inserted sequence. Cloning sites are sometimes called polylinkers or multiple cloning sites, particularly if they are elaborate and provide a choice of many different unique restriction endonuclease sites. Where a polylinker has many restriction endonuclease sites, one or two might be used to direct insertion of a sequence, and the remaining flanking sites might be used for restriction mapping of the plasmid candidate to check the size, identity, and orientation of the inserted sequence. If the restriction endonuclease sites that were originally used to insert the sequence are lost during cloning, these flanking sites might also be used to excise the sequence to move it to a different cloning vector. Some cloning sites have features that facilitate the insertion of foreign DNA sequences, such as *lox* sequences that allow for directed recombination using Cre recombinase, or tethered topoiso-merase enzymes that improve the kinetics of DNA insertion.

Cloning sites may have other features that are useful in the molecular lab, such as surrounding reference, coding, or regulatory sequences. It is common to find plasmid vectors that have several short segments of M13 bacteriophage DNA called the 'universal forward' and 'universal reverse' primer sequences. These forward and reverse primer sequences, respectively 5'-GTAAAACGACGGCCAG-3' and 5'-CAGGAAACAGCTATGAC-3', are engineered into many plasmid vectors so that they can be used as a common priming site for DNA sequencing. Many plasmids also carry flanking bacteriophage RNA polymerase promoter sequences, such as those of SP6, T3, or T7. The many uses of these promoters are discussed in more detail in Chapter 6, but suffice it to say that these are sites at which transcription can be initiated *in vitro*, using the specific bacteriophage enzyme. Two such promoter sequences often flank a cloning site so that transcription of the inserted sequence can be directed along either strand.

A cloning site may also interrupt a coding sequence in a plasmid, with the result that a gene product is rendered nonfunctional by an insertion, or extended as a fusion protein by in-frame insertion of additional sequence. The interruption of a gene can be used to help in the screening of candidate colonies for plasmids bearing insertions. Alternatively, two coding sequences can be combined to make a coding sequence that encodes a chimeric or fusion protein, provided that the sequences share the same triplet reading frame. Both of these matters are discussed in more detail later in this chapter.

Finally, a cloning site may be placed strategically next to a regulatory sequence, so that the inserted sequence is under its control. For example, the plasmid pET-3a (Novagen®) contains an *Nde*I cloning site immediately downstream of an inducible transcriptional promoter and ribosome binding site, so that a researcher may express high levels of a foreign protein in *E. coli* using the natural translational start of the foreign gene (3). The use of regulatory sequences in plasmids is also discussed in more detail later in this chapter.

Plasmid capacity

How much DNA can a plasmid carry? There is no physical limit to the size of a circular piece of DNA, but large plasmids are harder to handle *in vitro* without shearing. In addition, they have lower transformation efficiencies when we try to put them into competent cells. When the overall size of a plasmid with a ColE1 origin of replication exceeds 10–15 kbp, for example, the cells carrying it may grow poorly. One reason for this may be that larger plasmids cannot easily be maintained at high copy number, and as a result have fewer copies of the selectable marker gene. The cells may grow more slowly because they cannot easily destroy the antibiotic in the medium. Using a lower concentration of antibiotic may reduce the selective pressure, as may a lower culture temperature to decrease the growth rate of the cells. Plasmids with large insertions of foreign DNA also have an increased chance of carrying sequences that interfere with the growth of the host cells, and this is exacerbated if the plasmid is replicated to a high copy number.

Cosmids are plasmids that carry one or more bacteriophage λ *cos* sites, a feature that is discussed in Section 4.3, but when cosmids are packaged in bacteriophage λ capsids, there is a viral packaging limit that prevents transduction of cosmids larger than about 52 kbp. P1-transducing vectors and M13-based vectors also have size limitations related to the packaging systems of the respective bacteriophage.

PAC and BAC plasmids are introduced into cells by transformation, rather than transduction, and are replicated to a low copy number in *E. coli*. These artificial chromosomes accept large insertions of foreign DNA, typically 100 to 250 kbp, as discussed in Section 4.3, but it is not clear what factors limit their sizes. The metabolic load on the cell may be increased if more DNA is being replicated, and it is interesting to note that the total mass of plasmid DNA is about the same in cells with large low-copy-number plasmids and cells with small high-copy-number plasmids.

Detection of inserts

During the preparation and cloning of recombinant plasmids, it is a common concern that plasmids re-close without a DNA insert during ligation. We could perform restriction mapping on the plasmid candidates to determine which ones carry an insert, but several methods have been devised to detect inserts without mapping. An early example of this was included in the plasmid pBR322, which had a *bla* gene conferring ampicillin resistance, and a *tet* gene conferring tetracycline resistance. The plasmid had a unique *Bam*HI restriction endonuclease site in the *tet* gene, so insertion of a sequence into that *Bam*HI site would prevent the recombinant plasmid from conferring tetracycline resistance to a host cell. A researcher would propagate the transformed cells in ampicillin, then test individual transformants for their resistance to tetracycline. A colony that was ampicillin-resistant and tetracycline-sensitive was likely to carry a version of the plasmid that had an insertion in the *Bam*HI site.

A different approach towards insert detection is adopted in the plasmid pZErO®-1 (Invitrogen™), which has a multiple cloning site in a fused *lacZa-ccdB* gene that encodes a toxic protein (see *Figure 4.3A*). Ligation of the pZErO® plasmid with a DNA insert generates a plasmid in which the *lacZa-ccdB* gene is disrupted, and so cells transformed with the recombinant plasmid do not express a functional ccdB protein domain. Those cells survive for two reasons: the plasmid they carry confers resistance to the antibiotic Zeocin™, which is added to the selective medium, and the cells are not expressing a toxic ccdB protein domain. That is, the culture is under positive selection for cells that carry an insertion in *lacZa-ccdB* (see *Figure 4.3C*). Cells that carry a re-closed pZErO® plasmid without a DNA insert express a functional ccdB protein domain, and notwithstanding the fact that the cells are Zeocin™ resistant, they still die because the ccdB protein interferes with topoisomerase II function (4) (see *Figure 4.3B*).

A related type of positive selection can be performed by separation, or lack of separation, of promoter sequences and toxic coding sequences. For example, in the BAC vector pBACe3.6 (Roswell Park Cancer Institute), a toxin-producing enzyme encoded by the gene *sacBII* is only expressed if the vector is uninterrupted, such that there is no DNA insert at the cloning

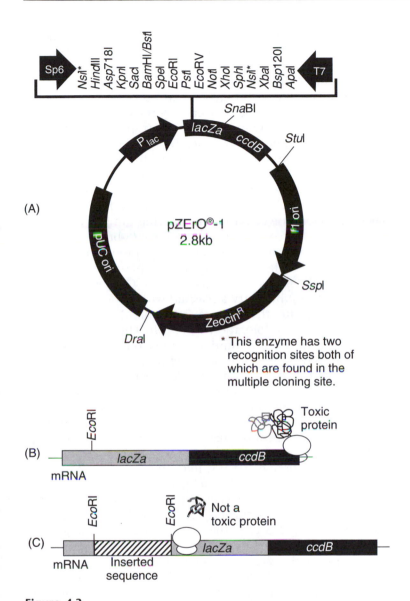

(A)

* This enzyme has two recognition sites both of which are found in the multiple cloning site.

Figure 4.3

Positive selection for DNA inserts, using the vector pZErO®-1. (A) Map of the plasmid pZErO®-1. Copyright 2006, Invitrogen Corporation. Used with permission. (B) When the *lacZa-ccdB* gene is not interrupted, a toxic product is made. Diagram of mRNA generated from unmodified pZErO®-1 plasmid, including *lacZa* sequence (gray) and *ccdB* sequence (black). On the right, a ribosome has completed synthesis of a toxic fusion protein between a β-galactosidase (*lacZa*) domain and a ccdB protein. (C) When the *lacZa-ccdB* gene is disrupted by a cloned DNA, a toxic product is not made. The inserted sequence, an *Eco*RI fragment, is shown as a hatched rectangle. The ribosome is shown having completed synthesis of a portion of the disrupted fusion gene, lacking the *ccdB* domain (with the assumption that an open reading frame was not maintained and the ribosome reached a stop codon).

site, and the *sacBII* promoter abuts the *sacBII* gene. The *sacBII* gene product is sucrose synthase, which converts sucrose into a polymer of fructose called levan. Levan is toxic to *E. coli*, and so when transformed bacteria are grown on sucrose-containing medium, the cells carrying an interruption at the cloning site lack the sacBII enzyme and survive.

Insertions in plasmids are sometimes detected by screening, rather than selection, and the blue-white screening test for interruption of a cloning site in the *lacZ* gene is a common technique. The method is as follows. The plasmid has a multiple cloning site that lies in the coding sequence for the β-galactosidase gene, called *lacZ*. If the plasmid has no insertion at the site, or has only a very small in-frame insertion that does not interfere with the function of the expressed product, the cell carrying the plasmid will have β-galactosidase activity. This enzyme can be detected by the cell's ability to convert the colorless substrate X-gal (5-bromo-4-chloro-3-indolyl-β-D-galactoside) to an insoluble blue-colored product on a bacteriological plate. Bacterial colonies that carry such a plasmid will appear blue, particularly if the plates have been cooled to 4°C after incubation and growth at 37°C. Colonies that carry a plasmid with a substantial insertion of DNA at the cloning site, that is to say, the colonies we want, will show the usual white or cream coloration of *E. coli*. The *lacZ* sequences are transcribed, but the sequence interruption at the cloning site keeps the translated protein from having β-galactosidase activity.

If the promoter driving *lacZ* expression is the *lac* operon promoter, then the promoter is induced by the addition of the lactose analog IPTG (isopropyl-β-D-thiogalactopyranoside). IPTG binds to the lac repressor, the gene product of the *lacI* gene, and prevents the repressor from binding to the operator DNA. With the operator free of bound repressor, transcription can proceed through the *lacZ* gene and the sequences inserted at the cloning site. For this regulation to be possible, the *E. coli* strain or the plasmid itself must carry and express the *lacI* gene.

One final clarification is that the *lacZ* gene is often split into two parts, a *lacZα* gene that is carried by the plasmid and encodes the N-terminal portion of β-galactosidase, and a *lacZΔM15* gene that is expressed from the genome and encodes the remaining sequences in the enzyme. pZErO®-1 is an example of a plasmid in which only the *lacZα* sequences are included (see segment denoted *lacZa*, in *Figure 4.3*). The *lacZα* and *lacZΔM15* protein products bind together to give a functional β-galactosidase enzyme, in a process called α-complementation. Why is the gene split up into two parts? The *lacZα* sequence is much smaller than the full-length *lacZ* gene, and this allows the plasmid to be kept at a smaller size. However, the blue-white screening will only work if the *E. coli* strain carries the remaining portion *lacZΔM15*.

The blue-white screening method in a plasmid therefore depends on the following chemical additives and genetic elements:

(i) growth medium containing appropriate antibiotic selection, IPTG, and X-gal;
(ii) a plasmid carrying the *lacZα* gene, with an internal cloning site;
(iii) an *E. coli* host that carries the *lacI* and *lacZΔM15* genes.

The researcher picks the white colonies off of a plate, which are likely to have an insertion in their plasmids, and does not pick the blue colonies.

There is a closely related application of blue-white screening in many viral vectors, as described in Section 4.2.

Fusion proteins, and chimeric genes

Many plasmids are designed to express an inserted DNA sequence as an 'add-on' to a gene, generating a translated protein that is chimeric, or a fusion between sequences. The inserted sequences may be added to the N-terminus or the C-terminus of the fusion protein, and the principle in either case is that the ribosome passes from the plasmid-encoded transcript sequence to the inserted sequence, and possibly to plasmid sequences again, while continuing translation of the mRNA in a single reading frame.

Expressing a foreign sequence as part of a fusion protein can have several benefits. The fusion 'partner', the consistent part that is encoded by the plasmid sequence, can be a polypeptide whose expression is traceable using an immunological test such as a Western blot or ELISA analysis. The fusion partner may help in purifying the expressed protein, by allowing the fusion protein to be bound to a matrix during affinity chromatography, or by increasing the mass of the fusion protein to aid in its detection by polyacrylamide gel electrophoresis. Many fusion protein expression systems allow for separation of the fusion partner from the expressed sequence of interest, using a specific protease cleavage site. Fusion proteins are often more stable in *E. coli* than foreign proteins expressed without a fusion partner, and more likely to be folded correctly.

Glutathione-S-transferase fusions

An example of a system for expressing foreign sequences as fusion proteins is the series of plasmid vectors called pGEX vectors (GE Healthcare). The pGEX vectors encode a glutathione-S-transferase (GST) gene from *Schistosoma japonicum*, and expression from their promoter (P_{tac}) is induced by addition of IPTG. The plasmids have a cloning site at the end of the GST gene that permits the insertion of extra foreign sequences, for the purposes of creating a fusion protein gene (see *Figure 4.4A*) (5). If the sequences inserted into the cloning site have an open ribosomal reading frame, and the cloning is done in a way that a ribosome can pass from the GST sequence into the foreign sequence while maintaining the correct reading frame, then the protein product will consist of GST followed by a C-terminal extension of foreign sequence. The GST portion of the protein is approximately 30 kDa, so if 10 kDa of extra foreign protein sequence is added to that fusion partner, the overall size will be 40 kDa.

The pGEX vectors are often available in three forms, representing plasmids with multiple cloning sites in the three possible reading frames. This is particularly useful if a researcher is using restriction enzymes to transfer a DNA sequence into the cloning site, and is otherwise limited in possibilities for changing the ends of the DNA. For example, the enzyme *Eco*RI might be used in cloning a DNA fragment into pGEX, and an *Eco*RI site is represented by the nucleotide sequence GAATTC. If an *Eco*RI site is in the gene sequence being transferred into the pGEX vector, the ribosome reading through that sequence in the native gene might read the sequence

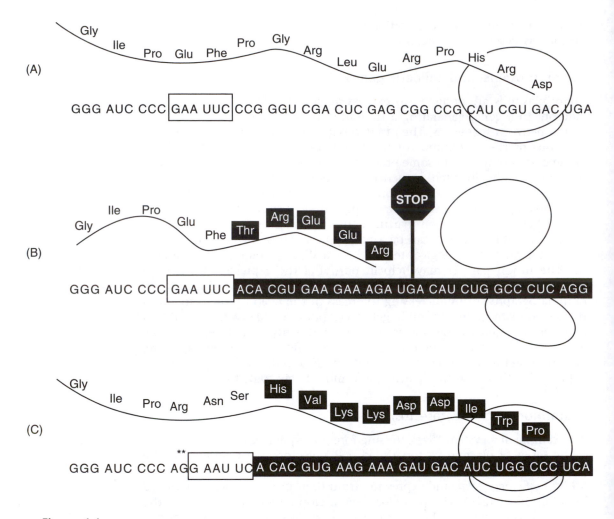

Figure 4.4

Engineering a fusion protein gene requires a consideration of reading frames. (A) Schematic representation of the mRNA sequence corresponding to the cloning site of pGEX-5X-1, with the sequence grouped as triplet codons. The sequence corresponding to the *Eco*RI cloning site is boxed. On the right, a ribosome is completing synthesis of the GST gene. (B) A human DNA sequence from the CFTR gene (GenBank NM_000492) is introduced into the EcoRI cloning site of pGEX-5X-1, shown as white lettering on a black background. The ribosome enters the new sequence in the wrong reading frame and translates nonsense, before terminating on a UGA stop codon. (C) The same human DNA sequence is introduced into pGEX-5X-3, a vector that is identical to pGEX-5X-1 except for the addition of two nucleotides to the left of the *Eco*RI site (marked by asterisks). With the frameshift of two nucleotides, the ribosome enters the CFTR gene sequence in the correct reading frame and generates a fusion protein between GST and CFTR.

in any of three reading frames; the first three nucleotides might be treated as a codon (GAA, encoding glutamic acid), or the ribosome might be shifted by one or two nucleotides to the right, so as to read AAT as a codon (for asparagine) or ATT as a codon (for isoleucine). The reading frames of the GST gene and inserted segment must match, or the ribosome will read the inserted sequence in the wrong frame and produce a nonsense protein (see *Figure 4.4B*). Depending on how the ribosome reads through the *Eco*RI in the native gene, the correct pGEX vector can be selected so that the ribosome reads into the inserted sequence using the correct reading frame (see *Figure 4.4C*).

Fusion proteins with GST can be purified by affinity chromatography, because GST binds to the substrate glutathione. For example, a column can be made with glutathione-derivatized Sepharose® beads, and a crude lysate of recombinant bacteria applied. The GST fusion protein in the lysate will bind to the glutathione on the column and be retained during washing (see *Figure 4.5A*). After the column has been washed free of contaminants,

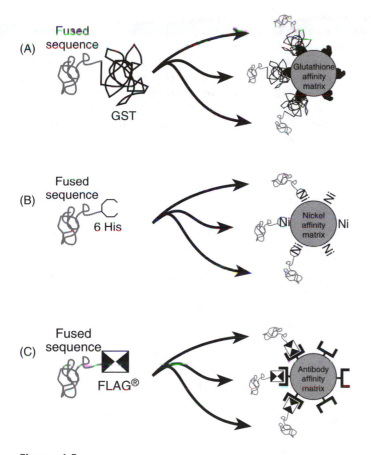

Figure 4.5

Affinity matrix capture of fusion proteins. (A) On the left, a schematic diagram of a protein of interest (gray) expressed as a fusion protein with GST (black). On the right, GST fusion proteins will bind to an affinity matrix having glutathione on the surface. (B) A schematic diagram of a protein of interest (gray) expressed as a fusion protein with a six-histidine tag (black). The six-histidine tag will bind to an affinity matrix having nickel on the surface. (C) A schematic diagram of a protein of interest (gray) expressed as a fusion protein with a FLAG® tag (black). The FLAG® tag will bind to an affinity matrix having anti-FLAG® antibodies on the surface.

the fusion protein can be eluted with glutathione, which binds to the fusion protein and dislodges it from the solid Sepharose® support.

The pGEX vectors also have specific protease sites encoded in their DNA, separating the GST gene from the multiple cloning sites. Examples of these proteases include the blood coagulation factors thrombin and factor Xa, which are commercially available. The fusion protein can therefore be divided after synthesis, to yield a foreign expressed protein and the GST fusion partner. The GST fusion partner can be removed by adsorption to a glutathione Sepharose® column, or separated by gel filtration chromatography or high performance (or high pressure) liquid chromatography (HPLC), leaving the purified foreign sequence without any appended sequences.

The glutathione–S-transferase is likely to be functional as part of a GST fusion protein, and it can be tracked during purification by a colorimetric enzymatic assay. The GST protein can also be detected using a commercially available antibody probe, in an ELISA or Western blot assay system. These techniques allow a researcher to track the progress of fusion protein purification to determine whether adequate amounts of fusion protein are being expressed in cells, and to look for evidence of proteolysis of the fusion protein by studying the apparent molecular weight of the product by chromatographic sizing or gel electrophoresis.

Polyhistidine tags

When a sequence of six to eight consecutive histidine amino acids are added to a protein, the polyhistidine sequence can often be used as a way of capturing the entire fusion protein on a nickel-derivatized column or solid support (see *Figure 4.5B*). The histidine amino acids wrap around the nickel ion and chelate it, and nonbinding contaminants can be washed away from the column and discarded. The fusion protein is eluted from the nickel column by adding imidazole, which has the same functional group as the amino acid histidine and competes for binding to the nickel sites.

The polyhistidine in a protein may function at either the N-terminus or C-terminus of a fusion protein, although each fusion protein is folded differently and this method may not work in some circumstances. An advantage of the polyhistidine is that it is a relatively short sequence to be adding to a foreign protein, and it is therefore less likely to perturb protein function than a larger fusion partner, such as GST. The addition of six to eight histidines is sometimes referred to as a protein tag, to convey the idea that the fused sequence is a small contribution to the overall product. The histidine tag can be tracked using a commercially available monoclonal antibody, and depending on the cloning vector, the tag may be separable from the rest of the expressed sequence using a specific protease.

FLAGs and other tags

There are several short sequences that are widely used to tag recombinant proteins, providing researchers with a way to purify them by affinity chromatography and detect them in a lysate. These short tag sequences have the advantage that they are not obtrusive and are unlikely to interfere with protein function. For example, the FLAG® sequence is the eight-amino-acid sequence DYKDDDDK, where D is aspartic acid, Y is tyrosine, and K is lysine. Monoclonal antibodies are available that bind to FLAG®, and several variants of the sequence, and these antibodies can be used in an affinity matrix to isolate FLAG®-bearing recombinant proteins (see *Figure 4.5C*). Other short tags include the HA-tag, derived from the hemagglutinin gene of influenza virus, the c-myc epitope tag, derived from the human *c-myc* gene, and the V5-tag, derived from the virus SV5. These tags, and many others that operate on the same principle, can be embedded in recombinant proteins and used for immunological detection and purification.

One that is slightly different is Strep-tag II, a nine-amino-acid streptavidin-binding peptide (NWSHPQFEK) (6). While it can be detected immuno logically, it can also bind to affinity matrices that involve streptavidin derivatives, and can be displaced and eluted from the streptavidin with biotin.

Fusion partners as positional reporters *in vivo*

One of the principal uses of tags is to report on the position of a recombinant protein *in vivo*, perhaps showing its site of synthesis or its movement and sequestration. The recombinant tagged protein can, in many cases, stand as a surrogate marker for an untagged wild-type protein, showing how its expression is regulated and how it behaves in a cell. The tags described thus far might be used to track an expressed protein in different biochemical fractions, or visually by indirect immunofluorescence in fixed cells using an immunological probe. However, one of the most commonly used fusion partners for proteins is green fluorescent protein (GFP), derived from the jellyfish *Aequorea victoria*. Fusions with GFP retain the ability to fluoresce, and can be used to localize the recombinant proteins in living cells, tissues, and whole animals.

Leaders for targeting or secretion

Many plasmids are available that permit researchers to express recombinant fusion proteins that contain leader sequences, allowing targeting or secretion of the protein. For example, the *ompA* signal peptide sequence from *E. coli* can be added to a recombinant protein to cause it to be secreted into the periplasm of *E. coli*. The α-factor leader peptide is used analogously in *S. cerevisiae* to engineer a protein that can be exported through the secretory pathway of the yeast cell. Sometimes plasmids have cloning sites that lie downstream of a series of short fused sequences: a signal sequence for secretion, a 6-His tag for purification, a FLAG® tag for tracking, and a thrombin protease site for cleavage. The series of tags can be cleanly removed from the protein by treatment with the specific protease, after it is secreted and purified.

Eukaryotes have specific signal sequences for import of proteins into organelles, such as mitochondria and chloroplasts, and these can be engineered into a fusion protein. A transgenic plant may carry a recombinant gene, for example, that resides and is transcribed in the nucleus, has an RNA that is translated on cytoplasmic ribosomes, and has a protein product that is imported into the chloroplast.

Combination of function

The engineering of recombinant proteins in a plasmid allows researchers to develop proteins with new combinations of functions. For example, the heavy and light immunoglobulin gene Fv domains might be fused, along with an intervening spacer, to make what is called a single chain Fv (ScFv) that retains the antibody binding characteristics of the original two-polypeptide protein. That ScFv might in turn be fused to a gene encoding an enzyme, to make a protein product that had an enzymatic activity that can be targeted to the antibody- (ScFv-) binding site.

Another common technique in biotechnology is called domain shuffling, in which the segments of DNA encoding protein domains are combined to make new versions of proteins. For example, a gene for a protein with domains A, B, and C could be recombined with a gene for a similar protein having domains a, b, and c, to make a collection of shuffled versions including aBC, AbC, aBc, and abC. The new combinations of domains in the recombinant proteins can result in new functions that were not part of the original set of domains. For example, domain shuffling has been used to generate new types of *cry* genes in *Bacillus thuringiensis* that encode protein products with novel insecticidal activities.

Regulatory sequences

Plasmids used in molecular labs often have elaborate systems for regulating the expression of genes of interest. For example, the aforementioned pET and pGEX vector systems permit regulated control of gene expression of foreign proteins in *E. coli*. Prokaryotic expression vectors have transcriptional promoters that may include adjacent binding sites for *trans*-acting activators or repressors. A ribosome binding site, also called a Shine–Dalgarno sequence, may be included if the inserted gene is to be translated (as well as transcribed). Translational stop codons in all reading frames may follow a multiple cloning site, to halt the ribosome if it reaches the end of the transcribed inserted sequence. Finally, transcription termination sites may be included to limit the progress of an RNA polymerase past the end of the inserted sequences. As was explained for the case of origins of DNA replication, regulatory sequences can be narrow range, working in only one or several species, or broad range.

Eukaryotic expression vectors also require regulatory sequences, such as transcriptional promoters, enhancers, splice sites, and polyadenylation sites. For example, when doing work in mammalian cells, the use of a multiple cloning site in the expression vector pBK-CMV (Stratagene®) allows researchers to express a foreign gene under the control of a human cytomegalovirus (CMV) promoter and an SV40 polyadenylation site. Vectors designed for the expression of recombinant proteins in plants are often based on the constitutive 35S promoter from cauliflower mosaic virus.

Ligation

The last step in handling a modified DNA vector, before it is introduced into a cell, is a ligation reaction to seal nicks in the phosphodiester backbone. This builds a covalent connection between the DNA ends, and joins strands into a contiguous piece that can be replicated by a host DNA polymerase. The usual product of a vector ligation is a closed, circular DNA.

Why is ligation important in the process? When DNA is engineered by annealing of cohesive overhanging ends, for example those left by some restriction endonucleases, nicks are left on each strand where the free 5′ and 3′ ends of the DNA meet. If the nicks are closely spaced on the opposite strands, the number of hydrogen bonds holding the DNA ends together may be too few to make a stable molecule at the temperatures used in DNA transformation. The base-pairing between these ends provides little strength; like two pieces of Post-it® note paper stuck together, the DNA fragments can be readily separated. However, ligation of the phosphodiester backbone makes the connection between the 'sticky ends' strong.

When blunt ends are being joined, there are no overhanging ends that could lend even a temporary connection by base-pairing. These ends can be joined only by a covalent connection, and so ligation becomes a critical part of the assembly process. Blunt end ligation is a little more challenging in the laboratory than cohesive end ligation, probably because having compatible overhanging ends allows a DNA to be loosely assembled prior to the arrival of the enzyme, like two pieces of fabric that are pinned together before a strong seam is sewn into them by a sewing machine.

The DNA ends must fit together

The most important principle underlying a successful ligation reaction is that the ends of the DNA have to fit together, either by having complementary (cohesive) overhanging ends, or by having ends lacking overhanging sequence (blunt ends). For example, these two fragments have overhanging ends that fit together by base-pairing like pieces of a puzzle, and they could be ligated together on both the top and bottom strand:

```
5′-...GGAGATCAGACTA        CGATCAGACAGCT...-3′
3′-...CCTCTAGTCTGA        TGCTAGTCTGTCGA...-5′
```

The product of ligation would have covalently joined strands:

```
5′-...GGAGATCAGACTACGATCAGACAGCT...-3′
3′-...CCTCTAGTCTGATGCTAGTCTGTCGA...-5′
```

However, if the overhanging ends had a mismatch in base-pairing or if they left a gap in base-pairing, then the enzyme T4 DNA ligase could not connect the strands by ligation:

```
5′-...GGAGATCAGACTC        CGATCAGACAGCT...-3′  (No base-pairing)
3′-...CCTCTAGTCTGA        TGCTAGTCTGTCGA...-5′

5′-...GGAGATCAGACTA        GATCAGACAGCT...-3′  (Gap would be left)
3′-...CCTCTAGTCTGA        TGCTAGTCTGTCGA...-5′
```

The details of engineering ends that fit together are discussed later in the context of restriction endonucleases (Chapter 5), DNA polymerases (Chapter 6), and polymerase chain reaction products (Chapter 7).

Many protocols for DNA ligation call for incubation of the reaction at 16°C, a temperature that is lower than the optimal temperature of T4 DNA ligase, but which favors the formation of hydrogen bonds between the cohesive ends. When blunt-end ligations are being conducted, there are no cohesive ends to join together and many researchers conduct those reactions at 25°C to 37°C to improve the performance of the enzyme.

The DNA ends must have a 5′ phosphate and a 3′ hydroxyl

The enzyme T4 DNA ligase requires a substrate with a 5′ phosphate group and a 3′ hydroxyl group (see *Figure 4.6A*). This is not usually a concern when restriction endonuclease fragments are being joined, because restriction endonucleases leave a 5′ phosphate and 3′ hydroxyl when the phosphodiester backbone is hydrolyzed. However, some methods of cloning involve the use of the enzyme alkaline phosphatase to remove the

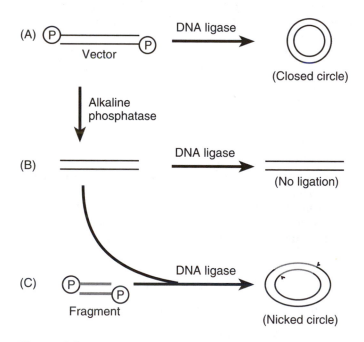

Figure 4.6

Using 5′ phosphate groups to control DNA ligase activity. (A) A linear vector DNA with compatible ends, and 5′ phosphates at each end can be converted to a covalently closed circle using T4 DNA ligase. (B) Removal of 5′ phosphates using the enzyme alkaline phosphatase results in a linear vector DNA that cannot be converted to a circular DNA using T4 DNA ligase. (C) If a DNA fragment having 5′ phosphates is provided in the ligation, it can be joined with the phosphatase-treated vector to make a nicked circular DNA. The product circle has one nick on each strand, corresponding to the locations of the 5′ phosphates that were removed from the vector.

5′ phosphates from vector DNA ends. Following removal of these 5′ phosphates, the vector cannot re-close on itself during ligation (see *Figure 4.6B*) unless a DNA insert that bears 5′ phosphate groups is included (see *Figure 4.6C*). It is important that the alkaline phosphatase be destroyed after treatment of a vector DNA, if it is used at all, because residual phosphatase activity could lead to the removal of phosphates from the DNA insert when the two samples are mixed. If that were to happen then the ligation would not work, because no 5′ phosphates would remain to serve as a substrate for the T4 DNA ligase.

How are these pieces of DNA connected if the vector DNA lacks phosphate groups at its two free 5′ ends? The answer is that there are two DNA strands that could be connected at each junction; if the vector is pre-treated with alkaline phosphatase then only one strand is connected, and the complementary strand is left unconnected. The product of this ligation has a single nick on each strand, corresponding to the positions of the two missing phosphate groups (see *Figure 4.6C*), but otherwise the DNA molecule is sufficiently strengthened by ligation that it can be introduced into cells by transformation. These two nicks are not a permanent flaw in the DNA, in any case. After introduction of the DNA into a bacterium, the host enzymes repair these nicks.

Many methods of cloning involve products generated through the polymerase chain reaction (see Chapter 7), and these products typically lack 5′ phosphate groups because the oligonucleotide primers used during synthesis are synthesized with 5′ hydroxyl groups. Can PCR products with 5′ hydroxyls be introduced into a vector by ligation? Yes, but the vector must provide the 5′ phosphates this time, and once again, the ligation product will have a nick on each strand corresponding to the positions of the missing phosphates. The ligation will not work at all if the PCR-generated DNA insert lacks 5′ phosphates and the vector has been treated with alkaline phosphatase to remove 5′ phosphates.

Adding and subtracting 5′ phosphates from DNA is easily managed using enzymes. Alkaline phosphatase can be used to remove a 5′ phosphate from a DNA end, leaving a 5′ hydroxyl, and T4 polynucleotide kinase can be used to transfer the γ phosphate from an ATP substrate to a 5′ hydroxyl.

The DNA ends must find each other in solution

What is the chance that two DNA ends will bump into each other in solution? This is an important question, because if the DNA ends are at such a low molar concentration that collisions between ends rarely occur, the ligation will not be efficient. The molar concentration of DNA ends can be calculated from the gram concentration of the DNA fragment, by using the DNA molecular weight. The average base pair in DNA is approximately 660 Da, so a 1-kbp DNA would be approximately 660 kDa, and a 5-kbp DNA would be approximately 3300 kDa. If we mix together 10 ng of a 5-kbp plasmid and 5 ng of a 1-kbp fragment, there will be 3 fmol of the plasmid DNA present (1×10^{-8} g ÷ 3.3×10^6 g mol^{-1} = 3×10^{-15} mol), and 7.5 fmol of the 1-kbp DNA (5×10^{-9} g ÷ 6.6×10^5 g mol^{-1} = 7.5×10^{-15} mol). If the DNA molecules are mixed in a volume of 5 μl, the concentration of each

DNA will be 0.6 nM (3×10^{-15} mol $\div 5 \times 10^{-6}$ l = 6×10^{-10} M). The molar ratio between the plasmid and 1-kbp DNA is 1:2.5 (3:7.5 fmol). Let's suppose that the 5-kbp plasmid DNA and 1-kbp DNA fragment are linear molecules, and have ends that are mutually compatible for ligation. The rate of ligation will be roughly proportional to the concentrations of each DNA, so if the ligation is performed in 50 μl instead of 5 μl, each concentration will be reduced 10-fold, and the rate of ligation reduced by 10^2 or 100-fold. A schematic representation of a 10-fold dilution is shown in *Figure 4.7B* and *C*.

If a plasmid vector has ends that can re-circularize, for example if a DNA vector is digested with the restriction endonuclease *Eco*RI (as discussed in Chapter 5), then the joining of the two ends of the plasmid without a DNA fragment insertion may be an unwanted side reaction during ligation (see the unimolecular reaction in *Figure 4.7A*). How likely is that to occur? It depends on how often the different ends bump into each other in solution. DNA behaves as a random coil in solution, and the compactness of the coil depends on ionic strength, temperature, and DNA sequence characteristics. The ends of two different DNA fragments will find each other at a rate that is proportional to their individual concentrations, but a DNA fragment will find its own 'other end' at a rate that is independent of concentration, the other end being tethered through the length of the DNA fragment.

When we want to bring two different DNA fragments together, it is important to perform the ligation in as concentrated a solution as we can manage so that the rate is favorable (see *Figure 4.7B*). For example, a ligation reaction between a DNA fragment and a plasmid might be established in 2.0 μl, using 0.5 μl of a DNA fragment, 0.5 μl of a DNA plasmid vector, and 1.0 μl of a $2 \times$ ligase buffer solution that has been prepared from a $10 \times$ buffer, and contains the enzyme T4 DNA ligase. The reaction can be set up as a small spherical droplet in a tube, under a drop of mineral oil, which coats the surface of the tube and keeps the reaction volume from adhering to the plastic. The pipetting device and pipette tips must be of high quality to deliver small volumes of 0.5 μl accurately. The products of the ligation can be recovered from the small droplet by addition of additional aqueous volume, removal of traces of the mineral oil by transfer to ParaFilm® (as discussed in Chapter 2), and ethanol precipitation. We may add polyethylene glycol (e.g., 15% w/v PEG-8000) to a ligation reaction to increase the rate of the joining reaction. As explained in Chapter 3, the effect of the PEG is to exclude solution volume and increase the activity of the DNA ends.

Bringing two different DNA molecules together also depends on the relative molar concentrations of the DNA molecules. We set up the ligation reaction using a molar ratio of plasmid DNA to DNA insert of approximately 1:2 or 1:3, so that insertion of DNA into the plasmid is favored over re-closure of the plasmid without a DNA insert. We do not usually wish to recover clones with insertions of multiple copies of the DNA fragment, so we do not use more extreme ratios such as 1:5 or 1:10.

Sometimes we don't want to bring two different DNA molecules together and we only want to re-close a plasmid DNA so that it is circular. For example, following mutagenesis using an inverse PCR method

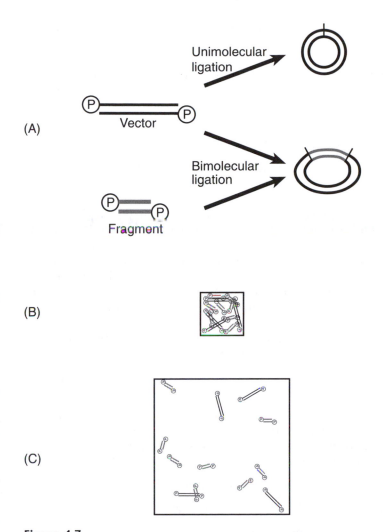

Figure 4.7

Ligation of a vector and a DNA fragment generates a mixture of unimolecular and bimolecular reaction products. (A) The closure of the vector with a DNA insert is usually desired, but depends on two DNA molecules coming together in a bimolecular reaction (bottom). (B) Schematic representation of a mixture of DNA insert and vector in a concentrated solution. A concentrated solution results in a greater frequency of collisions between different DNA molecules (bimolecular). (C) The same number of schematic molecules, distributed over a 10-fold greater area. Diluting a solution reduces the collisions between different DNA molecules (bimolecular), but does not affect the rate at which the two ends of a single molecule collide (unimolecular).

(Chapter 7), we may have a linear DNA that simply needs to be ligated between its own two ends. In that case we would treat the DNA with T4 polynucleotide kinase, to add 5′ phosphate groups to the inverse PCR product, and conduct the ligation in a dilute solution of perhaps 20 to 50 μl (see *Figure 4.7C*). In this case, the ligation reaction is between two ends that are tethered to each other through the length of the DNA, and so there is no point in setting up the reaction in a small volume such as 2 μl. In fact, diluting the sample into a large volume of a clean buffer makes side reactions involving insertion of contaminating fragments less likely, and dilutes out the effects of inhibitors that may accompany the DNA sample.

Ligations produce a mixture of products

When we conduct a ligation reaction, we often expect to obtain a mixture of desired products and products of side reactions. For example, if we digest a plasmid vector with a restriction endonuclease such as *Bam*HI, and prepare a DNA fragment for insertion using the enzyme *Bgl*II, the two types of ends can be ligated because they each have 5′-GATC overhanging ends (see Chapter 5). However, we might obtain a ligation product in which the plasmid vector re-closes at its *Bam*HI sites without insertion of a *Bgl*II fragment, or we might obtain a product in which several *Bgl*II fragments had joined together to make a larger DNA insertion. We can also expect to have many examples of *Bgl*II fragments closing on themselves to form circles, although these will not be detected in *E. coli* after transformation because they lack the origin of replication and drug resistance marker gene necessary for stable maintenance.

We can sometimes take steps to prevent side reactions, for example pre-treating the plasmid vector digested with *Bam*HI with the enzyme alkaline phosphatase to remove the 5′ phosphate groups and prevent closure without an insert (see *Figure 4.6*). We may also take steps to eliminate the products of side reactions; in the case of ligating the ends generated by *Bam*HI and *Bgl*II, we may digest the products of ligation with *Bam*HI to open plasmids that had re-closed without an inserted DNA.

```
                      BamHI  plus  BamHI  ends
NNNNNG        GATCCNNNNN   →   NNNNNGGATCCNNNNN
NNNNNCCTAG        GNNNNN       NNNNNCCTAGGNNNNN

                      BglII  plus  BglII  ends
NNNNNA        GATCTNNNNN   →   NNNNNAGATCTNNNNN
NNNNNTCTAG        ANNNNN       NNNNNTCTAGANNNNN

                      BglII  plus  BamHI  ends
NNNNNA        GATCCNNNNN   →   NNNNNAGATCCNNNNN
NNNNNTCTAG        GNNNNN       NNNNNTCTAGGNNNNN
```

The ends generated with *Bam*HI and *Bgl*II-generated ends are cohesive and fit together during ligation, but the fused ends make neither a *Bam*HI nor a *Bgl*II site after joining. Provided that there are no *Bam*HI sites in our inserted sequence, treatment with *Bam*HI should cause no ill effects to the ligation product that we want, but it should linearize the re-closed vector that we don't want and make it less effective in transformation.

Plasmid vectors are unlikely to re-close without a cloned DNA insertion if their ends are not compatible. For example a DNA vector digested with the two enzymes *Eco*RI and *Bam*HI would have the incompatible overhanging ends 5′-AATT and 5′-GATC. If a DNA fragment were prepared that also had *Eco*RI and *Bam*HI ends, then it might be inserted into the plasmid ends by ligation with *Eco*RI end joining with *Eco*RI end, and *Bam*HI end joining with *Bam*HI end. However, even in this case we may find a few unexpected side reaction products. The *Eco*RI and *Bam*HI ends of the plasmid vector could be spanned by not one, but three tandem DNA inserts, joined in the order *Eco*RI to *Bam*HI, *Bam*HI to *Eco*RI (i.e., back-to-back), and *Eco*RI to *Bam*HI. This might happen readily if the molar ratio of DNA insert to plasmid vector were very high during the ligation reaction. The calculation of molar concentration is important, particularly when the DNA insert is much smaller in size than the plasmid and might be overloaded in the reaction. For example, if we are cloning a 100-bp DNA insert into a 5000-bp plasmid, the plasmid is 50 times larger than the DNA insert. If we set up a ligation with 10 ng of each DNA, the mass ratio of DNA insert to plasmid is 1:1 but the molar ratio is 50:1. We would be inviting multiple inserts during ligation if we set up the reaction with that ratio.

Another problem we might face in this example is inefficiency in one or both of the endonuclease reactions used to prepare the DNA fragments. For example, if we conducted our *Eco*RI and *Bam*HI enzyme reactions in the same buffer and the *Eco*RI enzyme was inefficient, many of our plasmid vectors might have only been digested with *Bam*HI and not *Eco*RI. These might be readily re-closed during ligation to form the original plasmid, or they might collect a pair of back-to-back DNA fragment inserts *Bam*HI to *Eco*RI and *Eco*RI to *Bam*HI, joined in the middle at an *Eco*RI site.

If we design a ligation reaction so that it involves asymmetrical cohesive ends, for example 5′-GGAC in a plasmid DNA paired with 5′-GTCC in a DNA insert, and 5′-ACTA in a DNA insert paired with 5′-TAGT in a plasmid vector, then we have a greatly reduced chance of side reactions involving re-closure of plasmids or back-to-back insertion of DNA inserts. Some restriction endonucleases leave asymmetric ends, and these enzymes can be employed to do very specific cloning work; however, the easiest way of engineering asymmetric ends is to use the 3′ to 5′ exonuclease in a DNA polymerase such as Klenow enzyme to conduct a controlled digestion of ends. The ends of a DNA fragment generated by PCR can be controlled by the design of oligonucleotides, so a DNA fragment and plasmid vector can be engineered to fit together perfectly (see Chapter 7). This method may also be taken to the extreme of ligation-independent cloning, in which the cohesive ends are so lengthy that they are stable by their base-pairing, and ligation is unnecessary as a prerequisite for transformation.

If we cannot prevent side reactions from occurring during a ligation, we must find a way of identifying the one product we want from all of the unwanted products. If the insertion of a DNA fragment causes a signifi-cant change in size for a plasmid vector, it is relatively easy to rule out candidate plasmids that have re-closed without a DNA insert, or accepted a DNA insert of the wrong size. In a very simple screening procedure, the supercoiled plasmid DNA sizes can be compared by gel electrophoresis and plasmids lacking an insert detected (see *Figure 4.8A*). The plasmid may also be digested with one or more restriction endonucleases to determine the overall size, structure, and orientation of the inserted DNA (see *Figure 4.8B, C*). Finally, a DNA sequencing reaction may be conducted from each end of the multiple cloning site to check that the insertion was made correctly,

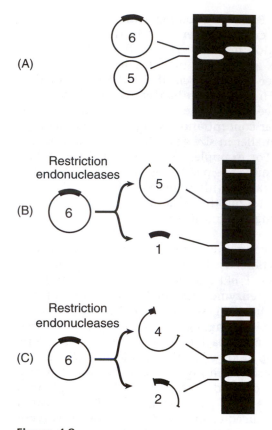

Figure 4.8

Diagnostic tests for plasmid structure. (A) Plasmids can be compared by size, as supercoiled molecules. A 5-kbp vector and a 6-kbp vector (containing a 1-kbp DNA insert) are compared by agarose gel electrophoresis. (B) The DNA insert can be excised to measure its size on a gel. This may not provide information on the orientation of the insert, or whether multiple inserts are present. (C) Restriction endonucleases that digest the insert sequence can be used to determine the orientation of the insert, and to confirm that only one DNA fragment is inserted.

and that no additional small DNA fragments were brought in at the junctions.

TOPO® technology

Vaccinia virus topoisomerase I is an enzyme that can break and re-build the phosphodiester backbone of a DNA, and can be applied in some cases as a substitute for T4 DNA ligase. Vaccinia virus topoisomerase recognizes the cleavage site CCCTT^, where the '^' symbol indicates the site of scission, and cuts one strand. It can be configured in a way that different types of overhanging ends are joined specifically (7). A significant difference between the two enzymes DNA ligase and vaccinia topoisomerase is that the topoisomerase enzyme acts through a DNA-(3′-phosphotyrosyl)-enzyme intermediate. Unlike the enzyme DNA ligase that requires a 5′ phosphate in the substrate, the vaccinia virus topoisomerase eschews the 5′ phosphate and rebuilds the phosphodiester backbone by transferring the 3′ phosphate to a 5′ hydroxyl. PCR-generated fragments generally lack 5′ phosphates, because that is the way most oligonucleotide primers are supplied, but for other applications the 5′ phosphates may be removed by use of the enzyme alkaline phosphatase.

One widely used application is the TOPO® (Invitrogen™) cloning system for inserting PCR-generated fragments into a vector. In this very rapid method of DNA joining, a plasmid vector is linearized and sold as a DNA-(3′-phosphotyrosyl)-topoisomerase intermediate, and the enzyme completes the joining reaction when a suitably configured DNA fragment arrives. Since the enzyme is tethered to the DNA in an activated form, the kinetics are very fast and the joining reaction can be completed in 5 min.

Gateway® technology

The bacteriophage λ genome integrates into the E. coli genome during the process of lysogeny, and the specific recombinase enzyme and sites involved in this joining reaction have been developed into a system for transfer of cloned sequences between plasmid vectors. In applying this Gateway® (Invitrogen™) technology, a researcher initially clones a gene of interest into an entry vector, perhaps by DNA ligation or use of a TOPO® technology. With that initial clone in hand, the DNA insert may be transferred to a variety of different destination vector systems that have the cognate λ recombination sequences. The transfer is accomplished in vitro, using a recombinase enzyme mixture that causes recombination between attL and attR sites, and between attB and attP sites. The recombined destination vector may then be introduced into E. coli cells by transformation.

Transformation and transfection

When we take a plasmid from the 'in vitro workbench' described in Chapter 1 and move it into living cells, that process is called transformation in bacteria, and transfection in eukaryotes. Why is one word used for prokaryotes and a different one for eukaryotes? By the time DNA could be effectively moved into eukaryotic cells in the 1980s, the word 'transformation' had already taken on other meanings among eukaryote biologists, for example transformation as a step in carcinogenesis. That does not entirely resolve the language issues, because the prokaryote biologists use the word transfection to mean a transformation of cells with a bacteriophage nucleic acid, leading to viral production and propagation of the phage.

Transformation

Bacterial cells that are ready to take up DNA from their surroundings, or be transformed, are said to be competent cells. A growing culture of E. coli may not have many competent cells in it, and so to increase the efficiency of transformation in an experiment we use a method to make more of the cells competent. These methods depend in part on whether the DNA will be introduced by heat shock or electroporation of the cells. In the heat-shock method, competent cells are incubated with DNA on ice then subjected to a sudden increase in temperature to 42°C for a brief period of time, perhaps 30 to 45 s. Following the heat shock, the cells are quickly cooled. In the electroporation method, competent cells are incubated with DNA and placed in a cuvette that has stainless steel plates serving as electrodes. The cuvette is placed in an electroporation instrument that delivers a pulse of electricity through the cells in the cuvette (8).

How do these methods work? The cellular details are not entirely clear, but both heat shock and electroporation involve destabilization of the membrane of the cell, and that event probably leads to the uptake of surrounding liquid that includes the nucleic acid. In the case of the heat-shock method, a common way of making the cells competent is to chill them on ice and wash them repeatedly with a 100-mM $CaCl_2$ solution. This treatment may lead to adjustments in the composition of lipids in the cellular membranes to preserve membrane fluidity at lower temperatures. When the cells are suddenly transferred to 42°C, their membranes may be too fluid for the higher temperature, and the cells take up the surrounding DNA as a result. The heat shock is very brief, because cells in this delicate situation do not survive lengthy treatment at high temperatures. In the electroporation method, the delivery of an electric charge to the cells leads to capacitance effects across the membrane and to its momentary breakdown, allowing DNA entry. Electroporation is performed using bacteria in a very low-ionic-strength solution so that the rate of release of charge from the capacitors in the instrument is controlled.

An instrument used for electroporation typically allows for the setting of the total electrical capacitance, and the voltage across the capacitors. With bacterial cells, high electrical field strengths of approximately 1700 $V\,mm^{-1}$ are used, and a capacitance of approximately 25 μF. These parameters must be optimized for each application.

We may think of water raised to a height behind a dam as an analogy for electrical charge. The difference in height between the surface of the water and the point of release of the water is analogous to voltage, and the total volume of water that is backed up behind the dam is analogous to capacitance. The release of water from a dam would start out quickly and gradually slow down, as the height of the water decreased. With the release of electrical charge from a capacitor, the amount of time it takes depends on the capacitance and the resistance of the circuit, and the pattern is again one of starting out quickly and gradually slowing down. The time constant (τ) of a capacitor, which in an electroporation instrument might be called the pulse time, is the number of milliseconds that it takes for the voltage in the charged capacitor to drop to 63.2%, or $1-1/e$ of its starting value. The pulse time is usually measured and reported back to the researcher after completion of a pulse; it is not controllable, except indirectly through the setting of the capacitance and the resistance in the instrument.

After heat shock or electroporation, the cells are chilled briefly and allowed to recover for 30 to 60 min in a rich growth medium lacking antibiotic or other forms of selection. This growth period after transformation allows the cells time to express the selectable marker genes that may have been transferred to them, so that when they are placed under selection they are prepared to survive. The cells are then collected and spread on a bacteriological plate containing the selective medium in an agar base. The population of transformed cells may be plated on two plates, with 1% of the transformation being spread on one and 99% on the other. If the transformation yields thousands of colonies, too many to pick individually, the plate with 1% of the transformed bacteria will have a more manageable tens of colonies. In spreading cells on a plate, it is common for researchers to alcohol-flame a glass spreading tool, let it cool, and use the sterile spreading tool to coat the surface of the plate evenly with cells. This spreading tool must be re-sterilized between uses to avoid cross-contamination. The plates may be stored upright for a few minutes to allow absorption of the liquid, then incubated overnight at 37°C with the lids facing down. This upside-down incubation tends to limit the formation of condensation on the lid, prevents condensed water from raining onto the surface of the agar medium, and prevents colonies from running together.

Transfection

DNA is introduced into eukaryotic cells by a variety of methods, depending on the type of cell and the methods of culture. Several of the early successful methods used to introduce DNA into mammalian cells included aggregation of the DNA with DEAE dextran, or co-precipitation of the DNA with calcium phosphate salts. In both cases, the electrostatic repulsion between the negatively charged cell surface and the phosphodiester backbone is assuaged by cationic shielding, and the DNA can approach the cell surface and be taken into the cytoplasm. From the cytoplasm, the DNA may make its way to the nucleus, or an organelle.

A related approach is to aggregate the DNA with a mixture of cationic and neutral lipids, the typical 'transfection reagent' that is sold as a proprietary formula. The DNA-lipid aggregate is introduced into cells by fusion with the plasma membrane, possibly in the form of a DNA-filled liposome. The cells are exposed to the transfection reagent for only a limited period of time, on the order of 30 min to a few hours, because transfection reagents can have some cell toxicity. Besides the cell exposure time, the overall amount of DNA and the charge ratio of cationic lipid to DNA are important parameters to optimize.

DNA can be introduced into cells by microinjection, but this is a slow and difficult method. The work is done under a microscope, which has to be placed on a vibration-free surface. The cells must be adherent to the surface, or otherwise held tight so that they can be pierced by the very fine pipette that delivers the DNA. The movement of the pipette tip is controlled by a micromanipulator mounted next to the microscope, an instrument that reduces the pipette movements to microscopic distances. The researcher injects the DNA directly into the nucleus, and then withdraws the pipette and moves on to the next cell. The total number of cells that can be injected in a day is limited, so this is a method that is typically used in limited applications such as oocyte injections.

Electroporation is widely used for transfection of eukaryotic cells, but the method is different than that described for bacterial cells and must be optimized for each situation. Eukaryotic cells are larger than bacteria, and are more susceptible to killing during the delivery of electrical charge. The typical method for electroporation of animal cells involves a lower voltage compared to bacteria methods, approximately 300 V, in conjunction with a higher capacitance such as 500 µF. In adjusting these parameters, the researcher may have to choose between having a small number of surviving cells that are very efficiently transfected, and having a larger number of surviving cells that are less efficiently transfected.

Plant cells have a cell wall that interferes with the entry of DNA in many transfection protocols. Plant cells may be converted to protoplasts in some situations, and then transfected by electroporation, but there are other approaches. One solution is to use a gene gun to force DNA-coated microprojectiles into the cells, a method that is also called biolistics. The microprojectiles may be particles of gold or tungsten, accelerated in the gene gun using a compressed gas such as helium. These microprojectiles carry the DNA through the cell wall and plasma membrane, and release the DNA into the cytoplasm. Biolistic gene guns can be used with whole plant tissues, such as an intact leaf, and are also sometimes used with whole animals, for example in DNA immunization through the skin. *A. tumifaciens*-mediated transfer of DNA into plants is a widely used approach that is described below, in the context of a method of conjugation between cells.

Conjugation

Conjugation is a method by which fragments of DNA can be transferred from one bacterium to another. Unlike transformation, it is not a general method that works with any DNA, but it is highly efficient when it is feasible. A natural example of conjugation is the fertility factor F in *E. coli*. F^+ bacteria are 'male', and capable of conjugation with an F− 'female'. Conjugation depends on the assembly of a conjugation bridge that connects the two 'mating' cells. The DNA that is to be transferred must bear specific *cis*-acting sequences that allow it to be mobilized or moved into the recipient. At the start of transfer the DNA is nicked at the *oriT* (origin of transfer) locus, and a single strand is extruded from its free 5' end. The 3' end remains double-stranded, and is extended by a DNA polymerase to fill the gap as the extruded single strand continues to be pulled out and transferred through the conjugation bridge to the recipient (see *Figure 4.9*).

Researchers may split the transfer functions between two plasmids; one plasmid bears the *oriT* locus, and is intended for mobilization and transfer to the recipient, and the second plasmid is a helper plasmid that provides the *trans*-acting transfer functions encoded by the *tra* genes. The helper plasmid stays behind in the donor and is not transferred, because it lacks *oriT*. Both the mobilized plasmid and the helper DNA are maintained stably in the donor cell by a plasmid origin of replication, which is different than *oriT*. The researcher may have a strain of bacteria that carries the helper plasmid only, but is ready to receive a mobilizable plasmid by DNA transformation.

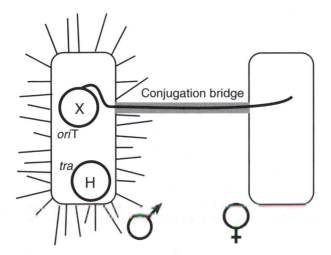

Figure 4.9

Transfer of a plasmid between bacteria by conjugation. Schematic diagram of a conjugation bridge between two *E. coli* bacteria, with transfer of a single stranded 'X' DNA from the pili-coated male (left) to female (right). The *trans*-acting factors that mobilize the plasmid 'X' are shown as genes (*tra*) on a separate helper plasmid 'H'. The plasmid 'X' has a *cis*-acting factor (*ori*T) from the F episome that is required for transfer. The male bacterium retains a double-stranded copy of the plasmid 'X'. DNA entering the female may recombine with the chromosome (not shown) or be retained as a replicon if transfer is complete.

Agrobacterium-mediated DNA transfer to plants

Strictly speaking, conjugation refers to the transfer of DNA between two bacteria; however, a conjugation-like transfer is possible across Kingdom lines: the formation of crown galls in plants is a natural example. Crown galls form as a result of the transfer of a piece of DNA called T-DNA from the soil bacterium *A. tumefaciens* to a plant. The T-DNA becomes integrated into the genome of the plant and carries genes that are expressed in the plant, including some that lead to hormone production and plant tissue overgrowth in the gall. The T-DNA is a part of a plasmid called Ti (tumor inducing), which is stably maintained in the *A. tumefaciens* cells. The *cis*-acting sequences that control T-DNA mobilization are the right-border and left-border sequences, which as their name suggests, flank the T-DNA. The right-border sequence is the point at which transfer is initiated. The *trans*-acting factors required for the conjugative-like transfer of DNA from bacteria to plants, are encoded by the *vir* (virulence) genes on the Ti plasmid.

This natural process of DNA transfer can be put to use in the laboratory by replacing the natural T-DNA in a Ti plasmid with an artificial sequence that is of interest to the researcher. This new T-DNA can be transferred into plants by an *Agrobacterium*, provided that the T-DNA carries the right- and left-border sequences, and provided that the *vir* genes are expressed in the cell. The new T-DNA may be integrated into the plant genome, and introduce a selectable marker gene and a gene of interest, both under the

control of plant transcriptional regulatory sequences. Several types of vectors are used for Ti-mediated gene transfer to plants, and the details of their composition are discussed in Section 4.3.

Shuttle vectors

When we want to stably maintain a plasmid in more than one species of host cell, we use what is called a shuttle vector. An example of a shuttle vector includes the plasmids that are used in Ti-mediated gene transfer to plants, which are first propagated in *E. coli*, and then transferred to *A. tumifaciens*. Shuttle vectors are usually designed to work in *E. coli*, and one other host; for example, *E. coli* and *S. cerevisiae* or *E. coli* and a cultured human cell line.

Why is *E. coli* always involved? *E. coli* is used as a starting point for a lot of work because we can obtain highly competent cells for transformation. Furthermore, we can obtain *E. coli* strains that are mutated to remove unwanted cellular activities, such as recombinase or mismatch repair. Once we have a plasmid established in *E. coli*, we can grow up enough of the plasmid to transform or transfect other species that might not take up DNA as easily, or use a system of conjugation if that is more efficient.

Problems

The solutions to these problems are provided at the end of the chapter.

(i) Which of these double-stranded linear fragments can be circularized by ligation, making use of base-pairing at the ends? (The '...' indicates a long internal sequence.)

```
A.  5'-GATCATTATACCGC...AGGTACAGAAGATC-3'
    3'-     TAATATGGCG...TCCATGTCTT    -5'

B.  5'-CTAGGTAAGTCACT...GACGATTCAG    -3'
    3'-     CATTCAGTGA...CTGCTAAGTCGATC-5'

C.  5'-     ACGGATACGT...AGGCATACAT    -3'
    3'-CTAGTGCCTATGCA...TCCGTATGTAGATC-5'

D.  5'-     CAACGTACAG...ACAAAGTTCTGATC-3'
    3'-CTAGGTTGCATGTC...TGTTTCAAGA    -5'

E.  5'-TCAAGATCCATCAG...AGGCACCTCTTT    -3'
    3'-     TTCTAGGTAGTC...TCCGTGGAGAAAAG-5'
```

(ii) Looking at the same collection of linear fragments, which ends can be brought together in bimolecular ligation reactions?

(iii) Suppose that we want to set up a ligation reaction using 25 ng of vector DNA, where the vector is a 3.5-kbp plasmid, inserting a DNA fragment that is 1 kbp. We want to set up the ligation so that the molar ratio of DNA fragment to vector is 2:1.

   A. If the reaction is conducted in 2.5 μl, what are the gram and molar concentrations of each DNA? (Reminder: the average base pair in DNA is approximately 660 Da.)

   B. If reaction is conducted in 25 μl, what are the gram and molar concentrations of each DNA?

(iv) Suppose we want to set up a ligation reaction in 2.5 µl, using the same 3.5-kbp vector but inserting a 250-bp fragment instead of a 1-kbp fragment. How would we establish a 2:1 molar ratio between DNA fragment and vector in the reaction, if we use 25 ng of vector?

4.2 Viruses

Viruses can be harnessed to carry nucleic acids into cells, a method that is called transduction when it is performed in a prokaryotic organism. It is a straightforward idea: the nucleic acid is packaged into a viral particle, either *in vitro* or *in vivo*, and the viral particle infects the target cell. Transduction complements two methods of introducing DNA into cells that were discussed previously, transformation (or transfection) and conjugation. We use the term bacteriophage to describe a virus that infects bacteria, but it is not improper to use the word virus when talking about bacteriophage. Viral vectors are also used with eukaryotic organisms, particularly in cases in which electroporation would not be feasible, such as when tissues or whole organisms are the intended recipients.

Some viral vectors are replication competent; they replicate in the target cell, spreading recombinant sequences to other cells. Other vectors are simple delivery vehicles that are not capable of repackaging themselves into viral particles and being spread. There is also variation in the amount of nucleic acid that can be packed into viruses, with some viruses holding strict packaging limits and others being able to accommodate a broader range of fragment sizes. Genetic elements from viruses are widely used in cloning vectors, whether or not the vector is packaged into a viral particle when it is used. The applications of viral elements in chimeric vectors are discussed in Section 4.3.

Handling and storage

Bacteriophage and other viruses have differing optimal storage conditions. For example, bacteriophage λ particles are stored at 4°C in a buffered solution containing 10 mM magnesium ion, which is critical for bacteriophage viability. A common bacteriophage λ storage solution is called 'SM buffer', and consists of 50 mM Tris-Cl, pH 7.5, 100 mM NaCl, 10 mM $MgSO_4$, and 2% (w/v) gelatin. The titer of a bacteriophage stock does drop with time at 4°C, and long-term storage is best done at -70°C in a solution containing 7% dimethylsulfoxide (DMSO).

A crude stock of a bacteriophage or virus may amount to nothing more than a cell-free supernatant from a culture. Bacteriophage and viruses are sufficiently small that they are not brought out of suspension by moderate centrifugation. For example, centrifugation at $10\,000 \times g$ for 20 min would pellet most cellular debris but would leave viral particles in liquid suspension. Most viruses will also pass through a 0.45-µm filter, provided that the filter is low protein binding or that there is sufficient carrier protein in the solution to block binding. Why do virus stocks lose infectivity over time? Cell supernatants often have proteases released from dying cells, and these may attack the surface proteins on a virus and reduce its infectivity. A highly purified stock of virus, for example a bacteriophage λ stock that has been purified by CsCl-density-gradient centrifugation, will tend to retain its titer longer than a crude stock.

When working with bacteriophage and other viruses, it is important to disinfect virus-containing waste solutions, and disinfect the laboratory equipment and disposable plasticware that comes in contact with the viruses. Some viruses, such as bacteriophage M13, can spread through cultures in a laboratory by contaminating aerosol droplets, which may form during procedures that lead to liquid shear, for example agitation of a culture flask, vortexing, or sonication. Pipetting can also lead to aerosol droplets, particularly if a stream of liquid is allowed to drop into a receiving flask from some height, forming bubbles in the solution that pop, or if the last drop of a dispensed liquid is pushed out of the pipette by air pressure and allowed to pop on the pipette tip. With pathogenic viruses, the contamination of aerosol droplets may be a health and safety concern as well as an experimental concern. Many viruses are only handled in biological containment hoods with laminar flow, to prevent accidental contamination.

Plaque purification

It is important always to start a viral growth from a pure stock that is genetically uniform. The process of isolating a uniform stock of virus or bacteriophage is called plaque purification, and it involves a repetitive process of plating the virus on host cells in such a way that isolated viral plaques can be 'picked' and re-grown. This may be done in semisolid medium, for example agar-based plates used with bacteriophage, but the problem with any system is that the virus may diffuse from one plaque to another and cross-contaminate a sample.

In a typical process of plaque purification, a virus or bacteriophage is plated at low density so that plaques will be widely separated. The virus or bacteriophage isolated from a single plaque is re-plated, again at very low density, so that a single plaque can be isolated without cross-contamination. A third round of plaque purification may then be performed, and the single plaque isolated is assumed to be homogeneous.

What's wrong with having a little bit of contamination in a viral stock? The real problem is that bacteriophage and viral variants may be propagated at different rates; if a heterogeneous stock is used to start an infection, it may change over time. The predominant virus, the one the researcher wants to grow, may be quickly overtaken by faster-growing variants. Even if it is not overtaken during growth, experimental results may be ascribed to the predominant virus in a population, when in fact the results are an effect of a contaminating virus that is unknown to the researcher.

Multiplicity of infection

When it is time to propagate a bacteriophage or other virus, a stock of the virus is mixed with host cells and the cells are then grown in a nutrient medium to allow productive infection. In some protocols the virus and host cells are mixed together in a small volume first, perhaps being incubated on ice for several minutes to allow for adsorption of the virus, then diluted into a larger volume of growth medium. This adsorption step is not always required, but the higher initial concentration does improve the chances of the virus and host cell finding each other in solution.

The initial ratio between number of infectious viruses and number of host cells is called the multiplicity of infection (MOI), and this is an important parameter to calculate when setting up a viral growth. The number of viruses in a stock can be determined by a process called titering, in which small amounts of a virus stock are mixed with cells and the number of infective centers or plaques is counted on a plate. It is assumed that each plaque represents the virus propagated from a single infection event, so the number of plaque-forming units per μl of virus stock is a measure of infectious particles per μl. The number of viral particles in a culture may exceed the number of infectious units, if some of the particles have been inactivated or degraded.

The number of host cells in a prokaryotic culture can be estimated by optical density at 600 nm. For example, an *E. coli* culture with an optical density of 1.0 at 600 nm will have approximately 5×10^8 cells per ml, although the viability of the cells will decrease as they age. In the case of infection with bacteriophage λ, the bacterial stock is usually grown in medium supplemented with maltose to induce expression of the bacteriophage receptor on the cell surface. A stock of bacteria used in infection should be grown from a single colony on a fresh plate to minimize the number of variants in the culture. Maltose is not added to the growth medium of an infected culture because it competes with the bacteriophage for binding to the cell.

When bacteriophage λ are being propagated in *E. coli* to make a high titer stock, either for the purposes of long-term bacteriophage storage or for DNA preparation, it is common for a researcher to use an MOI of approximately 0.0005 to 0.005: a ratio of 1 plaque forming unit (pfu) for every 200 to 2000 bacteria. That may not seem like much, but the bacteriophage grow much more quickly than the bacteria. The highest yield of bacteriophage will be obtained if the uninfected bacteria continue to grow, and are only overtaken by virus at about the time when the bacteria have exhausted the medium anyway. If too much virus is added the bacteria will be killed off before they can make a suitable amount of viral progeny, and if not enough virus is added the bacteria will exhaust the growth medium before they have a chance to be infected. In the latter case, an overgrown flask of λ-infected *E. coli* can sometimes be made productive for virus if fresh medium is added and the period of growth extended. A perfectly timed infection results in a flask that is opalescent and has suspended aggregates of lysed cell debris. Lysis may sometimes be improved in a flask by chilling it on ice and adding a few drops of chloroform. The bacterial debris should be promptly separated from the λ in solution by centrifugation, to prevent loss of virus by adsorption to the debris.

Multiplicity of infection is also important in productive infections with eukaryotic viruses. Eukaryotic cell numbers are usually estimated by counting a small representative sample in a hemocytometer under a microscope. The central 'bullseye' of a hemocytometer is a 5×5 grid of squares, representing a volume of 0.1 μl, or 1×10^{-4} ml. A hemocytometer facilitates the estimation of eukaryotic cell numbers, because if 100 cells are found to lie in the central (5×5) area of the hemocytometer, the concentration of cells is 100 per 1×10^{-4} ml, or equivalently 1×10^6 per ml. The dimensions of the central (5×5) area of a hemocytometer is $1 \times 1 \times 0.1$ mm, and the entire counting surface of the hemocytometer is a 3×3 array of these. It is common practice to count five of the $1 \times 1 \times 0.1$ mm squares, and average the results. When a culture is so concentrated that the cells are overlapping on the grid, or otherwise too numerous to count, a dilution may be made to bring the number of cells per grid down to a countable number. The dilution factor must then be included in the calculation of cell concentration in the original, undiluted stock.

Extracting viral genomes

All viruses have a capsid assembled from coat proteins that protect the nucleic acid genome. Some viruses have an integumentary layer and viral envelope as well. A researcher may add DNase and RNase to a preparation of virus to destroy host nucleic acids that contaminate a viral preparation, without fear that the enzymes will also destroy the viral nucleic acid. Following that pre-treatment with nuclease, most viruses can be opened up through the addition of detergent, protease, and mild heat.

For example, the bacteriophage λ capsid can be digested through the addition of sodium dodecyl sulfate (SDS) to a concentration of 0.5% (w/v), proteinase K to 100 μg ml^{-1}, and incubation at 37°C for approximately 1 h. The released viral DNA can be separated from the protein fragments by phenol extraction, and concentrated by ethanol precipitation. The viral DNA from bacteriophage λ is approximately 50 kbp, and susceptible to shearing if handled roughly.

Bacteriophage M13 is a filamentous bacteriophage that causes a persistent infection of *E. coli*. The cells are not killed by the virus, but grow more slowly than uninfected cells and shed virus continually. The virus is propagated in liquid culture by mixing virus with uninfected cells, and growing the cells in fresh medium for a period of a few hours. The infected cells can be separated from the supernatant by centrifugation, and the supernatant saved as a stock of virus. The replicative form of the viral genome can be isolated from the collected cells if a researcher wishes to preserve a double-stranded version for *in vitro* work with restriction endonucleases. The virus can be concentrated from the clarified supernatant by precipitation, using 0.5 M NaCl and 5% (w/v) polyethylene glycol (PEG-6000 or PEG-8000), and centrifugation for 20 min at $12\,000 \times g$. The pellet containing the virus can be drained of all residual PEG, then re-suspended in a 10 mM Tris-Cl, pH 8.0, 1 mM EDTA buffer, and extracted with phenol to remove the lipids and protein. The aqueous fraction containing the viral DNA can then be extracted with chloroform and the single-stranded DNA genome precipitated with isopropanol. Alternatively, the re-suspended viral pellet can be treated with 4 M NaI for approximately 5 min, which will solubilize the proteins, and the single-stranded DNA can be precipitated with ethanol.

Cloning sites and bacteriophage arms

How do we genetically modify viral vectors to insert a new piece of DNA? A common way of working with viral genomes is to use restriction endonucleases to open a viral DNA at a multiple cloning site, and insert a new piece of DNA using the enzyme DNA ligase. When we insert a piece of DNA into a vector based on the filamentous bacteriophage M13, we work with the double-stranded replicative form (RF) that is extracted from persistently infected *E. coli* cells. The RF is a circular DNA molecule, and so inserting a DNA fragment to make a recombinant molecule involves the very same steps that would be used with a plasmid cloning vector (see Section 4.1). On the other hand, when we use a vector based on bacteriophage λ, we work with a double-stranded linear DNA that is extracted from viral particles. The ends of the λ vector are the cohesive ends, or *cos* sites, that promote circularization of the vector after infection. When the λ vector is treated with a restriction endonuclease that digests the DNA at a cloning site, the DNA is separated into linear 'arms' that must be reassembled with the insert to make a recombinant (see *Figure 4.10*). That is, the 'left arm' and 'right arm' of the bacteriophage hold the inserted DNA between them, and must both be present for the vector to be functional.

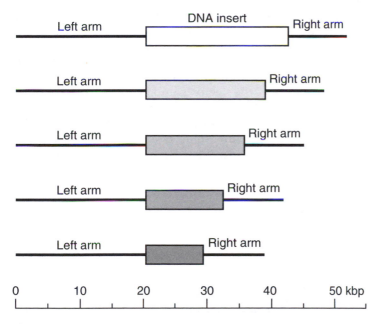

Figure 4.10

Bacteriophage λ vectors allow the cloning of DNA inserts in a range of sizes. The DNA insert is cloned between the left and right arms of bacteriophage λ vectors, and the range of allowable insert sizes is shown schematically against a scale of kilobase pairs (bottom). In the case of Lambda FIX® II (Stratagene®), the left arm is approximately 20 kbp and the right arm is approximately 9.1 kbp.

Some vectors have a piece of DNA called a stuffer that is typically removed and replaced during cloning. The stuffer DNA typically serves only one simple function and that is to take up space. For example, some viruses have size limits on both ends of the scale, being unable to package or propagate viral DNA that is either too small or too large. A stuffer DNA may be placed in such a vector to allow propagation of the viral vector before it is used in DNA cloning. The stuffer maintains the genome at a viable size, and it is replaced during cloning with a fragment of interest.

When vectors carrying a stuffer DNA are used in cloning, they must be digested with restriction endonucleases that cut on both sides of the stuffer fragment, separating it from the rest of the circular DNA or viral DNA arms. Some vectors have symmetric multiple cloning sites flanking the stuffer DNA, so that a single restriction endonuclease can be used to excise the stuffer. For example, the vector Lambda FIX® II has a multiple cloning site on either side of the 14-kbp stuffer DNA (see *Figure 4.11*). On the left arm the order of endonuclease recognition sequences is *Xba*I, *Sac*I, *Not*I, *Sac*I (again), *Sal*I, *Xho*I, *Eco*RI, and *Xba*I (again), whereas on the right arm the order is (exactly reversed) *Xba*I, *Eco*RI, *Xho*I, *Sal*I, *Sac*I, *Not*I, *Sac*I (again) and *Xba*I (again). A researcher might use *Eco*RI to excise the stuffer and prepare the vector for receipt of a new DNA fragment, and after cloning, the distal enzymes *Xba*I, *Sac*I, *Not*I, *Sal*I, and *Xho*I would still be present in the multiple cloning sites, and useful for mapping or excision of the DNA insert.

What keeps the stuffer DNA from going right back into the vector during cloning? There are a couple of ways of keeping the stuffer DNA from competing with the DNA insert during ligation, or selecting against the stuffer DNA during propagation of the candidate bacteriophage. With bacteriophage λ vector systems, one simple way is to purify the arms by gel electrophoresis. In the case of the vector Lambda FIX® II, the left arm, stuffer, and right arm are respectively 20 kbp, 12.8 kbp, and 9.1 kbp (see *Figure 4.11*). If a researcher wished to purify the two arms, then the 20-kbp and 9.1-kbp fragments of digestion would be saved and the 12.8-kbp fragment discarded. Some bacteriophage λ systems, for example the EMBL

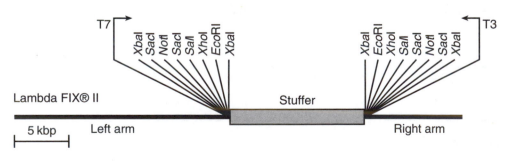

Figure 4.11

Map of the Lambda FIX® II replacement vector. The vector contains two polylinker regions, with flanking T7 and T3 RNA polymerase promoters. The DNA separating the polylinkers is the stuffer, which is removed during cloning and replaced with a foreign DNA insert.

vectors, have stuffer DNA that can be broken into small fragments using restriction endonucleases, and that can simplify the purification of arms by their size.

Many bacteriophage λ vectors allow for selection against the re-ligation of stuffer fragments, by making use of a 'sensitive to P2 inhibition' or 'sensitive to P2 interference' (Spi) phenotype. Vectors such as Lambda FIX® II (see *Figure 4.11*) and the EMBL vectors include the *red* and *gam* genes in their stuffer sequences, and these genes make the bacteriophage unable to replicate lytically in *E. coli* lysogenic for bacteriophage P2. When the stuffer is excised and replaced with a DNA insert, then the *red* and *gam* genes are lost and the bacteriophage can replicate on *E. coli* lysogenic for P2. So, in a mixed population of λ bacteriophage, some with leftover stuffers and others with replacement inserts, only the bacteriophage with replacement inserts will be propagated on *E. coli* lysogenic for P2.

An alternative way of preventing the stuffer fragment from re-entering the λ bacteriophage arms is to change the ends so they are incompatible during ligation. The enzyme alkaline phosphatase may be used to cleave the 5′ phosphate groups from the arms and the stuffer fragment prior to ligation. The replacement insert is not treated with alkaline phosphatase, so it still has its 5′ phosphate groups intact. When the arms, stuffer, and replacement insert are mixed together in a ligation reaction, the arms and the replacement insert may be covalently connected using the 5′ phosphate groups from the insert. The stuffer cannot be reconnected to the arms because, between the two, there are no 5′ phosphate groups that can be used to rebuild the phosphodiester backbone.

Another method of preventing the stuffer from re-inserting into the arms is to separate the arms from the stuffer using the enzyme *Xho*I, and then fill in two of the four nucleotides in the overhanging end using Klenow DNA polymerase. This method of partial fill-in of a restriction site is described in Chapter 6; however, for this discussion the two important points are that the treated stuffer and arms are no longer compatible with each other during ligation, but the ends of a replacement insert can be made to be compatible with the arms. The partial fill-in of the bacteriophage λ arms must be conducted after the *cos* ends of the arms are ligated, to prevent damage to the cohesive ends by the enzyme Klenow.

Bacteriophage P1 vectors may also carry a stuffer fragment to reserve space in the genome for an inserted DNA, but in this case the stuffer may not be completely excised during cloning work. Bacteriophage P1 has a packaging limit of approximately 100 kbp, and in a vector that has a packaging limit of about 80 kbp of DNA there may still be considerable variability in phage genome size. After insertion of a DNA fragment into the P1 vector, the stuffer fragment remains with the vector and serves the purpose of filling out the overall size so that it is at least 100 kbp. During packaging of the DNA into bacteriophage P1 particles, the excess stuffer sequence beyond this 100 kbp limit is removed. So, in this case, the stuffer serves the purpose of 'packing material' to fill out the inside of the P1 particle. A more general approach to using the P1 replicon for making an artificial chromosome does not rely on bacteriophage packaging, and is discussed in Section 4.3.

Bacteriophage packaging limits and capacity

A packaging extract is used to load DNA into a bacteriophage capsid *in vitro*, so that it may be transduced into a cell. The packaging extract consists of all of the proteins involved in capsid assembly, and when the researcher adds DNA capable of being packaged then the reaction proceeds and capsids are filled. Bacteriophage have different requirements for the *cis*-acting sequences that initiate packaging, and the example of bacteriophage λ will be primarily discussed here.

Bacteriophage λ initiates packaging of DNA at a *cos* site, and terminates packaging at the next *cos* site encountered in a DNA fragment. Wild-type lambda is approximately 48.5 kbp, and the range of sizes that can be successfully packaged is approximately 79% to 106% of the wild-type size, or 38-52 kbp. Prior to ligation with a DNA insert, the arms of bacteriophage λ may be pre-ligated at their *cos* sites. Then, the bacteriophage λ is digested with restriction enzymes to prepare a cloning site for a DNA insert. During ligation with a DNA insert, a long concatemer of bacteriophage λ genomes may be assembled, and this becomes the substrate for the packaging reaction (see *Figure 4.12*). Within the concatemer will be genome monomer units consisting of a *cos* site, bacteriophage right arm, DNA insert, bacteriophage left arm, and a second *cos* site. The bacteriophage λ capsid is filled, starting at a *cos* site, and terminating at the next *cos* site at the other end of the monomer genome. Another empty bacteriophage λ capsid will pick up where the previous (now filled) one left off, packaging the next genome in the concatemer. The process continues, and leads to a collection of bacteriophage λ capsids that are filled with unit length recombinant genomes.

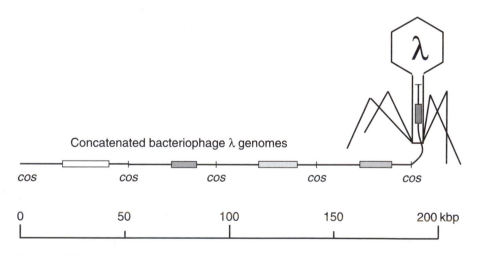

Figure 4.12

Schematic representation of a bacteriophage λ packaging reaction.
Concatenated bacteriophage λ genomes are the substrate for packaging. Empty λ capsids are filled with DNA, starting and ending at a *cos* site. When the capsid on the right is filled, another empty capsid will be used to package the next λ genome in line.

The distance between the *cos* sites must be 38 to 52 kbp if the packaging is to be successful, and this explains why large capacity bacteriophage λ vectors have stuffer fragments. Vectors such as Lambda FIX® II and the EMBL vectors have arms that combine to make 29 kbp, leaving a shortfall of 9 kbp. The stuffer fills the shortfall, and can be replaced with a DNA insert of between at least 9 kbp and at most 23 kbp, to yield the overall required size range of 38 to 52 kbp. Some bacteriophage λ vectors are designed to carry small DNA inserts, and they have no need of a stuffer. For example Lambda ZAP® II is 40.8 kbp, and its starting size exceeds the lower limit of 38 kbp even without an inserted DNA. It can accept DNA inserts of up to approximately 10 kbp without exceeding the packaging limit of the bacteriophage capsid.

How do bacteriophage λ packaging extracts work? There are several types of extracts that are commercially available, or that can be made in a lab, but they share the feature of providing all of the protein factors required for bacteriophage assembly. One type of packaging system is constructed by mixing two extracts; each is deficient in a different protein required for assembly. When these two extracts are mixed together with a DNA substrate provided by the researcher, the assembly of functional viral particles is enabled by complementation between the extracts. The endogenous λ genomes are inactivated by UV irradiation, and so are not packaged to produce viable contaminating bacteriophage. A different type of extract is generated from cells that express all of the required protein factors, but are unable to package their endogenous bacteriophage λ genomes because of a mutation in their *cos* sites. This single-tube extract is missing only the *cis*-acting sequences for packaging, and so can package a properly configured (*cos*-containing) DNA substrate added by a researcher.

The only *cis*-acting sequences from bacteriophage λ that are required for packaging are the *cos* sites, and it is possible to package a plasmid that contains *cos* sites and no other bacteriophage λ sequences. This type of vector is called a cosmid, the word being derived from '*cos*' and 'plasmid'. The important principle of cosmids is that they are plasmids that are passed to a bacterium by transduction in a bacteriophage λ capsid. They are not virulent and cannot spread from cell to cell by infection since they contain no genes from bacteriophage λ, and they can only be packaged if the cosmid carries a sufficient amount of inserted DNA so that the *cos* sites are 39 to 52 kbp apart (see Section 4.3).

Bacteriophage P1-packaging extracts initiate packaging from a *cis*-acting sequence called *pac*, and continue filling the capsid until the packaging limit of approximately 100 kbp is achieved. A P1 vector may be designed to have *lox* sites flanking the stuffer so that the stuffer can be removed by Cre-mediated site-specific recombination, once inserted into an *E. coli* cell.

Bacteriophage M13 vectors such as mp18 and mp19 are introduced into cells by transformation, so there is no need for an *in vitro* packaging extract. There is no exact packaging limit for bacteriophage M13, which has a wild-type size of 6.4 kb, but the replication cycle may disfavor genomes with large insertions and lead to rearrangements and loss of sequence. The M13 vectors mp18 and mp19 are 7.2 kb, and are typically used in cloning inserts of approximately 1 kbp. This limitation of mp18 and mp19 may be more a function of the persistent infection than the packaging step, because large DNA sequences of at least 13 kbp can be packaged into filamentous bacteriophage particles *in vivo*, using phagemid technologies described in Section 4.3. Unlike the M13 cloning vectors mp18 and mp19, phagemids are simply packaged, single-stranded versions of plasmids.

Helper viruses and packaging cell lines

In thinking about how bacteriophage and viruses are produced, we distinguish between genetic elements that are required in *cis*, and those that are required in *trans*. An example of a required *cis*-acting sequence is the *cos* site of bacteriophage λ, and an example of a *trans*-acting factor is a protein factor used in the construction of the viral particle. In bacteriophage packaging extracts, the extract provides the *trans*-acting factors and the DNA substrate provides the *cis*-acting factors. However, the same principle can apply *in vivo*, if *trans*-acting factors are provided by a 'helper' virus or by a specialized packaging cell line.

For example, Lambda ZAP® II (Stratagene®) has an intergenic region from the filamentous bacteriophage f1, and this region contains an origin of replication that can be activated by co-infection with an M13 filamentous bacteriophage as a 'helper'. As described in Section 4.3, the portion of the Lambda ZAP® II carrying the DNA insert is packaged in M13 capsids as a phagemid and secreted from the co-infected cells. The phagemid inside the M13 capsid is not virulent and cannot be spread from cell to cell by infection; it contains a single-stranded copy of a plasmid pBluescript II® (Stratagene®).

To take an example from animal viruses, the Ψ element of murine leukemia virus (MLV) is a *cis*-acting sequence required for packaging of a retroviral RNA genome. An RNA sequence carrying the Ψ element may be packaged in a cell line that expresses the specific retroviral *gag*, *pol*, and *env* genes, which encode *trans*-acting viral proteins. To prevent the development of replication competent retroviruses by recombination, these retroviral genes may be distributed in the chromosome of the packaging cell line. MLV-based vectors are limited by the overall packaging size of the viral genome, approximately 10 kb. They are used to target nucleic acids to dividing cell populations, but not to quiescent cells.

Herpes simplex virus (HSV) vectors carry an HSV *pac* element and can be packaged by a cell line expressing the *trans*-acting protein factors from a 'helper' genome. Cells are transfected with a vector that provides the *cis*-acting sequence, and a DNA containing multiple copies of the vector may be packaged into a single virion. For example, ten tandem copies of a 15-kbp sequence may be packaged into a virion that would normally hold a 150-kbp genome. The 'helper' genes may be integrated into the packaging cell-line chromosome and broken into several independent segments to prevent recombination events that might lead to the building of a replication-competent viral particle.

Adenoviral vectors are available for introducing DNA sequences into a variety of eukaryotic cells, and these are packaged in a human embryonic kidney (HEK 293) cell line. The cell line expresses the E1 genes that are required for propagation of the virus; however, the E1 genes are not included in the vector DNA. The generated virus may be used to carry a gene of interest into a quiescent or proliferating target cell, but they are replication incompetent and do not spread from cell to cell after gene delivery.

Adeno-associated viruses (AAV) are small DNA viruses that depend on an adenovirus or herpes virus as a helper virus during propagation. They are considered important vectors for gene therapy because they are nonpathogenic. To establish a packaging cell line for recombinant AAV, the *rep* and *cap* open reading frames of AAV are expressed from a separate vector, and the cells are infected with adenovirus. Alternatively, a replication-incompetent adenovirus may be used to limit the contamination of recombinant AAV.

Viral regulatory sequences

Cloning vectors that are based on viruses fall into two principal categories; those that carry enough of the viral genome to allow productive infection and spread of the virus, and those that do not. Bacteriophage λ vectors such as Lambda FIX® II, the EMBL vectors, and Lambda ZAP® II, as well as M13 vectors such as mp18 and mp19, are examples of the first type, because a DNA insert is cloned into the context of a working viral genome. Viral-based vectors that do not lead to productive infection include cosmids and phagemids, and retroviral vectors carrying a Ψ sequence but no retroviral genes. Many of these are chimeric vectors that combine plasmid and viral elements, and these will be discussed in more detail in Section 4.3.

The life cycle of bacteriophage λ

Bacteriophage λ is the source of many important regulatory elements used in recombinant vectors, so a brief introduction to its life cycle is in order. When an *E. coli* cell is infected with bacteriophage λ, it may either enter a lytic cycle and generate new viral particles, or become lysogenic and integrate the bacteriophage λ DNA into the cell's chromosome. During lysogenic growth, the viral DNA is propagated along with the host chromosome. At some later time the cell may emerge from its lysogenic state and enter a productive lytic cycle. The selection of one or the other path is based on a regulatory switch involving several viral and host proteins, as well as several key viral transcriptional promoters.

While the story of bacteriophage λ regulation is a bit too complicated to explain in any detail here, a key point is that the lysogenic state is maintained by expression of a viral protein repressor called cI (pronounced 'see one'). The *cI* gene product represses the expression of a gene called *cro*, which if expressed would open up the rightward and leftward transcription of the genome, leading to lytic growth. The cI repressor prevents expression of cro protein by binding to the operator region of a promoter called P_R. If a sufficient amount of cI repressor is expressed shortly after infection, a previously uninfected cell may become lysogenic for bacteriophage λ. However, a host protein called hfl tends to decrease the expression of cI repressor in freshly infected cells, and so in rapidly dividing cells expressing high levels of hfl, there is an increased likelihood of establishing lytic growth. Cells that are lysogenic for bacteriophage λ are immune to new infection with bacteriophage λ, because cI repressor expressed from the resident bacteriophage λ DNA also represses the newly arrived bacteriophage λ DNA and prevents it from causing a lytic infection.

When wild-type bacteriophage λ infects *E. coli* on a bacteriological plate, the plaques are turbid or cloudy because the infection leads to a mixed outcome. Some cells are engaged in lytic viral growth, which causes virus release and cell death, but there is also some overgrowth of lysogenic cells in the presence of the virus and these cloud the plaque. A bacteriophage λ that does not produce a functional cI repressor cannot repress lytic growth, and so produces no lysogeny upon infection. These *cI* mutants produce clear plaques instead of turbid plaques, because there is only lytic growth and cell killing.

Many bacteriophage λ vectors are designed to carry large fragments of foreign DNA into an *E. coli* cell, and to make room for this DNA the vectors generally only carry the essential genes for lytic growth. Vectors such as Lambda FIX® II and the EMBL vectors lack the *cI* gene and all of the other genetic elements associated with lysogeny, and produce only clear plaques when plated on wild-type *E. coli* cells. In these vectors, the behavior of bacteriophage λ is simplified; they will only grow lytically, and not lysogenically.

There are other bacteriophage λ vectors, for example Lambda ZAP® II and λgt11, that carry smaller DNA inserts and have not lost the ability to grow lysogenically. These λ vectors generate turbid plaques on a plate of *E. coli* cells because they carry a *cI* gene and all of the other necessary genetic elements for lysogeny. These vectors make use of a temperature-sensitive allele of the *cI* gene, called *cI857*, that produces a cI repressor protein that is functional at 32°C and nonfunctional at 39°C. A bacteriophage with the *cI857* allele supports only lytic growth at the higher temperature, and permits a mixture of lytic growth and lysogeny at the lower temperature.

Bacteriophage and viral insert detection

Although infrequently used, the bacteriophage λgt10 is an interesting example of a vector in that it has a cloning site that interrupts the *cI* gene (9). If the λgt10 vector carries no DNA insert, then the *cI* gene is functional and the bacteriophage vector produces turbid plaques on a plate of *E. coli*. However, if the λgt10 vector carries a DNA insert that interrupts and interferes with the *cI* gene, then the bacteriophage vector produces clear plaques because only lytic growth can occur. Furthermore, a researcher could plate a mixed population of λgt10 candidate phage on an *hfl* strain of *E. coli*, which makes no hfl protein, and only λgt10 candidates carrying DNA inserts will grow lytically on the *hfl* strain. Lysogeny would be favored in those λgt10 without a DNA insert because *cI* is intact and expressed in the absence of hfl protein. That's a nice way of selecting for success during a cloning experiment.

When we use large-capacity bacteriophage λ vectors that require a stuffer fragment, we don't need to worry about propagating bacteriophage in which the two arms have come together during DNA ligation, and without a DNA insert between them. These 'arms only' bacteriophage would be about 29 kbp and too small to be propagated. However, we do need to take some steps to prevent reinsertion of the stuffer DNA, as discussed earlier, and to avoid having multiple small DNA inserts come together during ligation to make a large chimeric insert. DNA that is going to be inserted into a bacteriophage such as Lambda FIX® II or the EMBL vectors may be size-selected before ligation, so that only DNA fragments in the size range 14 to 23 kbp are included. A single DNA fragment in this range may be ligated together with bacteriophage arms to produce a bacteriophage genome that is appropriately sized, but if two or more of them come together then the ligated DNA is too large to be packaged. Multiple insertions are not likely if the bacteriophage arms were prepared by partial fill-in of a *Xho*I site, because the DNA inserts will similarly have a 2-bp overhanging end and will not be compatible with other potential inserts during ligation. For example, when the DNA inserts are prepared by limited digestion with *Sau*3AI, leaving a 5'-GATC overhanging end, the complementary strand can be partially filled in with Klenow enzyme and the substrates dGTP and dATP to leave a 5'-GA overhanging end. The 5'-GA ends of the DNA inserts are compatible with the 5'-TC over-hanging ends of the arms, but not compatible with the 5'-GA overhanging ends of other DNA inserts. This method is discussed in more detail in Chapter 5.

We do worry about re-closure of bacteriophage arms without a DNA insert when using a vector such as Lambda ZAP® II, because these bacteriophage can be propagated without any stuffer or inserted DNA. The previous example of selecting for DNA inserts in λgt10 is not applicable to Lambda ZAP® II and λgt11, because the *cI* gene remains intact during cloning. However, the cloning sites in Lambda ZAP® II and λgt11 are within the coding sequence of a β-galactosidase gene and that allows for blue-white screening using the colorimetric substrate X-gal, as described in Section 4.1. In both Lambda ZAP® II and λgt11, the β-galactosidase gene is under control of the *lac* operon promoter, and synthesis of β-galactosidase is induced by the addition of IPTG. Lambda ZAP® II carries only the *lacZα* portion of the gene, alternatively called *lacZ'*, and must be plated on a complementing strain of *E. coli* that carries the *lacZΔM15* gene. On the other hand, λgt11 carries the entire *lacZ* gene and does not require a *lacZΔM15* gene in the host cell.

When a mixed population of these bacteriophage, some with DNA inserts and some without, are plated with *E. coli* in the presence of X-gal and IPTG, two types of plaques will be observed. The blue plaques are caused by infection with bacteriophage expressing β-galactosidase, and therefore have no DNA insertion in *lacZ*, and the colorless plaques are caused by infection with bacteriophage carrying a DNA insert in *lacZ*. The colorless plaques are an indication of successful cloning. When foreign DNA is inserted into Lambda ZAP® II or λgt11, addition of IPTG to the growing culture causes expression of the inserted sequence as a fusion protein with a fragment of β-galactosidase. This allows for an important application of these bacteriophage in immunological screening, as discussed in Section 4.4.

The M13-based vectors mp18 and mp19 can be screened for inserts using blue-white screening, because the cloning site lies within a *lacZα* gene. When these vectors are used in DNA cloning and transformed into *E. coli* (*lacZΔM15*) cells, the transformed cells become persistently infected with the M13 vector and produce plaques. If the bacteriological plates contain X-gal and IPTG, the plaques are colorless if a DNA insert disrupted the *lacZα* gene, and blue if the *lacZα* gene is uninterrupted.

The Bac-N-Blue™ system for generating recombinant baculoviruses also makes use of a *lacZ* and X-gal mediated colorimetric assay, but the blue color in infected cells is indicative of a successful cloning experiment rather than an unsuccessful one. The recombinant virus is made by co-transfection of Sf9 insect cells with a linearized viral genome and a transfer vector containing a gene of interest for expression in eukaryotic cells. Successful recombination between the baculovirus genome and a transfer vector leads to rebuilding of the *lacZ* gene, and β-galactosidase can then be expressed and detected as a reporter gene. No IPTG is added to this system, because the *lacZ* gene is expressed from a constitutive baculovirus 'early to late' (ETL) promoter that is not regulated by IPTG. Furthermore, cells infected with nonrecombinant virus express the baculovirus polyhedrin gene, a characteristic used for insert detection. Polyhedrin gene expression leads to the formation of occlusion bodies in infected cells, so a lack of occlusion body formation in infected cells helps to confirm that the cell is infected with a recombinant virus.

Difficult-to-clone sequences

DNA fragments are not always stable in a prokaryotic host such as *E. coli*, which can be a problem if a researcher is trying to construct a representative library of genomic DNA. For example, some bacterial strains have restriction systems that cause digestion of DNA with 5-methyl cytosine bases in CpG sequences, a common modification of eukaryotic DNA. Methyladenine is a modified base that can also be important in a restriction system. Specific *E. coli* strains that are 'restriction minus' are used to improve the maintenance of methylated DNA; *mcrA*− *mcrBC*− strains are used when cytosines might be methylated, and *mrr*− strains are used when adenines or cytosines might be methylated. Conversely, some bacterial strains have restriction systems such as *Eco*KI, that cause digestion of DNA not modified with methyladenine at specific sites, and strains of *E. coli* that are *hsdR*− are used to prevent digestion of these unmodified sequences.

Foreign DNA fragments may also be rearranged or subjected to recombination in a bacterial host, leading to difficulty in maintaining direct or inverted repeats in foreign DNA sequences. Strains of *E. coli* that are deficient in the genes *recA*, *recB*, and *recC* are widely used as host cells for bacteriophage libraries, to try to minimize the rearrangement or loss of DNA inserts. Mismatch repair can also be a problem, for example if a researcher is attempting to perform site-directed mutagenesis and has transformed *E. coli* cells with a heteroduplex. The cell may erase the mutation in strand that was synthesized *in vitro* and replace it with a sequence complementary to the parental strand. This mismatch repair is triggered by bonding of the MutS protein of *E. coli* to a mismatched base, and methylation-directed mismatch repair can be limited by using a strain of *E. coli* that is *mutS* (10).

The protein products of many sequences are toxic to *E. coli* cells if expressed; they are difficult to clone because they poison the cell that carries them. Many plasmid sequences may be expressed at low levels from a cryptic promoter, or from a promoter that is repressed but still slightly active or 'leaky'. Several approaches may be taken to help resolve this type of problem. A researcher may select a plasmid that has a very low level of background transcription in *E. coli*, or a plasmid that can be maintained at a very low copy number in cells.

Problems

The solutions to these problems are provided at the end of the chapter.

(v) Suppose we have 250 ml of a solution containing a bacteriophage lambda clone, at a concentration of 10^9 pfu per ml. How many µg of phage DNA could be extracted from the 250 ml of stock, assuming that the phage genome is 50 kbp and the average base pair in DNA is approximately 660 Da?

(vi) Suppose you have a growing bacterial stock that has an optical density at 600 nm of 0.6, and a stock of bacteriophage λ that is 10^6 pfu per ml. How would you dilute and add bacteriophage λ to 10 µl of the bacterial stock, to obtain an MOI of 0.0005? Assume that 1 A_{600} unit represents 5×10^8 cells per ml.

4.3 Chimeras and artificial chromosomes

Many cloning vectors used in molecular biology are not easily categorized, as they contain parts from many different systems and bring together elements from plasmids, viruses, and genomic DNA. For example, a cosmid is a plasmid that can be packaged into a bacteriophage λ capsid, and a phagemid is a plasmid that can be replicated and packaged into a filamentous bacteriophage capsid. They are more 'plasmid' than 'virus', but how do we categorize a bacteriophage λ vector that carries an internal phagemid? These chimeras have been largely saved for the discussion below, so that the principles of plasmid and viral cloning vectors can be considered at the same time.

Phagemids, cosmids, PACs, BACs, and fosmids

Phagemids

Phagemids provide an easy way of generating a single-stranded copy of a DNA for use in preparation of DNA probes, or as a template for site-directed mutagenesis. Phagemids are plasmids that carry an f1 origin of replication from a filamentous bacteriophage, in addition to their own plasmid origin of replication. Upon activation of the f1 origin of replication by infection with an M13 helper bacteriophage, the single-stranded DNA is packaged into an M13 capsid and secreted into the medium of growing *E. coli* cells. The viral particles can be isolated from the medium by removal of the cells, and concentrated by polyethylene glycol (PEG) precipitation.

Lambda ZAP® II is a bacteriophage λ vector that contains a phagemid at its core, a plasmid sequence that can be packaged in an M13 capsid (11). The plasmid inside Lambda ZAP® II is called pBluescript II®, and it carries the intergenic region of bacteriophage f1, including an origin of DNA replication that is activated by an M13 helper bacteriophage. If a researcher has cloned a DNA insert into Lambda ZAP® II, the virus can be grown in cells that are co-infected with the M13 helper virus, and the medium will primarily contain two types of viral particles: the original Lambda ZAP® II recombinant and the M13 capsids carrying the pBluescript II® phagemid sequence. E. coli cells infected (transduced) by the M13 particle carrying a phagemid will propagate the packaged genome as a plasmid (see *Figure 4.13*).

Cosmids

A cosmid is a plasmid with a pair of *cos* (cohesive end) sites from bacteriophage λ. If a researcher inserts enough DNA into a cosmid to bring its size into the range of 39 to 52 kbp, then it can be packaged into a bacteriophage λ capsid and introduced into E. coli by transduction. It may seem odd that we would want to take this extra step of packaging; why not just transform the same plasmid into E. coli? For one thing, by packaging the DNA into λ, it is possible to erase the background level of plasmids that did not take up a DNA insert, or only took up a very small insertion. Those unwanted cosmids would be too small to be packaged. Furthermore, the packaged DNA is not susceptible to shearing or nuclease attack, and is efficiently transferred to E. coli by transduction.

PACs

P1 artificial chromosomes (PACs) are circular plasmids that carry an origin of replication from bacteriophage P1. They are maintained in E. coli cells at a low copy number of approximately one to five copies per cell, and are appropriate for cloning DNA fragments of approximately 100 to 200 kbp. PACs are widely used for physical mapping studies of DNA, for curating genomic libraries, and for isolation of large eukaryotic genes and regulatory regions. The PAC is introduced into E. coli by DNA transformation, and carries a selectable marker gene conferring resistance to an antibiotic such as kanamycin or chloramphenicol. This allows the researcher to select against bacteria that do not take up and maintain the PAC DNA.

Some PAC vectors also allow for positive selection of PACs that have taken up a DNA insertion. For example, PAC vectors carrying an insertion at a cloning site in the *sacBII* gene will not express sucrose synthase, and will not produce the toxic polymer levan in the presence of sucrose. On the other hand, PAC vectors lacking a DNA insertion would have an uninterrupted *sacBII* gene and would be selected against by their expression of a functional sucrose synthase. In this scenario the PAC vector would carry a stuffer fragment as it is being prepared for use in cloning, to prevent *sacBII* expression. The stuffer may also carry an alternate origin of replication, one that allows for high-copy-number production of

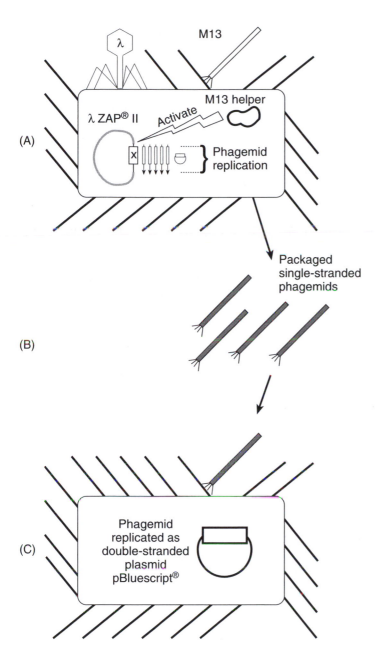

Figure 4.13

Excision of phagemid from Lambda ZAP® II bacteriophage clone. (A) A male *E. coli* cell (with pili) is co-infected with a Lambda ZAP® II clone (on the left) and an M13 helper phage (on the right). The M13 phage expresses *trans*-acting proteins that activate the f1 origin of replication in the Lambda ZAP® II clone, leading to the generation of single-stranded DNA copies of the phagemid region (downward-pointing arrows). The phagemid region includes the foreign DNA (rectangle marked 'X') that is cloned into Lambda ZAP® II. The single-stranded copies are circularized and packaged into M13 virions. (B) The M13 virions carrying single-stranded copies of the phagemid are secreted from the infected cells and accumulate in the growth medium. (C) Infection (transduction) of *E. coli* with the M13 virions carrying phagemid DNA leads to replication of the phagemid as a double-stranded plasmid, pBluescript®.

plasmid in *E. coli*. This alternate origin of replication is typically used for preparation of the vector when it is small in size, prior to insertion of a large DNA, and is not present in the recombinant PAC in which the stuffer is replaced.

PACs often carry *lox* sites that allow a researcher to make use of the Cre-*lox* system of recombination from bacteriophage P1. For example, a PAC might carry a *loxP* site that allows targeted recombination with a cognate *loxP* site in a chromosome. The *loxP* site in the chromosome would have been prepared in advance, as a target. Upon expression of the Cre recombinase enzyme, the PAC will integrate into the chromosomal *loxP* site, and the result will be a fully incorporated PAC that is flanked by a pair of *loxP* sites. The Cre recombinase is expressed only briefly when this method is used, because the continued presence of Cre in cells may cause the PAC integration to be reversed through recombination of the flanking *loxP* sites. There are different types of *lox* sites, and they can be used with some level of specificity (12). For example, Cre will cause recombination between two *loxP* sites, or two *lox511* sites, but not between a *loxP* and a *lox511* site.

BACs

Bacterial artificial chromosomes (BACs) are similar to PACs, in that both allow for the low-copy-number maintenance in *E. coli* of DNA fragments of approximately 100 to 200 kbp. BACs are introduced into *E. coli* by DNA transformation, and are maintained under antibiotic selection, as are PACs. The principal difference between BACs and PACs is in their origin of replication, with BACs being based on an origin of replication from the *E. coli* F factor.

The F (fertility) factor is an episome that causes *E. coli* to synthesize long cell-surface extensions called pili. In nature these pili help the *E. coli* adhere to surfaces such as intestinal cells, and are an important factor in virulence. The F factor is spread from F+ 'male' bacteria to F− 'female' bacteria in a population by conjugation. Upon receipt of a complete F factor, a female will express pili and become a male. Bacteria carrying BACs only have a small portion of the F factor replicon, and are not necessarily male. However, some BACs may be passed from one bacterium to another by conjugation, provided that the necessary *cis*- and *trans*-acting factors are provided.

BAC vectors may be engineered similarly to PAC vectors, having a *sacBII* gene for positive selection of DNA inserts during cloning, and a stuffer fragment to interrupt the gene and prevent cell killing when the vector is being prepared for cloning work. This stuffer may contain an alternate origin of replication for high-copy-number production, one that is excised and discarded along with the stuffer. Alternatively, BACs may contain two origins of replication, one being the F factor origin that allows for low-copy-number maintenance, and the second being an inducible *oriV* origin that allows amplification of the DNA to a higher copy number (13). This DNA copy number increase only occurs when a *trans*-acting factor from the *trfA* gene is expressed, for example as part of the CopyControl™ system from Epicentre® Biotechnologies.

Fosmids

Fosmids are cosmids that carry the replicon of an F factor, so they are plasmids that can be maintained at a low copy number in *E. coli*, and can be packaged into bacteriophage λ capsids if they are in the range of 39 to 52 kbp. A fosmid would be expected to carry a selectable marker gene, perhaps conferring resistance to an antibiotic, as well as *cos* sites and the F factor origin of replication. It may seem odd that a researcher would want to keep a moderately sized plasmid at a low copy number, but some DNA sequences carry genes that are toxic if expressed at a high level in *E. coli*. By keeping the plasmid copy number low, the overall expression of the toxic gene is also lower in the cell. A fosmid can be used in place of a cosmid if a low plasmid copy number is desirable.

YACs, Ti plasmids, and baculovirus vectors

YACs

Yeast artificial chromosomes (YACs) are vectors that can be maintained in the yeast *S. cerevisiae*, and typically maintain DNA fragments of approximately 250 to 2000 kbp. While YACs have an exceptionally large capacity for foreign DNA, there are occasional problems with sequence instability and recombination of the cloned sequences with the yeast genomic DNA. The YAC vector is linear in yeast, and the ends carry telomere sequences to maintain the integrity of the ends during DNA replication. The YAC also carries an autonomously replicating sequence (ARS), which is effectively an origin of DNA replication in yeast, and a centromere (CEN) function. The vector is prepared in *E. coli* as a circular DNA plasmid, and divided *in vitro* into two arms by restriction endonucleases. One of the restriction endonuclease sites lies between a pair of telomere sequences, and the other lies in the multiple cloning site. The ligation of the YAC arms to DNA inserts results in a linear chromosome that is introduced into yeast by transfection, and each arm of the YAC carries a yeast selectable marker such as *URA3* or *TRP1*.

Binary and cointegrate Ti plasmids

In Ti-mediated transfer of DNA from *A. tumefaciens* into a plant cell, a plasmid might be constructed first in *E. coli* where it can be easily introduced and characterized, then transferred to *A. tumefaciens* by transformation or conjugation. Such a shuttle vector would need to have an origin of replication that was functional in *E. coli*, because it needs to be stably maintained in those cells. If the plasmid is also expected to be stably maintained without recombination in *A. tumefaciens*, then it would need to have a second origin of replication that would work in those cells. A single selectable marker gene might be used in both hosts, driven by a broad-host-range transcriptional promoter.

One way of organizing the sequences involved in T-DNA transfer is to use a strain of *A. tumefaciens* carrying a helper plasmid that expresses the *vir* genes. This helper plasmid is said to be disarmed because it lacks a right-border sequence. The cells also carry a second, mobilizable plasmid, carrying a right-border sequence and genes of interest, which may then be transferred to plants. This is called a binary vector system because there are two plasmids involved, and both plasmids carry origins of replication that function in *A. tumefaciens*.

A second way of organizing the sequences is to have a disarmed helper plasmid in *A. tumefaciens* that becomes a target of homologous recombination with a mobilizable plasmid sequence. This recombination takes place in *A. tumefaciens* and is directed by a segment of matching DNA sequence between the helper plasmid and the second introduced plasmid. Recombination between the plasmids builds a larger, Ti-like plasmid that carries both T-DNA sequence and *vir* genes. This is called a cointegrate vector system because the two plasmids become recombined into one larger one. The helper plasmid carries an origin of replication that functions in *A. tumefaciens*, but the second plasmid in the cointegrate system does not have an origin of replication; instead, the second plasmid carries a selectable marker gene that is effective in *A. tumefaciens*. The *Agrobacterium* is placed under drug selection and survives if the selectable marker gene is maintained and expressed. This is most easily accomplished by recombination of the plasmids to bring together the origin of replication and selectable marker gene in a single plasmid.

In both the binary and cointegrate approaches, a segment of T-DNA becomes mobilizable and can be transferred from *A. tumefaciens* to plant cells in culture. These plant cells may then be placed under selection for uptake, expression, and stable maintenance of the recombinant T-DNA.

Baculovirus transfer vectors

Baculoviruses infect invertebrate cells, and vectors based on the virus are widely used to introduce DNA into insect cells, such as the *Spodoptera* cell lines Sf9 or Sf21. One way of putting foreign DNA into a baculovirus is to use a transfer vector, a small plasmid that is prepared and cloned in *E. coli* and introduced into insect cells by transfection, along with the full length viral DNA. Recombination between the transfer vector and the full-length viral sequence leads to a recombinant baculovirus carrying the DNA insert. This is similar to the method previously described for preparation of cointegrate Ti plasmids, in the sense that homologous recombination is used to build the resulting vector.

The Bac-N-Blue™ system from Invitrogen® and the pPolh vectors from Sigma-Aldrich Co. are examples of transfer vectors. They contain a multiple-cloning-site sequence flanked by segments of DNA from baculovirus, which directs a double crossover event. In the case of the pPolh vectors, the DNA recombination is directed by the ORF 603 and ORF 1629 sequences of the baculovirus. In the case of the Bac-N-Blue™ system, the recombination is directed by ORF 1629 and a segment of the *E. coli lacZ* gene that is shared between the transfer vector and specialized Bac-N-Blue™ baculovirus DNA. Upon successful recombination, the *lacZ* fragments are brought together to encode a functional β-galactosidase. This can be detected colorimetrically using the substrate X-gal; hence the 'Blue' in the name 'Bac-N-Blue™'.

4.4 Common applications

Gene cloning

Cloning vectors are used to capture DNA sequences for a variety of purposes, including physical DNA mapping and sequencing, gene expression, probe preparation, and site-directed mutagenesis. In most cases, the vector is joined with a DNA insert *in vitro*, using a joining enzyme such as DNA ligase, and then transferred to a recipient *E. coli* cell by DNA transformation or transduction. The process of preparing the ends of a DNA vector and DNA insert so they can be joined may involve several techniques, including polymerase chain reaction (Chapter 7), restriction endonuclease treatment (Chapter 5), and DNA polymerase treatment (Chapter 6). There may be some damaged parts of the plasmid that are repaired by *E. coli*, for example mismatched nucleotides, nicks, or gaps in the double-stranded DNA. The plasmid DNA may also have methylation patterns that result in damage by a host restriction system. The efficiency of transformation will typically be lower with damaged DNA than intact DNA, but efficiency is not a concern if the right clone can still be obtained. Once identified and isolated in a pure state, the correct candidate clone can be stably maintained in *E. coli* as a plasmid or stored as a bacteriophage stock.

Mapping and sequencing

When a researcher wants to do a bit of fine-structure mapping of a gene, he or she may clone a segment of the gene into a plasmid or bacteriophage λ vector that has a substantial multiple-cloning site, with many choices of restriction endonuclease sites that are unique or rare in the vector. The researcher may initially digest the DNA of the clone with these corresponding restriction enzymes, gathering information about the presence of restriction enzyme sites in the DNA insert.

For example, suppose a plasmid vector is 3.5 kbp and has a single *Eco*RI recognition sequence in the multiple cloning site, and this *Eco*RI site is next to a *Bam*HI recognition sequence that is used in cloning a DNA fragment. If a recombinant plasmid is treated with *Eco*RI and the researcher sees two restriction fragments, one that is 4.0 kbp and one that is 0.7 kbp, then it is reasonable to conclude that the cloned DNA fragment has an *Eco*RI recognition sequence that lies 0.7 kbp from the *Eco*RI site in the multiple cloning site (see *Figure 4.14*). Furthermore, since the larger fragment is 4.0 kbp and not 3.5 kbp, there is an additional 0.5 kbp of DNA that lies beyond this internal *Eco*RI site. This is a very little bit of mapping information that helps to establish the size and orientation of the DNA insert. Additional restriction endonuclease digestions with different enzymes would add to this base of information, and could be used to build a coherent restriction map of the DNA insert. Enzymes are most useful if they yield a simple pattern for the base vector, either a single band or several predictable bands that do not obscure the pattern of bands related to the DNA insert.

In mapping large DNA inserts, such as one taken from a bacteriophage λ clone, a researcher may use a restriction enzyme that cuts at the multiple cloning site so that the DNA can be end-labeled with a ^{32}P phosphate, a

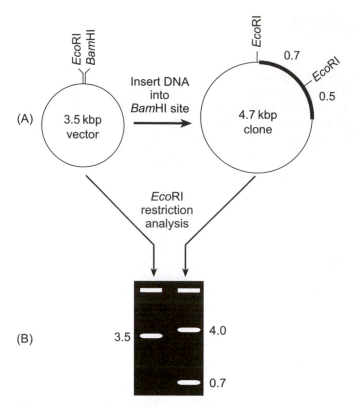

Figure 4.14

Simple restriction analysis of a plasmid. (A) A 3.5-kbp vector is modified by the insertion of a DNA fragment into the *Bam*HI site, generating a 4.7-kbp clone (on the right). The inserted DNA sequence is shown as a thick line. The location of an *Eco*RI site in the inserted DNA is inferred from restriction analysis using the enzyme *Eco*RI, and the knowledge that the original vector has an *Eco*RI site next to the site of insertion. (B) Schematic diagram of results of gel electrophoresis on *Eco*RI-digested DNA products of the original vector (left) and modified vector (right). The apparent sizes (kbp) of DNA products are indicated next to the gel.

marking that allows the end to be more easily tracked during restriction analysis. Another approach is to use Southern blotting to assist in the analysis, and to use end probes developed from phage RNA polymerase promoter sequences that are often placed strategically in the flanking regions of multiple cloning sites. A large fragment may also be cloned in smaller segments, a process called subcloning, to simplify the process of mapping.

DNA sequencing is often conducted using a cloned DNA template, and many plasmid vectors have universal sequencing primer sites (discussed in Section 4.1) to assist with the sequencing of an unknown DNA fragment. If the DNA insert is large and cannot be readily sequenced in a single reaction, a researcher may use restriction endonucleases that have recognition sequences in the multiple cloning site and the DNA insert to delete portions of the insert. This deletion can bring the universal sequencing primer sites closer to segments of the DNA insert that have not been adequately sequenced, and allow more information to be collected from the interior parts of the DNA insert.

DNA sequencing can be performed on a double-stranded DNA template, provided that it is pure and not too large. It is challenging to sequence a DNA that is embedded in a large vector, for example bacteriophage λ, because the increased formula weight of the template necessitates more micrograms of DNA in the reaction. A researcher will usually subclone a large DNA insert into plasmid vectors prior to DNA sequencing. If a single-stranded template is desirable for a sequencing reaction, then a researcher may use a phagemid vector. These vectors contain a filamentous bacteriophage (f1) origin that allows for the production and isolation of a single-stranded template, packaged and secreted in M13 virions (discussed in Section 4.3).

Gene expression

Cloning vectors are sometimes engineered so that they allow expression of cloned DNA sequences; this may be for the purpose of purifying a recombinant RNA or protein product or studying the expression of a recombinant gene in a host cell. An important principle of genetic engineering is that the expression system has to be concordant with the regulatory sequences included in the DNA clone. For example, if a gene is to be expressed in a mammalian cell then it must be flanked by a promoter sequence that will function in the cell of choice. Sometimes regulatory sequences allow for regulated expression, or tissue-specific expression, and other times these sequences are designed for constitutive expression. A cloning vector designed for gene expression will have a cloning site embedded in host regulatory sequences. It may also encode protein tags such as V5 or FLAG® (discussed in Section 4.1), which may be used to monitor protein expression or localization. When a cloned sequence is expressed as a fusion protein, it is important that it be cloned in-frame with the triplet reading frame used by the ribosome (see *Figure 4.4*). Most of these engineering issues are easily resolved if a PCR product is being cloned, since in that case the oligonucleotides can be designed to establish the appropriate junction sequence.

The rest of the expression vector may contain the usual elements required for plasmid maintenance and selection in *E. coli*, and perhaps selection in a different expression host such as plants, yeast, or animal cells. If the vector is to be stably maintained in the expression host as a plasmid, then a host-specific origin of replication is required. Alternatively, if a portion of the vector is to be integrated into the host genomic DNA, then a sequence directing homologous recombination or targeted Cre-*lox* recombination may be present.

Probe preparation

A cloned DNA may be radioactively or chemically labeled for use as a probe in Southern or Northern blot analysis. A protocol such as nick translation or random oligo primed synthesis does not require any special cloning vector features; however a researcher may choose to gel-purify just the DNA insert for labeling and not include the vector sequences. In this case, having flanking restriction enzyme recognition sequences in a multiple cloning site is a convenience. Nick translation and random oligo-primed synthesis from a double-stranded template produces labeled DNA from both strands. A single DNA strand can be labeled specifically if the template is single-stranded, for example if a phagemid is induced to produce packaged genomes (see *Figure 4.13*).

RNA probes may be prepared using flanking bacteriophage RNA polymerase promoter sequences, if these are present in the vector. RNA probes are limited to the single strand that is transcribed, and the DNA template can be eliminated from the reaction products by treatment with RNase-free DNase. However, RNA end probes prepared in this way tend to be short because of the constraints of high-specific-activity labeling (as discussed in Chapter 6).

It is worth cloning a gene to make a probe, because a DNA fragment that isn't cloned may be contaminated with other unwanted DNA fragments from the host. For example, a PCR product may look pure on a gel, but if used to make a probe then a minor contaminant may become labeled and produce an experimental artifact. Not realizing that the probe is impure, a researcher may reach an erroneous conclusion and compromise the validity of an experimental result. On the other hand if the gene is cloned before a probe is made, it is isolated from contaminating host factors and there are fewer opportunities for mistaken interpretations of Southern and Northern analyses.

Site-directed mutagenesis

Strategies for site-directed mutagenesis do not usually depend on a specific vector, as discussed in Chapters 6 and 7. However, if the strategy depends on starting with a single-stranded DNA template as the 'parental' strand, it may be helpful to use a phagemid that allows production of M13 virions containing a copy of the plasmid. One plasmid that is specifically designed for site-directed mutagenesis is the Altered Sites® plasmid (Promega®), that facilitates selection for an altered 'daughter' strand. The Altered Sites® plasmid has selectable marker genes for ampicillin and tetracycline resistance, but one of the two is mutagenized and nonfunctional. The mutagenesis of a cloned gene of interest is performed simultaneously with mutagenesis that corrects of one of the selectable marker genes. Selection of transformed cells with that specific antibiotic leads to isolation of plasmids that have a corrected marker gene, and these are likely also to be mutated in the gene of interest (14).

More generally, PCR can be used to perform site-directed mutagenesis of a sequence that is cloned in a plasmid vector. A pair of oligonucleotides can be designed to introduce a mutation in a cloned sequence and regenerate the entire plasmid by inverse PCR, as described in Chapter 7. The plasmid is converted to a linear DNA by inverse PCR, and can be re-closed into a circular DNA by ligation prior to transformation into *E. coli*.

Libraries

A 'library' is a collection, and when we talk about libraries in molecular biology we mean a collection of recombinant DNA molecules, bacteriophage, or cells. Cloning vectors are used to preserve libraries of DNA fragments that may come from genomic DNA, cDNA, or synthetic materials. The goal is to have a representative library, one that has at least one copy of each desired sequence, so the ligation reaction must be highly efficient in cloning a wide variety of DNA fragments.

Genomic libraries

When a researcher makes a genomic DNA library, he or she must give careful consideration to the size of the genome and the average size of the DNA inserted into the cloning vector. A large genome will need many independent clones to represent it, particularly if small DNA fragments are inserted into the cloning vector. Making a genomic DNA library is a bit like painting a wall, the number of brushstrokes we have to use to cover the wall completely depends on the size of the wall and the size of our paintbrush. Of course it is a bit more complicated than this, because the best painters will overlap their brushstrokes a bit and will apply several coats of paint overall. When we are making a library from random DNA fragments, there is no way to make sure that every sequence has been cloned at least once in our library. Some sequences in the genome may be represented five times in the library, others only once, and still others not at all. That is like employing a blind painter who makes random brushstrokes on the wall; some parts of the wall may be painted and repainted many times before the whole wall is adequately covered.

Restriction endonucleases are typically used to cut genomic DNA into fragments that are compatible with a cloning site in a vector. As discussed in Chapter 5, restriction endonuclease recognition sequences are randomly distributed in a genomic DNA at a frequency that depends on recognition sequence composition and length. The choice of restriction enzymes is important, because it can affect the representation of DNA fragments in a library.

For example, suppose we completely digest a genomic DNA with the enzyme BamHI and size-select DNA fragments that are in the range of 10 to 20 kbp for cloning into a vector such as Lambda FIX® II or EMBL3. Would every DNA sequence in the genome be represented in that size-selected collection of DNA fragments? No, because many, and in fact most BamHI-cut fragments would probably be excluded because they were smaller than 10 kbp, and some would be larger than 20 kbp. On the other hand, suppose we digested a genomic DNA with the enzyme Sau3AI, which cuts most DNA sequences into small fragments less than 1 kbp. If we only add enough enzyme so that the digestion is partially completed, perhaps yielding DNA fragments with an average size of 10 to 20 kbp, then the Sau3AI fragments will be overlapping and will tend to represent the entire genome well. The Sau3AI ends are compatible with the ends left by BamHI in a cloning vector, a matter discussed in Chapter 5, and this is a common approach used in preparing a genomic library of random, overlapping fragments.

When genomic libraries are made in large-capacity vectors, such as BACs and PACs, it takes fewer independent clones to represent a complete genome than when libraries are made in small plasmids or bacteriophage λ vectors. For example, 50 000 independent BAC clones could carry up to

10^7 kbp, if the average DNA insert is 200 kbp ($5 \times 10^4 \times 2 \times 10^5$ bp = 1×10^{10} bp). However, 50 000 independent Lambda FIX® II clones could only carry 10^6 kbp, with an average DNA insert of 20 kbp ($5 \times 10^4 \times 2 \times 10^4$ = 1×10^9 bp). How far would 50 000 independent clones take us, in terms of representation of the human genome? The human genome is approximately 3×10^6 kbp, so in the BAC library each human DNA sequence would be found an average of 3.3 times (10^7 kbp $\div$ 3×10^6 kbp), and in the Lambda FIX® II library each human DNA sequence would be found an average of 0.33 times (10^6 kbp $\div$ 3×10^6 kbp).

However, even with an average of 3.3 copies of each sequence in the BAC library there is no certainty that any given DNA sequence will be represented. The chance of representation of any sequence is determined by a Poisson distribution. For example, in these 50 000 element BAC and Lambda FIX® II libraries, the probability of representation of any given human sequence at least once is 96% and 28% respectively ($1 - 1/e^{3.3}$ = 0.96, and $1 - 1/e^{0.33}$ = 0.28). The 50 000 element BAC library does a better job of representing the DNA in the human genome, but remembering our analogy of the blind painter, we can think of the BAC library as being like a bigger paintbrush. With a big paintbrush, 50 000 random brushstrokes will go much farther towards finishing the job than 50 000 random brushstrokes with a small paintbrush.

cDNA libraries

Libraries made from cDNA do not contain particularly large DNA fragments, at least in comparison to most genomic DNA libraries, so cDNA libraries do not need to be prepared in high-capacity vectors. A bacteriophage λ vector with a capacity of 0 to 10 kbp, such as Lambda ZAP® II, has all the capacity that is needed for cDNA and does not select against the cloning of very small cDNA elements that might only be a few hundred base pairs. The preparation of cDNA inserts for a library is discussed in Chapter 6, but suffice it to say that it is usually more complicated than just digesting a DNA with a restriction endonuclease.

The origin of the RNA used in a cDNA library is particularly important, because RNA molecules are sometimes specific to a particular cell, tissue, or developmental stage. A human spleen cDNA library would have many elements in common with a human pancreatic cDNA library, but there would be many differences as well. For example, both would contain cDNA for tubulin, but only the pancreatic library would be expected to have cDNA for trypsinogen. Only the splenic library would be expected to have cDNA for immunoglobulins, aside from the fact that the pancreatic tissue might have been a bit bloody and infiltrated with B-lymphocytes at the time the RNA was extracted.

Some cloning vectors for cDNA libraries can be used for protein expression, allowing for an immunological screening with antibody probes instead of nucleic acid probes. In this method, a researcher starts with an antibody probe and searches for a cDNA clone that encodes a protein product that binds to the antibody. For example, the vectors λgt11 and Lambda ZAP® II allow for insertion of DNA within a *lacZ* coding sequence. If a cDNA for a gene of interest happens to be cloned in the right orientation, and in the same triplet reading frame as *lacZ*, then a portion of the cDNA may be expressed as a protein fused with β-galactosidase. The fusion protein will be expressed after induction of the transcriptional promoter with IPTG, and released into the surrounding medium of the bacteriological plate. Some of the released protein can be absorbed onto a nitrocellulose blotting membrane and detected with an antibody probe directed against a gene product of interest. Going back to the plaque that produced the immunologically detectable protein, a researcher has a candidate sequence that may be a cDNA for the gene of interest.

How many independent elements are needed for a cDNA library to be representative? It is important to consider that some RNA molecules are expressed at high-steady-state levels in a cell, and others are expressed at low-steady-state levels. If these proportions between RNA species are preserved during the process of cDNA generation, then our library will have to have many replicates of the common cDNAs so that the rare ones are also represented.

Libraries for cloning by complementation

The usual approach to cloning a gene by complementation is to start with a strain or cell line that has lost the ability to perform a specific function, due to a (probable) genetic lesion in a single gene. At the outset, we predict that if we introduced a fresh copy of that gene, one that was not mutated, we could restore the specific function to the cell. Once we see exactly which DNA restores the function, we have an idea what gene might have been mutated or lost its function in the original cells. We have to test a large number of possible genetic solutions, so we need a library of genes.

The point of genetic complementation is that the replacement gene must be expressed, and possibly even regulated appropriately. The replacement gene can only help the cell to overcome a simple loss-of-function, not a mutated allele with a dominant mode of action. If the loss of function can be resolved without replacing the mutated gene, perhaps by over-expressing a different gene, then the clones isolated by genetic complementation may include some spurious results that do not point to the mutated gene.

Vectors that are used for genetic complementation must be capable of being maintained in the host cell, either by plasmid maintenance or integration into the chromosome. The researcher might try to use a cDNA library for complementation, provided that the library is complete and full length, but since a cDNA contains no transcriptional promoter the gene cannot be regulated in its usual way. An alternative is to introduce a genomic DNA library for complementation, with the hope that the entire corrective gene might be contained in a single library clone and might be expressed and regulated appropriately from its own promoter.

When it is found that a specific clone does complement a defect in a cell line or strain, then the researcher can rescue the clone and study the DNA insert it contains. If the complementing clone is being maintained as a plasmid in a host cell, it can very likely be shuttled back into *E. coli* by DNA transformation. Alternatively, if the clone DNA has integrated into the host cell chromosome then the vector sequences will very likely 'tag' the integration site for recovery of the gene, either by cloning or polymerase chain reaction. In the end, the researcher has a gene sequence that is functionally related to the genetic defect in a cell line or strain, and may point to the site of the genetic defect.

Phage, bacterial, and yeast display libraries

Display libraries are used to express a collection of protein epitopes for selective binding. For example, a bacteriophage vector may have a cloning site in a coat protein gene, so that inserted sequences are expressed in the form of a fusion protein exposed on the surface. The library consists of variations in the sequences inserted, so that each bacteriophage candidate has a slightly different (possibly random) element displayed. A researcher takes a display library of bacteriophage, exposes the collection to a target surface such as a protein bound to an ELISA plate, and isolates the bacteriophage candidates that bind. After several sequential rounds of selection, perhaps with propagation and amplification of the binding clones between each round, the candidate clones are isolated and plaque-purified. Each candidate can be tested for binding affinity on the target surface, and the sequence of the DNA insertion in the coat protein gene can be determined. It is common for researchers to report convergence of predicted amino acid sequences in the DNA inserts, and to interpret these data as evidence of a consensus sequence for binding.

A similar type of repetitive selection is used for bacterial or yeast display libraries, which are based on expression of a foreign coding sequence on a cell-surface protein. The cloning vectors involved in these libraries contain a gene for a cell-surface protein, for example a flagellar or pilus protein in *E. coli* or the a-agglutinin receptor of *S. cerevisiae*, and allow insertion of a random DNA element within the coding sequence of the gene to generate a library of fusion proteins. Selection may be based on binding to a solid support, as in the example of bacteriophage display libraries, or binding of cells to a fluorescent target in solution with subsequent sorting of cells in a fluorescence-activated cell sorter (FACS).

Yeast two-hybrid system libraries

Yeast two-hybrid libraries are a different way of detecting protein–protein interactions; the binding of two proteins is detected in the nucleus of a living cell, rather than on the surface. Suppose we are interested in a human protein 'X' and we wonder what other human proteins might bind to it. A yeast two-hybrid library would be one way to identify proteins that bind to protein 'X'.

Explaining how these libraries work requires a brief discussion about transcriptional regulation. Genes are regulated in yeast by transcriptional activators, which are proteins that bind to a specific DNA sequence upstream of a gene, an upstream activation sequence (UAS), and encourage the assembly of a transcription complex. In a laboratory system, we might place a UAS upstream of a reporter gene, or a selectable marker gene, so that the gene is only expressed if a transcriptional activator protein binds to the UAS. However, the really interesting thing is that the transcriptional activator has two domains; a DNA-binding domain (DBD) that is responsible for binding the UAS, and an activation domain (AD) that is responsible for recruiting the RNA polymerase to the promoter. If we separate these two domains from each other and express them in separate polypeptides, the promoter is not activated. That is, a DBD bound to a UAS is not enough to cause promoter activation. The AD and DBD have to be physically associated for activation to occur.

The way that these domains can be brought together in a cell, restoring transcriptional activation, is by making each separate domain into a fusion protein. The DBD is expressed as a fusion with a 'bait' sequence, and the AD is expressed as a fusion with a 'fish' (or 'prey') sequence. The DBD-bait fusion does not change in an experiment, but many different AD-fish combinations are tested to see if they will interact. For example, AD-fish could be a library of human cDNA sequences, expressed as a fusion protein with AD. If DBD-bait and one of the AD-fish library elements can bind to each other, the AD will be localized to the UAS and will be able to help recruit RNA polymerase to the promoter. Both the DBD-bait and AD-fish fusion proteins include a nuclear localization domain, so they will be in the right compartment of the cell to activate transcription if a bait-fish interaction occurs.

Genome applications

How do we genetically modify an organism? We have several possible approaches to consider. We may introduce a vector that has no origin of DNA replication, intending to look for an immediate or transient effect of the DNA in the cell. No selection is applied in that case because the DNA is not maintained and the experiment is of short duration. When this is done to a eukaryotic cell, it is called a transient transfection. If we want the introduced DNA to be stable in the cell, it must either be integrated into the genome or replicate extrachromosomally using its own origin of replication. In plant cells there is also the possibility of integration and maintenance in the chloroplast genome. In any case we would need to apply selection, such as an antibiotic, to select for cells that had taken up and expressed the DNA. The selection might be stopped after a period of time, if the DNA is integrated into the host chromosome, because the integration is likely to be maintained in the absence of selection. However, if the introduced DNA is being maintained extrachromosomally then it is likely that selection must be continued indefinitely, unless the element has a centromere-like function that allows for stable maintenance.

The cloning vectors we use for each of these applications look slightly different, depending on the host cell and the intended use. A vector being used for transient transfection of mammalian cells would have a gene of interest flanked by mammalian transcriptional regulatory sequences. A vector being used for stable maintenance by integration into the host genome might have a targeting sequence that directs homologous recombination. The gene of interest is flanked by a pair of targeting sequences if a double crossover event is planned; in this case the vector would very likely be linearized before transfection. The gene could carry its own transcriptional regulatory sequences, or make use of the sequences that are resident in the chromosome at the targeting locus.

One common application is to knock out a gene by homologous recombination. In that case a selectable marker gene is embedded in targeting sequence from a genetic locus, and a double crossover event replaces the wild-type sequence with the selectable marker gene. A single crossover event with a circular plasmid results in integration of the entire plasmid sequence.

Integration can be targeted using the Cre-*lox* system from bacteriophage P1, and in that case the chromosome would already have been prepared so that it had a resident *lox* site. The plasmid introduced would also have a *lox* site, and brief expression of Cre recombinase would lead to integration. A pair of *lox* sites of different specificity could be used to flank the gene of interest, and the target in the chromosome, if a double recombination was desired.

Integration can also be nonhomologous, meaning that it does not become directed to a specific site in the chromosome. This is also called ectopic integration. The integration may not be completely random, because if the cells are placed under selection there is bias for chromosomal sites that are favorable to expression of the selectable marker gene.

RNA production

Many cloning vectors allow for RNA production, *in vivo* or *in vitro*, from a bacteriophage RNA polymerase promoter. These promoters are short sequences that can be embedded upstream or downstream of a multiple cloning site, to direct transcription from either strand of the cloned DNA. When a promoter is represented on a map with an arrow, the direction of the arrow is the 5′ to 3′ direction of the transcript (see *Figure 4.15*). If the cloned DNA has a coding sequence, we often use the expressions 'sense strand' and 'antisense strand' to refer to the two strands of DNA. The sense strand of DNA reads the same as an RNA transcribed for gene expression; for example, if the RNA transcript reads 5′-GUUCAA-3′, the sense DNA strand reads 5′-GTTCAA-3′ and the antisense DNA strand reads as the reverse complement, 5′-TTGAAC-3′ (see *Figure 4.15*).

Suppose a cDNA is cloned in a multiple cloning site, between flanking T7 and T3 promoter sequences, with the start of the gene next to the T7 promoter and the end of the gene next to the T3 promoter. The T7 promoter could be used to make a sense RNA, *in vitro* or *in vivo*, that could be translated to yield the protein product encoded by the cDNA. The T3 promoter could be used to make antisense RNA for genetic studies, or radiolabeled antisense RNA for use as a probe in Northern analysis (see *Figure 4.15*).

If these bacteriophage promoters are used *in vitro*, the respective bacteriophage RNA polymerase is added to the DNA template, along with ribonucleoside triphosphate substrates. Each polymerase is specific; for example, the T3 enzyme will not work with the T7 promoter.

Another type of cloning vector that directs RNA synthesis is one designed for RNAi production in a cell, for example a mammalian cell. A small double-stranded RNA may be produced as a short hairpin RNA from the RNA polymerase III promoter, as in the example of the pSIREN or

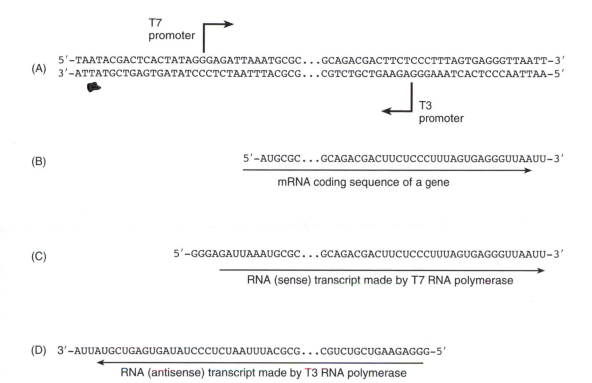

Figure 4.15

Transcription of sense and antisense RNA using bacteriophage RNA polymerases. (A) Double-stranded DNA representation of a gene sequence, cloned so that bacteriophage T7 and T3 RNA polymerase promoters are located at the left and right ends, respectively. The bent arrows indicate the start site for transcription, and the direction of transcription. (B) A segment of RNA that represents a cellular mRNA transcribed from the gene. (C) RNA transcribed using T7 RNA polymerase. This is a sense RNA because it matches the natural mRNA sequence. (D) RNA transcribed using T3 RNA polymerase. This is an antisense RNA because it is complementary to the natural mRNA sequence. Note that the 5′ end is shown on the right, and the 3′ end on the left.

siLentGene™-2 vectors that use the human U6 RNA promoter, or the GeneClip™ Cloning System that uses the human U1 RNA promoter.

Protein expression

Cloning vectors designed for recombinant protein expression are widespread, and used for a variety of applications in basic research and biotechnology. Sometimes the application involves expressing as much protein as possible, and in other cases the idea is to have the protein expressed in a controlled manner. Protein expression is primarily regulated indirectly, through the activity of the transcriptional promoter, although the sequence of the mRNA can also have profound effects on efficiency of translation. One example of a protein expression system described earlier are the pGEX® vectors (see *Figure 4.4*).

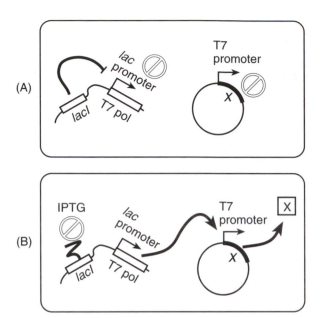

Figure 4.16

Regulated expression of genes using the pET plasmid system. (A) Schematic diagram of *E. coli* cell containing a pET plasmid with a cloned gene of interest (marked '*X*'). In the absence of inducer, the *lacI* gene product binds to the operator of a *lac* promoter and prevents expression of T7 RNA polymerase. (B) In the presence of the inducer IPTG, the *lacI* gene product binds to IPTG and the lac promoter is not repressed. T7 RNA polymerase is expressed, which activates transcription of gene *X*, causing expression of a gene product '*X*'.

The pET system of vectors (Novagen®) uses an indirect method for induction of expression, in that the gene of interest is cloned downstream from a bacteriophage T7 RNA polymerase promoter (see *Figure 4.16*). The T7 RNA polymerase gene is integrated into the *E. coli* genome, and is itself under the control of a *lac* promoter and operator region. The *lacI* gene is also expressed from a genomic copy, and so the addition of the inducer chemical IPTG causes a chain of events leading to protein expression; the IPTG binds to the lac repressor protein and causes expression of T7 RNA polymerase from the genome, and the T7 RNA polymerase activates transcription of the gene of interest from the pET plasmid. The added layer of regulation in the pET expression system keeps the gene of interest from being transcribed and translated until the inducer is added, which is helpful when trying to express products that are toxic to the *E. coli* cell.

Tet-Off and Tet-On

Another example is the BD™ Tet-Off and Tet-On system for regulation of protein expression in mammalian cells. In the Tet-Off system, a gene of interest can be maintained in a transcriptionally silent state by the addition of the antibiotic doxycycline. The way this works is that doxycycline binds to a Tet-controlled transactivator (tTA), and prevents it from activating transcription of a gene of interest by binding to an upstream tetracycline response element (TRE). The structure of tTA is reminiscent of the two-hybrid system in *S. cerevisiae*, in which a DNA-binding domain in the protein directs binding to a *cis*-acting sequence, and a separate activation domain directs transcriptional activation. The DNA-binding domain of tTA is derived from the *E. coli* tetracycline repressor, and TRE is its binding site. In the cloning vector, TRE is located upstream of a promoter such as the cytomegalovirus (CMV) promoter, or an alternative tissue-specific promoter. The activation domain of tTA is derived from a herpes simplex virus VP16 protein, and so when tTA binds to TRE it can help to initiate transcription at the downstream promoter. A gene of interest is kept transcriptionally silent by doxycycline, because the antibiotic binds to tTA and prevents it from activating transcription.

On the other hand, in the Tet-On system the tetracycline repressor portion of the fusion protein is mutated so that it binds to TRE in the presence of doxycycline, instead of in the absence of the antibiotic. This modified protein is called the 'reverse' Tet-controlled transactivator (rtTA), and it allows a gene of interest to be maintained in a transcriptionally silent state in the absence of doxycycline, and induced in the presence of doxycycline.

The Tet-Off and Tet-On systems have applications in transgenic animal research, as well as somatic cell cultures. A gene can be maintained in a transcriptionally active or inactive state in an animal until doxycyline is added to the diet. If the tTA or rtTA are expressed from a tissue-specific or developmentally regulated promoter, the inducible system can be used to provide additional insight into the effects of changing gene expression in tissues or whole organisms.

Problems

The solutions to these problems are provided at the end of the chapter.

(vii) Suppose a 4500-bp vector has an *Eco*RI site at nt 90, and a *Bam*HI site at nt 100, and three *Sma*I sites at nt 110, 1450, and 3810. You insert a 2-kbp fragment of DNA digested with *Bgl*II into the *Bam*HI site of the vector, to generate a 6500-bp clone. Digestion of the clone with *Eco*RI yields two bands on a gel, with sizes 6000 and 500 bp. Digestion of the clone with *Sma*I yields four bands, with sizes 2360, 2280, 1340, and 520 bp. Digestion of the clone with *Eco*RI and *Sma*I yields six bands on a gel, with sizes 2360, 1340, 1000, 780, 520, and 500 bp. What are the locations of the *Eco*RI and *Sma*I sites in the 2-kbp fragment?

(viii) Suppose that you prepare a genomic DNA library from an organism with a genome of 1×10^9 bp, using a PAC vector with random DNA inserts having an average size of 100 kbp. If you have 80 000 independent PAC clones in the library, what is the probability that a particular small segment of DNA is represented at least once in the library?

Further reading

Ausubel FW, Brent R, Kingston RE, Moore DD, Seidman JG, Smith JA and Struhl K (eds) (1987) *Current Protocols in Molecular Biology*. John Wiley & Sons. New York.

Glick BR, Pasternak JJ (2003) *Molecular Biotechnology*, 3rd Edn. ASM Press, Washington, DC.

Perbal B (1988) *A Practical Guide to Molecular Cloning*, 2nd Edn. John Wiley & Sons, New York.

Ptashne M (2004) *A Genetic Switch – Phage Lambda Revisited*, 3rd Edn. Cold Spring Harbor Laboratory Press, Cold Spring Harbor, NY.

Sambrook J and Russell D (2001) *Molecular Cloning: A Laboratory Manual*, 3rd Edn. Cold Spring Harbor Laboratory Press, Cold Spring Harbor, NY.

References

1. Birnboim HC, Doly J (1979) A rapid alkaline extraction procedure for screening recombinant plasmid DNA. *Nucleic Acids Res.* **7**: 1513–1523.
2. Clewell DB, Helinski DR (1969) Supercoiled circular DNA-protein complex in *Escherichia coli*: purification and induced conversion to an open circular DNA form. *Proc. Natl Acad. Sci. USA* **62**: 1159–1166.
3. Novagen (2006) *pET System Manual*, 11th Edn (Novagen® TB055 01/06). EMD Biosciences, La Jolla, CA.
4. Bernard P, Couturier M (1992) Cell killing by the F plasmid CcdB protein involves poisoning of DNA topoisomerase II complexes. *J. Mol. Biol.* **226**: 735–745.
5. Smith DB, Johnson KS (1988) Single-step purification of polypeptides expressed in *Escherichia coli* as fusions with glutathione S-transferase. *Gene* **67**: 31–40.
6. Schmidt TGM, Koepke J, Frank R, Skerra A (1996) Molecular interactions between the Strep-tag affinity peptide and its cognate target, streptavidin. *J. Mol. Biol.* **255**: 753–766.
7. Cheng C, Shuman S (2000) DNA strand transfer catalyzed by vaccinia topoisomerase: ligation of DNAs containing a 3′ mononucleotide overhang. *Nucleic Acids Res.* **28**: 1893–1898.
8 Siguret V, Ribba AS, Cherel G, Meyer D, Pietu G (1994) Effect of plasmid size on transformation efficiency by electroporation of *Escherichia coli* DH5 alpha. *BioTechniques* **16**: 422–426.
9. Huynh TV, Young RA, Davis RW (1985) Constructing and screening cDNA libraries in lambda gt10 and lambda gt11. *DNA Cloning* 1: 49–78.
10. Su SS, Modrich P (1986) *Escherichia coli mutS*-encoded protein binds to mismatched DNA base pairs. *Proc. Natl Acad. Sci. USA* **83**: 5057–5061.
11. Stratagene (2006) *Lambda ZAP® II Undigested Vector Kit Instruction Manual* (Catalog #236201, Revision #093002f). Stratagene®, La Jolla, CA.
12. Hoess RH, Wierzbicki A, Abremski K (1986) The role of the *loxP* spacer region in P1 site-specific recombination. *Nucleic Acids Res.* **14**: 2287–2300.
13. Wild J, Szybalski W, (2004) Copy-control tightly regulated expression vectors based on pBAC/oriV. *Methods Mol. Biol.* **267**: 155–167.
14. Promega (2005) *Altered Sites® II in Vitro Mutagenesis System: instructions for use of products Q6210, Q6090 AND Q6080* (Promega Part## TM001 Revised 4/05). Promega Corp., Madison, WI.

Solutions to problems

Solutions to problems posed in Section 4.1

(i)   There are three fragments that have ends that can be joined, to circularize the molecules: B, D, and E.

(ii)   The right end of A, the left end of C, and both the left and right ends of D are all mutually compatible for joining in bimolecular reactions. The right end of C could be joined to the right or left end of B.

(iii)   A. First we determine the formula weight of the vector DNA so we can calculate the number of moles of vector DNA in 25 ng. An average base pair in DNA is 660 g mol^{-1}, so the formula weight of a 3500-bp DNA is

$$3500 \times 660 = 2.31 \times 10^6 \text{ g mol}^{-1}.$$

Now we can calculate that

$$2.5 \times 10^{-8} \text{ g} \div 2.31 \times 10^6 \text{ g mol}^{-1} = 1.08 \times 10^{-14} \text{ mol (10.8 fmol) vector.}$$

If the reaction is conducted in 2.5×10^{-6} l, then the gram concentration of vector is:

$$2.5 \times 10^{-8} \text{ g} \div 2.5 \times 10^{-6} \text{ l} = 0.01 \text{ g l}^{-1}$$

or 10 mg l^{-1}, or 10 µg ml^{-1}. The molar concentration of vector is

$$1.08 \times 10^{-14} \text{ mol} \div 2.5 \times 10^{-6} \text{ l} = 4.33 \times 10^{-9} \text{ M}$$

or 4.33 nM.

We plan to have a 2:1 molar ratio of fragment to vector, so we know that the amount of fragment we need for the ligation reaction is twice the amount of vector, or 8.66 nM (2.16×10^{-14} mol in 2.5×10^{-6} l). To determine the gram concentration of DNA fragment, we first calculate the formula weight, which is 1000 bp $\times$ 660 g mol^{-1}, or 6.6×10^5 g mol^{-1}. The grams of DNA fragment used in the ligation is therefore

$$2.16 \times 10^{-14} \text{ mol} \times 6.6 \times 10^5 \text{ g mol}^{-1} = 1.43 \times 10^{-8} \text{ g}$$

or 14.3 ng. The gram concentration of DNA fragment is then

$$1.43 \times 10^{-8} \text{ g} \div 2.5 \times 10^{-6} \text{ l} = 5.71 \times 10^{-3} \text{ g l}^{-1}$$

or 5.71 mg l^{-1}, or 5.71 µg ml^{-1}.

We can check our calculation by reasoning as follows. The DNA fragment is 1/3.5 or 28.57% of the size of the vector. If we want a 1:1 molar ratio of fragment and vector, then the number of grams of DNA fragment would be

$$25 \text{ ng} \times 0.2857 = 7.14 \text{ ng}$$

However, we wanted a 2:1 molar ratio, so the amount of DNA fragment is doubled to 14.3 ng.

B. As the volume increases by a factor of 10, the concentrations of each solute decreases by a factor of 10, so the concentration of vector will be 0.433 nM (1 µg ml^{-1}) and the concentration of fragment will be 0.866 nM (0.571 µg ml^{-1}). The molar ratio is still 2:1 in this lower concentration solution, but the rate of the bimolecular reaction will be lower because the molecules will not collide as frequently.

(iv)   The amount of vector added is unchanged in this problem, compared to the previous one. The molar concentration of the 250-bp fragment is 8.66 nM, the same as in part A of the previous problem, because when we are using molar concentrations it doesn't matter that the fragment is smaller. However, 1 mol of the 250-bp fragment has 25% of the mass of 1 mol of the 1000-bp fragment in the previous problem, so the gram concentration of the 250-bp fragment will be one quarter of the answer in the previous problem, or

$$5.71 \text{ µg ml}^{-1} \times 0.25 = 1.43 \text{ µg ml}^{-1}$$

Solutions to problems posed in Section 4.2

(v)   First we need to determine the total number of bacteriophage in the 250 ml so we can calculate the number of moles of bacteriophage. We also need to calculate the formula weight of the bacteriophage from the length of the genome, so we can determine the number of µg of phage DNA from the moles of phage. The total number of bacteriophage in the solution is

$$250 \text{ ml} \times 10^9 \text{ pfu ml}^{-1} = 2.5 \times 10^{11} \text{ pfu}$$

and we use Avogadro's number to determine that

$$2.5 \times 10^{11} \text{ pfu} \div 6.02 \times 10^{23} \text{ mol}^{-1} = 4.15 \times 10^{-13} \text{ mole}$$

The formula weight of the bacteriophage genome is

$$5 \times 10^4 \times 660 \text{ g mol}^{-1} = 3.3 \times 10^7 \text{ g mol}^{-1}$$

so the number of grams of bacteriophage DNA in the solution (assuming that all viral particles are also plaque forming units) is

$$4.15 \times 10^{-13} \text{ mol} \times 3.3 \times 10^7 \text{ g mol}^{-1} = 1.37 \times 10^{-5} \text{ g} = 13.7 \text{ µg}$$

(vi)   The concentration of the bacterial stock is

$$0.6 \times 5 \times 10^8 \text{ cells ml}^{-1} = 3 \times 10^8 \text{ cells ml}^{-1}$$

We are starting with 10 µl of the bacterial stock, or 0.01 ml $\times 3 \times 10^8$ cells ml^{-1} = 3×10^6 cells. If we are to have an MOI of 0.0005, then we need to combine these bacteria with $0.0005 \times 3 \times 10^6 = 1.5 \times 10^3$ bacteriophage λ pfu. The concentration of bacteriophage λ in the stock solution is 10^6 pfu per ml, or 10^3 pfu per µl. Therefore, 1.5 µl of the stock will contain 1.5×10^3 bacteriophage. However, depending on the quality of our pipetting devices we might do better to dilute the bacteriophage first, for example taking 100 µl of the bacteriophage and adding 900 µl of phage storage buffer to make a 1:10 dilution. Then, the concentration of bacteriophage in the dilution would be 10^5 pfu ml^{-1}, and 15 µl would be 0.015 ml $\times 10^5$ pfu ml^{-1} = 1.5×10^3 bacteriophage.

Solutions to problems posed in Section 4.4

(vii) We may first consider the results of restriction digestion with *Eco*RI alone, the finding of a 6000-bp and 500-bp fragment. The 500-bp fragment must start at nt 90, the location of the original vector's sole *Eco*RI site, enter the DNA insert at nt 100, and extend an additional 490 bp into the DNA insert.

The 4500-bp vector has three *Sma*I sites, and would generate *Sma*I restriction fragments with sizes 2360, 1340, and 800 bp. The 2360- and 1340-bp fragments are shared between the 4500- and 6500-bp plasmids, and are unchanged by the 2-kbp insertion. The 800-bp *Sma*I fragment is swelled by the insertion of 2000 bp, but the total 2800 bp is divided into 2280 bp plus 520 bp by the extra *Sma*I site that is within the DNA insertion. The 520-bp fragment must have one *Sma*I end that comes from the DNA insertion, and one that comes from the original vector. We have two possibilities for the latter: either the *Sma*I site at nt 110 or the one at nt 3810 (using numbers appropriate to the original vector) must lie 520 bp from the new *Sma*I site in the DNA insertion. It must be the one at nt 110, because the other is more than 520 bp away from the *Bam*HI cloning

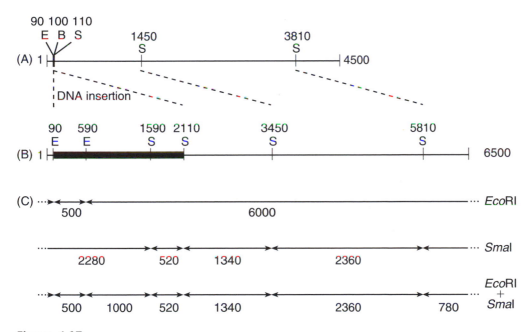

Figure 4.17

Solution to a restriction mapping problem. (A) Linear map of the 4500-bp vector, with index numbers indicating the locations of *Eco*RI (E), *Bam*HI (B), and *Sma*I (S) restriction sites. The vector is circular, and nt 1 (on the left) is joined with nt 4500 (on the right). (B) Linear map of the vector after insertion of a 2-kbp *Bgl*II DNA fragment at the *Bam*HI site. The correspondence of restriction sites in the 6500-bp plasmid is indicated by dashed lines over the map. (C) DNA fragment sizes generated by digestion of the 6500-bp vector with *Eco*RI (top), *Sma*I (middle), and both *Eco*RI and *Sma*I together (bottom).

site. Therefore, the new *Sma*I site is 510 bp from one end of the DNA insertion and 1490 bp from the other end.

We were able to solve this without consulting the double-digestion data (*Eco*RI and *Sma*I together), but it is often necessary to use double-digestion data to resolve complicated patterns. We can confirm that upon digestion with both *Eco*RI and *Sma*I, we release a 1000-bp fragment that is internal to the 2-kbp DNA insert, and represents the DNA between the newly inserted *Eco*RI and *Sma*I sites. The complete solution is shown in *Figure 4.17*.

(viii) We use the Poisson distribution to solve problems of this type. With 8×10^4 independent clones of 1×10^5 bp, we have a total product of 8×10^9 bp represented. The genome is 1×10^9 bp, so we have an average of 8 copies of each sequence. For any given sequence, the probability that it is represented at least once among the 80 000 clones is:

$$1 - 1/e^8 = 0.9997.$$

That is, we have a 99.97% chance that any given sequence is represented.

Restriction endonucleases

5

5.1 Recognition sequences

Nucleases are enzymes that break or 'digest' the phosphodiester backbone of DNA. Endonucleases are nucleases that can do their work at internal sites in a DNA sequence, meaning that they do not need to operate on a free 5′ or 3′ end. Restriction endonucleases operate near specific sequences in a DNA; they are endonucleases that are restricted or constrained to recognize specific sequences before they digest DNA (1–3). In particular the type II restriction endonucleases, that are the subject of this chapter, are practical for use in the recombinant DNA laboratory, because they cleave DNA in a highly predictable way after binding to their recognition sequence (4).

Names and their pronunciations

Restriction enzyme names come from the genus and species names from which the enzymes were first isolated. The enzyme *Sma*I, for example, is from *Serratia marcescens*, and the enzyme *Hpa*II is from *Haemophilus parainfluenzae*. The Roman numeral indicates the isolate number, so, for example, *Hpa*II is the second enzyme isolated from *Haemophilus parainfluenzae*, and *Hpa*II is a different enzyme from *Hpa*I. Some enzyme names include a strain descriptor, for example *Eco*RI and *Bam*HI are the first restriction endonucleases isolated from *Escherichia coli* R and *Bacillus amyloliquefaciens* H, respectively.

A word or two about pronunciation might be in order, since the names represent conventions of speech and a mispronunciation is immediately noticed by seasoned researchers. Many names are sounded out; for example, *Sma*I is universally pronounced with the 'S' and 'm' blended, as in the word 'smoke', and a short 'a' vowel that sounds like 'sm-aah'. The capital letter strain descriptors are pronounced as independent letters, and the Roman numerals are always pronounced as numbers, so an enzyme name such as *Eco*RI is read as 'eek-oh-are-one' or 'echo-are-one', not 'ee-core-eye'. Some enzyme names are pronounced as common words, for example *Bgl*II is 'bagel-two', *Sna*BI is 'snobby-one', *Bst*EII is 'beastie-two', and the 'Sau' in *Sau*3AI is 'sow' (rhymes with cow) or 'saw'. Others are broken into parts by convention, for example *Hind*III is 'hin-dee-three', not 'hind-three', but *Hinc*II is 'hink-two', not 'hin-see-two'. Perhaps the early researchers who set the style of pronunciation did not care for the indelicate sound of 'hind', although those were simpler times that preceded the discovery of *Sex*AI, *Bsm*I, and *Fok*I. Many names defy pronunciation in English and are read out as their letters, for example *Kpn*I, *Dpn*I, and *Nco*I.

Names such as *Sma*I, *Bam*HI, and *Eco*RI are so well known that they may be shortened in informal laboratory banter, for example, 'sma', 'bam', and 'R1'. In fact, laboratory speech must be very puzzling to a new researcher: 'I cut my mini DNA with R5 and kipped it' translates as 'I took DNA I prepared in a mini-lysate procedure, digested it with the restriction endonuclease *Eco*RV, and then treated it with calf intestinal alkaline phosphatase.'

Specificity

Restriction endonucleases recognize specific sequences in DNA (5). For example, the enzyme *Eco*RI binds to the sequence 5'-GAATTC-3' in double-stranded DNA, and the enzyme *Sau*3AI binds to the sequence 5'-GATC-3'. Some enzymes recognize more than one DNA sequence; for example, the enzyme *Hinf*I recognizes each of the sequences 5'-GAGTC-3', 5'-GAATC-3', 5'-GATTC-3', and 5'-GACTC-3'. These four sequences are all related, in that they start with a 5'-GA and end with a TC-3', but the nucleotide in the middle could be any of the four nucleotides G, A, T, or C. We say that such a sequence is partially degenerate to indicate that it is not completely specified. We use the letter 'N' to indicate the presence of any of the four nucleotides G, A, T, or C, so the *Hinf*I recognition sequence could be written in an abbreviated form as 5'-GANTC-3'.

Not all degenerate sequences allow any of the four nucleotides. For example, the recognition sequence for the enzyme *Hinc*II is written as 5'-GTYPAC-3'; where the 'Y' indicates a pyrimidine base, either C or T, and the 'P' indicates a purine base, either G or A. This recognition sequence is also sometimes written as 5'-GTPyPuAC-3', where Py and Pu indicate pyrimidines and purines, respectively, or 5'-GT(C/T)(A/G)AC-3', where the points of degeneracy are indicated by nucleotides in parentheses. *Hinc*II consequently has four related recognition sequences: 5'-GTCAAC-3', 5'-GTCGAC-3', 5'-GTTAAC-3', and 5'-GTTGAC-3'.

There is a standardized code for writing degenerate nucleotides; for example 'W' is used to indicate either of the nucleotide bases A or T, and 'S' is used to indicate either G or C. The letters are sometimes helpful mnemonics; for example W is an abbreviation for 'weak', since A/T base pairs are based on only two hydrogen bonds and S is an abbreviation for 'strong' since G/C base pairs are based on three. The restriction endonuclease *Bme*1580I is an example of an enzyme that has a degenerate recognition sequence 5'-G(G/T)GC(A/C)C-3', and in the one-letter code this could be written as 5'-GKGCMC-3'. The more esoteric parts of the code are not widely used, beyond the straightforward N, Y, and P letters. Not many researchers probably keep in mind, for example, that the letter V stands for any of the bases A, G, or C.

When we look at an example such as *Hinf*I, with a recognition sequence of 5'-GANTC-3', we can see that only four specific nucleotides define the site. The nucleotide in the middle does not contribute to the sequence specificity, except for the fact that it serves as a wild-card or placeholder. Many restriction endonucleases have interrupted recognition sequences, and a more extreme example is the enzyme *Xcm*I which recognizes the sequence 5'-CCANNNNNNNNNTGG-3'. The *Xcm*I recognition sequence is specific at only six nucleotide positions, and nine 'N' degenerate bases separate the two blocks of specific sequence. This sequence might be abbreviated as 5'-CCA(N9)TGG-3' in some references, where the 'N9' indicates a block of nine 'N' nucleotides.

Symmetry

Most recognition sequences for restriction endonucleases are symmetrical, or palindromic, in the sense that they read the same 5' to 3' on each strand. For example, the enzyme *Bam*HI has a recognition sequence of 5'-GGATCC-3', and the underlying complementary strand for this sequence is 3'-CCTAGG-5'. Reading this complementary strand right to left, or 5' to 3', it is clearly the same sequence: 5'-GGATCC-3'. Many enzymes bind to the DNA substrate as a homodimer, recognizing the two strands of the DNA simultaneously. The symmetry of the recognition sequence simply reflects the binding of a symmetrical pair of monomers. We've seen examples of degenerate sequences that were not strictly symmetrical at the nucleotide level, for example the 5'-GTYPAC-3' sequence of *Hinc*II allows for both symmetrical (5'-GTCGAC-3') and nonsymmetrical (5'-GTCAAC-3') restriction sites. However, the 5'-GTYPAC-3' site is symmetrical in the sense that Y and P are complementary sets of nucleotides.

There are recognition sequences for some restriction endonucleases that are highly specific and completely asymmetrical. For example, the enzyme *Bsa*I recognizes the sequence 5'-GGTCTC-3', which would read 5'-GAGACC-3' on the complementary strand. If we were searching for *Bsa*I sites in a printed sequence that shows only one DNA strand, we would need to look for both the 5'-GGTCTC-3' and 5'-GAGACC-3' versions.

Length

Most restriction endonucleases have recognition sequences of four, five, or six specific nucleotides. For example, the recognition sequence for *Bam*HI consists of six specific nucleotides (5'-GGATCC-3'), two more than the recognition sequence for *Sau*3AI (5'-GATC-3'). The internal four nucleotides of the *Bam*HI site are GATC, so every *Bam*HI recognition sequence contains a *Sau*3AI recognition sequence. The converse is not true; most *Sau*3AI sites are not also *Bam*HI sites.

Generally, the restriction enzymes with shorter recognition sequences will cut a DNA sequence more frequently because the specific site is present more often. If we imagine an idealized DNA sequence that consists of 25% A, 25% G, 25% C, and 25% T nucleotides, then the average separation between *Sau*3AI sites will be about 256 bp. This is because the probability of any nucleotide being a G is 0.25, and the probabilities of the next three sequential nucleotides being respectively A, T, and C, is 0.25 for each. At any starting nucleotide, the probability of finding a 5'-GATC-3' sequence is the product of the four individual probabilities: $(0.25)^4 = 1/256$.

By the same type of argument, at any point in an idealized sequence the probability of finding any four-nucleotide recognition sequence is 1/256, the probability of finding any five-nucleotide recognition sequence is 1/1024, and the probability of finding any six-nucleotide recognition sequence is 1/4096. We might expect fragments from a *Bam*HI digestion to average about 4.1 kbp, in this idealized situation. Restriction endonucleases such as *Not*I, with recognition sequence 5'-GCGGCCGC-3', or *Sfi*I, with recognition sequence 5'-GGCC(N5)GGCC-3', cut DNA infrequently, about every 65 kbp in the idealized sequence, because their recognition sequences have an unusually high number of specific nucleotides. That is not to say that there will not be fragments that are much smaller or much larger than the average, or that all DNA sequences will behave in this idealized way.

For example, the restriction endonuclease *Sph*I has a six-nucleotide recognition sequence 5'-GCATGC-3', having more G or C nucleotides than A or T nucleotides. In the human genome, which is 41% G or C nucleotides and 59% A or T nucleotides, the average fragment from an *Sph*I digestion would be predicted to be about 6,500 bp; larger than the idealized 4,096 bp because the *Sph*I sequence would be less frequently represented. However, these predictions based on individual nucleotide compositions can be very inaccurate. The restriction endonuclease *Pvu*I has a very similar recognition sequence, 5'-CGATCG-3', having the same nucleotide composition, but its recognition sequence would be found at a much lower frequency in human DNA than that of *Sph*I. One reason is that CG dinucleotides are rare in human DNA, and *Pvu*'I has two of them in its recognition sequence while *Sph*I has none. Furthermore, many CG dinucleotides are methylated in human DNA, and this methylation interferes with digestion by *Pvu*I. Methylation interference is discussed in more detail in Section 5.2.

Would an enzyme such as *Xcm*I be an extremely rare cutter of DNA, due to the length of its recognition sequence 5'-CCANNNNNNNNNTGG-3'? No, because the frequency of digestion in a DNA depends on the specific nucleotides in the recognition sequence, and the nine degenerate N nucleotides are non-specific. *Xcm*I would be predicted to cut a fragment of DNA at about the same frequency as *Nco*I, which has a recognition sequence of 5'-CCATGG-3'. The six-nucleotide recognition sequence of *Nco*I is related to that of *Xcm*I, but lacks the nine interrupting nonspecific nucleotides.

Problems

The solutions to these problems are provided at the end of the chapter.

(i) Determine how many recognition sequences for *Scr*FI (5'-CCNGG-3'), *Rca*I (5'-TCATGA-3'), and *Psp*GI (5'-CCWGG-3') are in the DNA sequence below (where N = G, A, T, or C; and W = A, or T):

```
5'-GAGTACTCCCGGACCGAGTTTACAGACCTGGGTCATGACCGAGCAGG-3'
3'-CTCATGAGGGCCTGGCTCAAATGTCTGGACCCAGTACTGGCTCGTCC-5'
```

(ii) Determine how many recognition sequences for *Bmg*BI (5'-CACGTC-3'), *Ear*I (5'-CTCTTC-3'), and *Aci*I (5'-CCGC-3') are in the DNA sequence below:

```
5'-ATTACAGAAGAGCCACGTCAGTACCGCTTTGCGGCATACAGTACAGT-3'
3'-TAATGTCTTCTCGGTGCAGTCATGGCGAAACGCCGTATGTCATGTCA-5'
```

5.2 Cleavage sites

Types of ends

The end of a double-stranded DNA molecule consists of two single-strand ends in close proximity, one being a free 5′ end and the other a free 3′ end. If two single strands are completely base-paired at an end, the structure they form is called a blunt end. When two DNA ends are blunt, they can always be brought together in a ligation reaction, which is to say that they are compatible. If these two single strands do not meet exactly at their ends, the strand that hangs out over the recessed end is called an overhanging end. Another way of looking at it is that the overhanging end has unpaired nucleotides and the recessed end does not. The overhanging end may also be called a cohesive end, or sticky end, these terms referring to the possible base-pairing of these ends to another complementary DNA. Either a 5′ or a 3′ end can play the role of an overhanging end. Overhanging ends are compatible with each other when they can be base-paired and ligated (6) (as discussed in Chapter 4).

The action of a restriction endonuclease is to break the phosphodiester backbone and leave a free 5′ phosphate and 3′ hydroxyl end. This nicking of the DNA occurs independently on each strand, and in a very precise location in the case of the type II restriction endonucleases. For example, in the case of *Eco*RI, the cleavage site on each strand is between the G and A nucleotide of the 5′-GAATTC-3′ sequence (see *Figure 5.1*). On the complementary strand, the cleavage site is also between the G and A nucleotide of the 3′-CTTAAG-5′, so the nicks are separated by four nucleotide pairs. The hydrogen bonds connecting those four nucleotide pairs are broken naturally by the ambient thermal energy, and the result is the creation of overhanging 5′-AATT ends. Recognition sequences are usually annotated with a slash or special mark to show the position of the nick; for example, the *Eco*RI recognition sequence

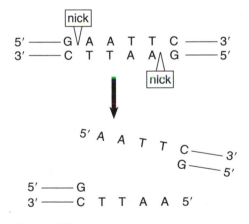

Figure 5.1

The action of restriction endonuclease *Eco*RI. The enzyme recognizes the specific sequence 5′-GAATTC-3′, and nicks the strands between the G and A nucleotides (top). The hydrogen bonding between the A and T base pairs is broken by thermal energy, and leads to separation of the cohesive ends (bottom).

may be represented as G/AATTC, or G^AATTC. By convention, a single-stranded sequence written from left to right can be assumed to be 5' to 3'.

Is this annotation sufficient for us to understand what the enzyme does? Yes, because the complementary strand has the same behavior. A researcher can take the information in the G/AATTC annotation, realize that the nick in the complementary strand must be similarly at 3'-CTTAA/G-5', and then see that when the ends are pulled apart each will have a four-nucleotide overhang.

It is important to emphasize that restriction endonucleases hydrolyze the covalent bonds in the phosphodiester backbone, and the hydrogen bonds in the base pairs fall apart by the effect of temperature. The hydrogen bonds may reform stably at low temperature, for example if the sample is placed on ice, but even in this case the covalent bonds will not spontaneously re-form. The overhanging ends left by *Eco*RI digestion can be brought together more permanently by ligation, whether or not they were ever once connected (as discussed in Chapter 4).

The example of *Eco*RI highlights the creation of a 5' overhanging end, one of the three types of ends that can be left by a restriction endonuclease. The other two types are the blunt end and the 3' overhanging end. The enzyme *Sma*I, which has a recognition sequence CCC/GGG, leaves a blunt end because the two nicks are directly across from each other in the double-stranded DNA. After cleavage there are no hydrogen bonds that need to be broken before the ends can be separated. The enzyme *Kpn*I, which has a recognition sequence GGTAC/C, leaves a 3' overhanging end of GTAC-3' after separation of the ends. In general, when the site of nicking is in the dead center of a symmetric recognition sequence the end will be blunt; when it is left of center it will leave a 5' overhanging end, and when it is right of center it will leave a 3' overhanging end.

Cleavage at asymmetric recognition sites

The cleavage sites discussed so far have all fallen within the recognition sequence of the enzyme, but this is not universally the case. For example, the enzyme *Sap*I recognizes the sequence 5'-GCTCTTC-3', but nicks the DNA between the first and second nucleotide to the right of this sequence (5'-GCTCTTCN/N-3'). The enzyme recognizes an asymmetric site, and on the complementary strand the nick occurs three nucleotides further to the right (3'-CGAGAAGNNNN/N-5'). When taken together, the two nicks caused by *Sap*I result in a 5' overhanging end of three 'N' nucleotides. The overhanging sequences are not consistent with *Sap*I, since they are created from 'N' nucleotides.

There is a common way of annotating cleavage sites of this type, and that is to count the number of 'N' bases to the cut site on each strand and represent it as a ratio in parenthesis. For example, the *Sap*I enzyme leaves an end with the following structure:

```
...GCTCTTCN    3'
...CGAGAAGNNNN 5'
```

Its recognition sequence could be written as GCTCTTC(1/4), to indicate nicking after one nucleotide on the strand with the GCTCTTC, and after four on the complementary strand. It is purely a convention to show the cutting to the 'right' of the sequence; it would mean the same thing to base the recognition sequence description on the complementary strand, which in the case of *Sap*I would be (4/1)GAAGAGC. The simplified annotation is welcome when we consider enzymes such as *Eco*P15I with its recognition sequence of CAGCAG(25/27), or *Mme*I with its recognition sequence of TCCRAC(20/18). *Eco*P15I will leave a 5′ overhanging end of two nucleotides, and *Mme*I will leave a 3′ overhanging end of two nucleotides. We must simply bear in mind that (27/25)CTGCTG and (18/20)GTYGGA are alternate descriptions of these two recognition sequences when we are studying their cleavage sites in a sequence.

Not all asymmetric recognition sequences involve cleavage sites that lie in degenerate nucleotides. For example, the enzyme *Bbv*CI has a recognition sequence of 5′-CC/TCAGC-3′ on one strand, and 3′-GGAGT/CG-5′ on the complementary strand. This can be written as CCTCAGC(−5/−2), where the negative numbers mean that we find the cut sites by counting nucleotides from the right end of the sequence, going to the left.

Some enzymes will digest a DNA on both sides of a recognition sequence; for example, *Bsa*XI has a recognition sequence of (9/12)AC(N5)CTCC(10/7) and excises a small piece of DNA that is 30 nucleotides on each strand, with a 3′ overhanging end of three nucleotides at each end. The product of digestion will have the following structure:

```
5′–     NNNNNNNNNNACNNNNNCTCCNNNNNNNNNNNN    –3′
3′–  NNNNNNNNNNNNNT GNNNNNGAGGNNNNNNNN       –5′
```

That is, one strand excised by *Bsa*XI will be 5′-(N9)AC(N5)CTCC(N10)-3′ and the complementary strand will be 3′-(N12)TG(N5)GAGG(N7)-5′. Once again, we must bear in mind that (7/10)GGAG(N5)GT(12/9) is another way of describing the recognition sequence for *Bsa*XI.

Nicking endonucleases

Some restriction enzymes have been genetically engineered so that they only cleave one of the two asymmetric strands at a recognition sequence. This activity is called a nicking enzyme because it does not lead to full scission of the double-stranded DNA. For example, the nicking enzyme N.*Bbv*CIB, which is an altered form of the restriction endonuclease *Bbv*CI, has a recognition sequence 5′-CC/TCAGC-3′. There is no cleavage of the complementary strand. Nicking enzymes generate free 5′ and 3′ ends in a DNA sequence, and can cause relaxation of a supercoiled molecule, but they do not generate separate DNA fragments in the way that a fully functional restriction enzyme does.

Isoschizomers and neoschizomers

Sometimes the same restriction endonuclease activity is found in an enzyme from two different organisms, and these enzymes are then called isoschizomers. For example, *Hae*III from *Haemophilus aegptius* and *Pho*I from *Pyrococcos horikoshii* are isoschizomers because both digest DNA at the sequence GG/CC, leaving a blunt end. Sometimes there are slight differences between the activities of isoschizomers; for example, *Hae*III is not inhibited by sequence methylation but *Pho*I is inhibited by some combinations of overlapping CpG methylation. *Mbo*I and *Sau*3AI are also isoschizomers because both digest DNA at the sequence /GATC, but *Mbo*I will not digest the DNA if the adenosine is methylated and *Sau*3AI will digest the sequence whether the adenosine is methylated or not.

The previous examples included only enzymes that have the same recognition sequence and cut the DNA sequence in the same way, but what about enzymes that recognize the same sequence but cut it differently? For example, the enzymes *Kas*I (G/GCGCC), *Nar*I (GG/CGCC), *Ehe*I (GGC/GCC), and *Bbe*I (GGCGC/C) all recognize the sequence 5'-GGCGCC-3', but each leaves a different type of end: *Kas*I leaves a 5' overhanging end of four nucleotides, *Nar*I leaves a 5' overhanging end of two nucleotides, *Ehe*I leaves a blunt end, and *Bbe*I leaves a 3' overhanging end of four nucleotides. Some researchers use the name neoschizomers to describe enzymes that recognize the same site but cut it differently, and others continue to use the terminology isoschizomer for this type of phenomenon. Neoschizomers ostensibly recognize the same sequence but may have slightly different properties of methylation interference, just as was the case with isoschizomers.

Restriction endonucleases that yield compatible ends

When we consider whether two DNA ends are compatible, we are essentially asking whether the two ends can be brought together for ligation using T4 DNA ligase. This enzyme, which is discussed in Chapter 4, forms a covalent bond between adjacent 5' phosphate and 3' hydroxyl ends. Two blunt ends can be fit together, regardless of their sequence, and so blunt ends are always compatible. If two DNA fragments have overhanging ends that match in base-pairing, and allow the free 5' and 3' ends to abut each other, they will also be compatible. However, if there is a mismatch in base-pairing or if the free 5' and 3' ends do not abut, the ends are not compatible.

For example, the enzymes *Bam*HI (G/GATCC), *Bgl*II (A/GATCT), and *Sau*3AI (/GATC) all generate a 5'-GATC overhanging end, and so their ends are mutually compatible. It may seem odd that a fragment digested with one enzyme might be compatible with a fragment digested with another, but it is only the fit of the base-pairing that is being tested, not the source of the DNA. This compatibility is very useful in the preparation of genomic DNA libraries, as outlined in Chapter 4, because a genomic DNA can be partially digested with *Sau*3AI to yield large DNA fragments with 5'-GATC overhanging ends. These large *Sau*3AI fragments can be introduced into a unique *Bam*HI site in a cloning vector.

We might be tempted to think that a 5'-GATC end would be compatible with a 3'-GATC end, but 5' overhanging ends can never base-pair with 3' overhanging ends and are never compatible:

```
...NNNNN                        NNNNN...
...NNNNNCTAG -5'     3'- GATCNNNNN...
```

T4 DNA ligase requires a double-stranded substrate, so even if the 5' and 3' overhanging ends happen to collide in a reaction they cannot be ligated together as if they were blunt ends. In a similar way, the enzymes *Kas*I

(G/GCGCC) and *Nar*I (GG/CGCC) do not yield mutually compatible ends because base-pairing between a 5'-GCGC and a 5'-CG would leave a single-stranded gap:

```
...NNNNG              5'- CGCCNNNNN...
...NNNNCCGCG -5'          GGNNNNN...
```

A 5'-GATC end is also not compatible with a 5'-CTAG end; the ends do not base-pair:

```
...NNNNN              5'- CTAGNNNNN...
...NNNNNCTAG -5'      3'-     NNNNN...
```

Base-pairing requires that the DNA strands be antiparallel: if one strand is running 5' to 3', from left to right, the base-paired strand must run 5' to 3' from right to left.

When compatible ends are joined by ligation, they may or may not yield a sequence that can be re-cleaved by the same restriction enzyme. For

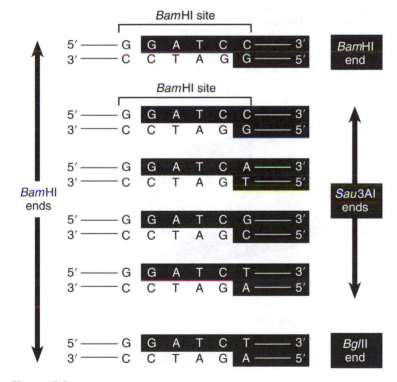

Figure 5.2

Pairing of cohesive ends generated by different enzymes. *Bam*HI ends (on the left) are paired with cohesive ends left by restriction endonucleases *Bam*HI, *Sau*3AI, and *Bgl*II (on the right, shown with white-on-black lettering). *Sau*3AI recognizes the specific sequence 5'-GATC-3', and so four possible nucleotides are shown for the position following the C nucleotide. *Bgl*II recognizes the specific sequence 5'-AGATCT-3'. The 5'-GATC overhanging ends are compatible among these three enzymes, but a *Bam*HI site 5'-GGATCC-3' is only created (or recreated) in the top two pairings of ends.

example, when a *Bam*HI-generated end is ligated to a *Bam*HI-generated end, it always yields a sequence that can be re-cleaved by *Bam*HI because the two halves of the recognition sequence (G/GATCC) are faithfully rejoined. If a *Bam*HI (G/GATCC) -generated end is ligated to a *Sau*3AI (/GATC) generated end, the resulting product will always be re-cleavable with *Sau*3AI, but will only be cleavable with *Bam*HI if the *Sau*3AI generated end contributes half of the *Bam*HI recognition sequence (see *Figure 5.2*). However, a *Bam*HI (G/GATCC) generated end ligated to a *Bgl*II (A/GATCT) generated end is never cleavable with either *Bam*HI or *Bgl*II, because the connection yields the mixed sequence AGATCC that contains neither of the recognition sequences (see *Figure 5.2*).

Sometimes the joining of ends by ligation can generate recognition sequences of unrelated enzymes. For example, if a DNA is digested with *Bgl*II (A/GATCT) and the 5' overhanging ends are filled in with Klenow enzyme (a technique that is discussed in Chapter 6), the now blunt ends can be joined by ligation to yield a sequence AGATCGATCT (see *Figure 5.3A, B*). While this does not regenerate a *Bgl*II recognition sequence, the junction of the ends creates the internal sequence ATCGAT that is the recognition sequence for the enzyme *Cla*I (AT/CGAT) (see *Figure 5.3C*).

Overhanging ends can be filled or erased using T4 DNA polymerase or Klenow enzyme which are useful for converting incompatible overhanging ends to compatible blunt ends (discussed in detail in Chapter 6).

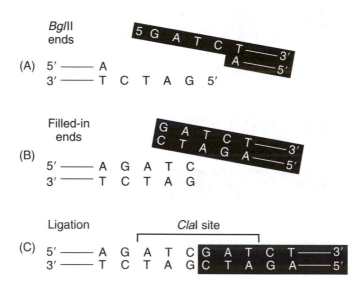

Figure 5.3

Generation of new restriction endonuclease recognition sequences at the site of ligation. (A) 5' overhanging ends generated by the enzyme *Bgl*II (5'-A/GATCT-3'). The end on the right is shown with white-on-black lettering to distinguish it from the one on the left. (B) The same *Bgl*II ends, with the overhanging ends filled in with a 5' to 3' DNA polymerase activity. (C) The two filled-in ends joined by ligation. The recognition sequence for the enzyme *Cla*I (5'-ATCGAT-3') is generated by the fusion of ends. A *Cla*I site is similarly generated by ligation of filled in *Bam*HI or *Sau*3AI ends.

However, the ligation of blunt ends is not as easily accomplished as ligation of compatible cohesive ends, as explained in Chapter 4.

Cleavage near a DNA end

Can a restriction endonuclease operate on a piece of DNA if the restriction site is very close to a DNA end? This is often an important question in cloning work, because some enzymes perform well at the ends of DNA and others do not. A multiple cloning site might be digested with several different restriction enzymes, and prior cleavage at one might affect the efficiency of cleavage at another. Furthermore, recognition sequences for restriction endonucleases may be included at the ends of DNA primers used in polymerase chain reactions (PCRs), but they need to be 'padded' with extra 5' nucleotides if the restriction enzymes are to digest the PCR products (see Chapter 7).

The enzyme *Eco*RI (G/AATTC), for example, will demonstrate 100% cleavage at a recognition sequence if the site is at least one base pair from the end of the fragment. That is, the double-stranded sequence

```
5'-NGAATTCNNNNN...
3'-NCTTAAGNNNNN...
```

should be digested efficiently with *Eco*RI. On the other hand, *Hind*III (A/AGCTT) does not efficiently cleave a DNA such as

```
5'-NAAGCTTNNNNN...
3'-NTTCGAANNNNN...
```

that only has one extra nucleotide in front of the recognition sequence. *Hind*III will digest a sequence that has two or more additional 5' nucleotides, for example:

```
5'-NNAAGCTTNNNNN
3'-NNTTCGAANNNNN
```

Restriction enzyme vendors sometimes provide information on how well their enzymes work at the tips of a DNA fragment. If in doubt, adding two or three nucleotides to the end of a recognition sequence will help to ensure that an enzyme can bind and function properly.

Star activity

When a restriction endonuclease is used under unusual conditions, for example at the wrong salt or pH, or in the presence of more than about 5% (v/v) glycerol, the enzyme may exhibit additional (and usually unwanted) enzymatic activities. The classic example is *Eco*RI, which normally has the recognition sequence G/AATTC, but under conditions of low ionic strength may recognize the more general N/AATTN sequence. The problem is that there are usually many more AATT sites in a DNA sequence than GAATTC sites, depending on the sequence composition, so this additional enzymatic activity, called *Eco*RI* ('star'), will tend to disrupt cloning and mapping work.

Star activities are known for several enzymes, and probably have yet to be characterized for many others. The star activities can generally be avoided by following the recommendations of the enzyme manufacturer for use of the enzyme. For example, this can be accomplished by using the proper buffer conditions, adding an appropriate but not excessive amount of enzyme to a reaction, and conducting reactions in a sufficient volume so that the glycerol in the enzyme storage buffer is diluted at least 10- to 20-fold.

Methylation effects

DNA is often methylated in a host cell by enzymes that transfer a methyl group to a DNA base, at a specific recognition sequence. For example, *dam+ E. coli* strains have a Dam methylase that converts the adenosine base in the sequence GATC to an N^6-methyladenosine. Another common methylation system in *E. coli* is the Dcm methylase, that converts the internal cytosine base in CC(A/T)GG to a C^5-methylcytosine. In higher eukaryotes, CG (or 'CpG') dinucleotides have C^5-methylcytosine modifications caused by a CpG methylase. All of these examples involve symmetric or palindromic recognition sequences, so the sequence on the complementary strand is also methylated.

How are restriction endonucleases affected by methylated sequences? It depends on the enzyme, and the reader should consult the information provided by the manufacturer or look up the information in a published table (7). Endonuclease recognition sequences do not need to contain the complete methylase recognition sequence to be affected; it is enough for the sequences to overlap (see *Figure 5.4*).

Not all enzymes are tolerant of methylated DNA. If we consider the three isoschizomers and neoschizomers *Mbo*I (/GATC), *Sau*3AI (/GATC), and *Dpn*I (GA/TC), we find an interesting difference (see *Figure 5.4*). *Mbo*I will not digest DNA that is Dam-methylated, that is, where the internal adenosine is methylated, but *Sau*3AI and *Dpn*I will digest it. In fact, *Dpn*I will only digest the GATC sequence if the adenosine is methylated. *Dpn*I and *Mbo*I are not inhibited by C^5-methylation of the GATC, but *Sau*3AI is inhibited (see *Figure 5.4*).

It is important to note that these methylation effects are host-cell specific. If human DNA is cloned into a bacterial vector and grown in *E. coli*, the patterns of CpG methylation that were established in the human will be lost as the DNA is replicated in *E. coli*. Conversely if a Dam- or Dcm-methylated DNA is propagated in human cells it will gradually lose these patterns of methylation as the DNA is replicated. In fact, this is often used as a simple test for replication. If a shuttle vector is grown in a *dam+ E. coli* host before transfection into mammalian cells, the DNA is initially *Dpn*I-sensitive and *Mbo*I-resistant. If it goes through several rounds of replication in the mammalian cell, it will become *Dpn*I-resistant and *Mbo*I-sensitive (see *Figure 5.4*).

Problems

The solutions to these problems are provided at the end of the chapter.

(iii) For each of the following restriction enzymes, make up an example of a double-stranded DNA sequence that contains a recognition sequence, and show how the DNA would look after cleavage and

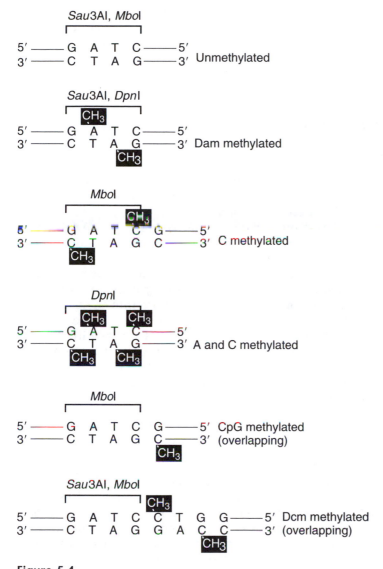

Figure 5.4

Effect of methylation of adenine or cytosine bases on recognition of the GATC sequence by three restriction endonucleases. Over each sequence description, the names of enzymes that will digest the DNA are indicated; for example, *Sau*3AI and *Mbo*I will both digest the unmethylated 5′-GATC-3′ sequence but *Dpn*I will not digest the unmethylated sequence (top). Methylated bases are indicated by a CH₃, with white-on-black lettering.

separation of ends: *Xba*I (T/CTAGA), *Sph*I (GCATG/C), and *Eco*RV (GAT/ATC).

(iv) Make up an example of a DNA sequence that can be digested with *Bsa*I (GGTCTC(1/5)) and *Bsg*I (GTGCAG(16/14)), leaving a 5′-GATC overhanging end at one end and a 3′-GC overhanging end at the other. After cleavage with these enzymes, the recognition sequences for these two enzymes should be present in a single large fragment.

(v) Which of these enzymes yield ends compatible with each other? *Bam*HI (G/GATCC), *Spe*I (A/CTAGT), *Ssp*I (AAT/ATT), *Nde*I (CA/TATG), *Nhe*I (G/CTAGC), *Bfa*I (C/TAG), *Bcl*I (T/GATCA), *Ase*I (AT/TAAT), and *Eco*RV (GAT/ATC).

5.3 Reaction conditions

A clean substrate

Restriction endonuclease reactions must be conducted under controlled conditions, and there are a few important things to consider when setting up such a reaction. We need to use DNA that is sufficiently pure that it can be digested, and this often requires the use of 'clean-up kits', spin columns, or DNA-binding resins that allow contaminants to be separated or washed away (as discussed in Chapter 4).

For example, there is a method of growing bacteriophage λ at high density on a bacteriological plate, overlaying the agar on the plate with phage storage buffer so the bacteriophage can diffuse into the buffer, and withdrawing the liquid to isolate the bacteriophage. If the DNA is extracted from this bacteriophage preparation, it is not readily digested with restriction endonucleases because there are inhibiting contaminants that leach out of the agar during isolation. Alcohol precipitation does not help to purify the DNA, because the contaminants are co-purified during precipitation. Use of a clean-up kit or spin column can separate the bacteriophage DNA from the restriction endonuclease inhibitors, and yield DNA that is suitable for most restriction enzymes.

Buffer and salt conditions

A second consideration is to use a buffer and salt condition that is optimal for enzyme function. Enzyme manufacturers often supply a $10 \times$ buffer that is optimized for use with their enzyme, as well as instructions for how compatible the enzyme is with other buffer conditions. In general, these buffer systems include a salt, for example NaCl, KOAc (potassium acetate), or KCl; a buffer, for example Tris-Cl, Tris-OAc, or Bis-Tris-Propane-Cl; a source of magnesium, for example $MgCl_2$ or $Mg(OAc)_2$; and a reducing agent, for example β-mercaptoethanol or dithiothreitol (DTT). When supplied as a $10 \times$ buffer, each component is 10 times more concentrated than it should be in the enzyme reaction. The researcher simply sets up the reaction so that 1/10 of the final volume is contributed by the $10 \times$ buffer, resulting in each component being diluted to its proper concentration. For example, if the reaction is to be conducted in 35 µl, then 3.5 µl of that volume should be $10 \times$ buffer. Information on how to prepare buffer solutions from scratch is discussed in Chapter 3.

The double digestion

The buffer condition is especially crucial when a researcher is performing a digestion of DNA with two enzymes simultaneously. Is there a middle ground at which both enzymes will function adequately, or must the reactions be run sequentially with a buffer change between them? There is no general answer to this question, other than to recommend that the reader consult a manufacturer's table that shows levels of compatibility. For example, suppose a researcher plans to conduct two double-digestion reactions, the first with *Eco*RI and *Sal*I, and the second with *Eco*RI and *Sac*I. The conditions recommended for these three enzymes, from three manufacturers, are shown in *Table 5.1*.

It is immediately obvious from *Table 5.1* that different manufacturers provide slightly different advice on how to use the same enzyme. The recommended conditions for both *Sal*I and *Eco*RI are high ionic strength, and we remember that for *Eco*RI this is critical to avoid *Eco*RI* activity. For a double-digestion with *Eco*RI and *Sal*I, Manufacturer N recommends that their *Eco*RI buffer be used with both enzymes in the same reaction solution. Manufacturer F recommends that their *Sal*I buffer be used with both enzymes in this situation, but also suggests that if the *Eco*RI buffer is used then a two-fold excess of *Sal*I enzyme should be added to compensate for lower activity. Manufacturer P indicates that *Eco*RI will be 50% to 75% active in the *Sal*I buffer, which again suggests that more enzyme should be added to compensate for the nonoptimal condition. These varying pieces of advice from the companies should be absorbed by the researcher as an indication that *Eco*RI and *Sal*I can function on a piece of DNA in a single buffer system. However, a slight increase in the amount of one of the enzymes might be appropriate, and extra time might be given to allow the reaction to reach completion.

The differences between the *Eco*RI buffer and the *Sac*I buffer are much more troubling; the *Sac*I enzyme seems to function best at low ionic strength, and the *Eco*RI enzyme does not function efficiently under those conditions. Indeed, Manufacturers N and F both recommend that the

Table 5.1 Recommended buffer conditions for three endonucleases

Endonuclease	Manufacturer N	Manufacturer P	Manufacturer F
*Eco*RI	100 mM Tris-Cl 50 mM NaCl 10 mM MgCl$_2$ 0.025% Triton X-100 pH 7.5 (at 25°C)	90 mM Tris-Cl 50 mM NaCl 10 mM MgCl$_2$ pH 7.5 (at 37°C)	50 mM Tris-Cl 100 mM NaCl 10 mM MgCl$_2$ 0.02% Triton X-100 100 µg ml^{-1} BSA (pH 7.5 at 37°C)
*Sal*I	50 mM Tris-Cl 100 mM NaCl 10 mM MgCl$_2$ 1 mM DTT 100 µg ml^{-1} BSA pH 7.9 (at 25°C)	6 mM Tris-Cl 150 mM NaCl 6 mM MgCl$_2$ 1 mM DTT pH 7.9 (at 37°C)	50 mM Tris-Cl 100 mM NaCl 10 mM MgCl$_2$ 100 µg ml^{-1} BSA pH 7.5 (at 37°C)
*Sac*I	10 mM Bis-Tris-Propane-Cl 10 mM MgCl$_2$ 1 mM DTT 100 µg ml^{-1} BSA pH 7.0 (at 25°C)	10 mM Tris-Cl 50 mM KCl 7 mM MgCl$_2$ 1 mM DTT pH 7.5 (at 37°C)	10 mM Bis-Tris Propane-Cl 10 mM MgCl$_2$ 100 µg ml^{-1} BSA pH 6.5 (at 37°C)

digestions be conducted sequentially, and Manufacturer P indicates that *Sac*I has <10% of its normal activity in the *Sal*I (high ionic strength) buffer. A sequential reaction can be conducted by incubating the DNA first with the enzyme that prefers lower ionic strength, in this case the enzyme *Sac*I in the recommended *Sac*I buffer. After the reaction is complete, a small amount of concentrated salt solution and buffer can be added to adjust the conditions, and the second enzyme can be added. The volume may also be increased to help dilute out the components of the first buffer, but this is not always necessary. For example, suppose a sample of DNA is digested with *Sac*I, in 50 μl of the *Sac*I buffer recommended by Manufacturer F. Upon completion of this first reaction, a researcher might add 6 μl of 1 M Tris-Cl, pH 7.5, 3 μl of 1 M NaCl, and after thorough mixing, 1 μl of *Eco*RI. The volume of this reaction would then be 60 μl, and it would be approximately 100 mM Tris-Cl, 50 mM NaCl, and 8 mM $MgCl_2$. The pH would be close to 7.5, since the Tris buffer would be much more concentrated than the original Bis-Tris-Propane-Cl. The buffer conditions would approximate the recommended *Eco*RI buffer from all three manufacturers. Extra *Eco*RI enzyme or extended reaction time could compensate for the nonoptimal reaction conditions.

When enzymes are added in sequential reactions, the enzyme added in an early step may retain some activity in later steps. In the previous example the *Sac*I enzyme may continue to function in the higher ionic strength buffer for *Eco*RI, albeit at a lower level. That isn't a problem in this case. However, had we decided to conduct the *Eco*RI reaction first, then to dilute the solution and conduct the *Sac*I reaction at a lower ionic strength, the residual *Eco*RI could have assumed a star activity and ruined the experiment.

Sometimes enzymes need to be heat-inactivated or removed by purification methods so that they do not interfere with subsequent steps in a procedure. If changing the salt conditions does not seem feasible in a double-digestion reaction, the DNA can be purified by ethanol precipitation, affinity matrix, or spin column to remove the salts, then brought into solution in a different buffer and salt conditions. There may be some loss of DNA during this purification process, but this might be tolerated if the researcher starts with an increased amount during the first step.

Unit definition

How much enzyme do we need to add when setting up a restriction endonuclease digestion? The answer to this question is a bit complicated, because it depends on the type of DNA being digested, the reaction conditions, and the length of time for the reaction incubation. Manufacturers provide us with tubes of enzyme at various concentrations, stated in terms of 'units' per μl. A unit of a restriction enzyme is the amount that is necessary to digest 1 μg of a standard DNA completely in a volume of 50 μl, in 1 hour, at the buffer and temperature conditions recommended. However, most researchers quickly learn that with normal variation in DNA purity and digestion preferences, it is safer to add 5 to 10 units of enzyme to a reaction containing 1 μg of DNA, or to extend the reaction time beyond 1 hour. Most enzymes do not work as well with supercoiled DNA as they do with a standard linear DNA, such as the bacteriophage λ genome, and a researcher may consult information provided by the manufacturer to make accommodations in terms of the number of enzymatic units to add. Some enzymes display site preferences that are difficult to explain or predict, as they relate to the sequence context surrounding the recognition sequence. These biases can cause some restriction sites to be easily digested while others are only incompletely digested.

Many enzymes survive well for a few hours under standard incubation temperatures, so a researcher might use 10 units of enzyme for 1 hour, or 2 units of enzyme for 5 hours, and achieve about the same result. Some manufacturers provide tables that indicate, for each enzyme they sell, the extent of survival of the enzyme in an overnight (e.g. 16-hour) digestion. These survival data are expected to vary between manufacturers, depending on the purity of their product. Overnight digestion may be helpful if a large quantity of DNA is being digested, but there is a risk that contaminating nucleases will have more time to damage the DNA. These contaminating nucleases could be present in the DNA sample or water, or as a contaminant in the commercial enzyme preparation. A bit of 'nibbling' on the overhanging ends of a DNA by a contaminating exonuclease may be inconsequential if the fragments are going to be used for a restriction analysis or Southern blot, but if the ends must remain intact for ligation and cloning experiments then the damage could be quite serious. In this case, adding a carefully determined amount of restriction enzyme and conducting the entire reaction in 1 hour is advisable.

Many manufacturers provide indirect information on the quality of their restriction endonucleases, by reporting the ability of DNA fragments to be re-ligated after over-digestion with their enzyme. For example, one manufacturer reports the following information for one of their enzymes, EcoRI (G/AATTC): 'After 50-fold over-digestion (3 units μg^{-1} DNA × 17 hours) with EcoRI, more than 95% of the DNA fragments can be ligated at a 5'-termini concentration of 0.02 μM.' The actual extent of over-digestion is not 50-fold, since the enzyme may only retain 10% to 20% of its original activity after 17 hours, but the information still provides reassurance that the enzyme preparation does not contain a significant amount of contaminating exonucleases. When the percentage of ligated fragments drops, or the extent of over-digestion is reduced, the researcher must take it as a cautionary note; for example, a manufacturer's description of their NaeI reads: 'After 10-fold over-digestion with NaeI, approximately 75% of the DNA fragments can be ligated.' The enzyme NaeI (GCC/GGC) leaves a blunt end after digestion, but blunt ends ought to be compatible with each other during ligation; this report is an indication that the enzyme might be inefficient in cloning work.

Some restriction enzymes do not leave consistent cohesive ends, an example being PflMI (CCANNNN/NTGG), which leaves a 3' overhanging end of three variable nucleotides. This sequence variation at the cleavage site means that most ends generated by the enzyme are not mutually compatible. Nonetheless, these types of enzymes can perform well in a ligation test. Compatible ends still exist among the degenerate sequences and can be brought together over time.

Setting up the reaction

When we are ready to set up a restriction endonuclease reaction, we have before us an empty tube, a sample of DNA, a restriction endonuclease (on ice, or in a cold block), a specific 10 × buffer (mixed well, if it was frozen and thawed), and some sterile distilled or deionized water. Does it matter how we put these components together? Yes, because the DNA and enzyme are very sensitive to the order of addition of components. If the DNA and 10 × buffer are added together first, the high concentration of Mg^{+2} ion may cause the DNA to precipitate. The best approach is to dilute the 10 × to the maximal extent possible before adding the DNA, and then mix the tube contents thoroughly before adding the enzyme. If the DNA and enzyme are mixed together first, the enzyme will not be buffered correctly and may begin to exhibit star activity while the reaction is being assembled. If the enzyme and 10 × buffer are mixed together first, the enzyme may be inactivated by the extreme salt condition.

Repetitive freezing and thawing of an enzyme tends to decrease its activity, and concentrated enzymes are usually distributed in a solution of 50% (v/v) glycerol to prevent freezing at −20°C. They may be shipped on dry ice in a frozen state, but once thawed to −20°C they remain in solution as a liquid and samples may be readily taken without further warming. It is difficult to pipette concentrated solutions of glycerol accurately using a regular air displacement pipetting device; pipette tips tend to pull up more volume than is indicated, wasting an expensive reagent. One solution is to use a positive displacement pipetting device, or to dilute a small sample of the enzyme for distribution to many independent reaction tubes. A concentrated enzyme may be diluted in 1 × reaction buffer, or a special diluent buffer; however, a diluted enzyme will gradually lose activity so this is only done with an amount that is needed for the reactions at hand.

A recipe for a typical reaction might be as follows, with the order of addition of components from top to bottom:

- 34 µl sterile deionized water
- 5 µl 10 × restriction endonuclease buffer 'X'
- (Mix well)
- 10 µl 100 µg ml^{-1} pQC-1 plasmid
- (Mix well)
- 1 µl *Eco*RI restriction enzyme (10 units µl^{-1})
- (Mix gently, but thoroughly. Incubate 1 hour at 37°C)

The total volume of the reaction is 50 µl, and so the 10 × buffer is diluted down to a concentration of 1 ×. The proposed final concentration of DNA in the reaction is 20 µg ml^{-1}, which is similar to the standard conditions under which the enzyme unit definitions are established. The DNA component of the reaction is variable in quality, and may contribute inhibitors such as traces of organic solvents, EDTA, and other small inhibitors. The enzyme may contain 50% (v/v) glycerol, and to avoid star activity it is important that it be diluted at least 10-fold into the reaction so that the glycerol concentration is less than 5% (v/v). If the reaction is conducted in a generous volume these inhibitors and preservatives will be diluted, and the enzymatic reaction may proceed more efficiently.

The details of technique are discussed more thoroughly in Chapters 2 and 3, but it bears repeating that the concentrated tube of enzyme is an expensive resource that must be guarded carefully. It should be stored in a cold block or on ice when it is being used, and a researcher should always change pipette tips before taking a sample from a stock tube. It is usually considered poor form to snatch a tube from the freezer and stand with the freezer door propped open while taking a sample of enzyme. This trick requires at least three hands, it allows the other enzymes in the freezer to warm up, it provides little opportunity for the researcher to study the label to confirm that the right tube has been selected, and it tends to indicate that the researcher plans to walk some distance with a loaded pipette tip.

Conducting the reaction

Mix and spin

It is important that an enzymatic reaction be thoroughly mixed, and this can be difficult for two reasons. When the enzyme in 50% (v/v) glycerol is added to a reaction, it will tend to sink to the bottom of the solution because of its high density and viscosity. Mixing the glycerol solution into the aqueous solution requires extra effort, but the researcher must perform this mixing with care because enzymes are also sensitive to excessive aeration and bubbling. A reaction can be drawn into a pipette tip repeatedly and flushed out again, to gently mix the components of a reaction without excessive aeration. However, if the DNA sample is high-molecular-weight, for example a genomic DNA with size greater than 50 kbp, then pipetting can lead to unwanted shearing of the DNA.

The thoroughly mixed reaction volume should be collected at the bottom of the tube by brief centrifugation in a microcentrifuge. This centrifugation can be accomplished in a few seconds, and it collects all of the droplets of liquid that may be adhering to the walls and cap of the tube. By collecting the reaction at the bottom, the researcher ensures uniform conditions for the reaction.

Incubation

The reaction can be incubated in a water bath, a dry block, or an air incubator; however, each has advantages and disadvantages. A water bath is highly conductive and will warm the tube to the correct temperature quickly. On the other hand, a tube in a water bath may be contaminated by dirty water if the cap comes in contact with the surface. A dry block may not heat a tube thoroughly if it is loose-fitting, and so the reaction may not achieve the necessary temperature for the enzyme. Both water baths and dry blocks tend to heat the bottom of the tube and leave the top exposed to the ambient air, and this can lead to the collection of condensed water in the cap. The condensed water is evaporated from the underlying reaction, so the concentration of solutes in the reaction is higher. This could lead to inconsistent results, depending on the sensitivity of the reaction to solute concentration. An air incubator will heat a tube uniformly, but the tube may be slow to heat up to the correct temperature.

Beware of enzyme names that start with the letter 'T'

Most, but not all restriction endonucleases work optimally at 37°C. However, an enzyme name that starts with the letter 'T' often comes from an organism that loves the heat: *Thermus*, *Thermococcus*, or *Thermophilus*. These organisms are extremophiles and their enzymes work optimally at higher temperatures such as 65°C to 75°C. There are also many examples of restriction endonucleases that work optimally at 50°C (e.g., *Sfi*I) or 25°C (e.g., *Sma*I). A researcher must be attentive to the conditions required by each enzyme, and not assume that all enzymes are used at 37°C.

Inactivating the enzyme

When a reaction is complete, it is common for a researcher to inactivate the enzyme so that it does not linger like an unwanted guest in a sample. For example, if a DNA fragment is prepared with *Bam*HI and a cloning vector is prepared with *Eco*RI, it would be a bad idea to mix those DNA samples while the enzymes were still active. The *Bam*HI might attack the cloning vector and the *Eco*RI might attack the DNA fragment, where the reason for conducting the reactions separately was to avoid just that.

One way to inactivate most enzymes is to add EDTA to a concentration of about 10 mM, because this chelates the Mg^{+2} ions that are necessary for enzymatic activity. However, if the DNA is subsequently purified by ethanol precipitation, it is important to dilute the 10 mM EDTA solution at least three-fold before adding the precipitating salts and ethanol. EDTA solutions of about 5 mM or higher will form a dense EDTA phase in an ethanolic solution. This liquid prevents the DNA pellet from sticking to the bottom of a centrifuge tube, and it may be lost when the tube is decanted.

Most enzymes can be heat-denatured at 65°C or 80°C, and manufacturers provide tables showing the appropriate temperature. There is variability between enzymes in terms of their heat sensitivity, and the researcher must again be wary of enzymes with names that start with the letter 'T'. If EDTA will not interfere with further work on the DNA, it is a good idea to add it to a concentration of at least 1 mM before heating the DNA to 65°C or 80°C. Heavy metal contaminants in the solution may catalyze damage to the nucleic acid at these high temperatures, and these contaminants are chelated by EDTA. Some researchers regularly heat restriction-digestions to 65°C after they are complete, to separate cohesive ends that are still joined by hydrogen bonding. This is probably not necessary because most short cohesive ends left by restriction endonucleases would not remain hydrogen-bonded at 37°C.

Very hardy enzymes can be extracted from a reaction by treatment with phenol and chloroform, as described in Chapter 3. If this is not sufficient to eliminate the enzymatic activity, the reaction can also be treated with $10\ \mu g\ ml^{-1}$ proteinase K and 0.2% (w/v) SDS at 50°C for 1 hour, prior to phenol and chloroform extraction. After extraction, the DNA can be purified by alcohol precipitation or binding to an affinity matrix. Traces of these solvents must be carefully removed if the DNA is to be used in subsequent enzymatic steps.

Partial digestion

When a restriction digestion reaction does not reach completion, we call that a 'partial' digestion. Partial digestions are sometimes planned, for example if a researcher uses the enzyme *Sau*3AI (/GATC) to generate 40-kbp fragments for a cosmid library, only about one out of every 160 *Sau*3AI recognition sequences is being cleaved. Another example of the use of planned partial digestion is when a researcher tries to discern the order of DNA fragments in a restriction map by seeing which smaller fragments fit together to make larger fragments. Digestions can be purposefully limited by adding fewer units of enzyme, by limiting the time of the reaction and inactivating the enzyme promptly, or by adding an inhibitor such as ethidium bromide to moderate the reaction.

Partial digestions are often an unwanted result if a researcher is trying to develop a restriction map or to prepare a DNA molecule for cloning. For example, suppose a researcher plans to prepare a cloning vector with the enzymes *Bam*HI and *Eco*RI, to receive a DNA fragment that is similarly prepared with these two enzymes. If the *Bam*HI and *Eco*RI sites are both in a multiple cloning site, it may be difficult to determine whether both enzymes have cut the vector to completion because the released *Bam*HI to *Eco*RI fragment may be just a few base pairs in length. If one of the two enzymes does not work efficiently then the vector will not have both *Bam*HI and *Eco*RI ends; it will not be ready to receive the fragment.

In situations in which a partial digestion is detectable by gel electrophoresis, it can often be recognized by the presence of faint or sub-molar quantities of DNA in individual stained bands, as explained in Chapter 2. These bands may be especially noticeable during Southern blotting, which is a highly sensitive method. However, there are other explanations for unexpected bands that must also be entertained. For instance, the sample could be completely digested but contain a contaminant DNA that was focused into a single band as a consequence of digestion. The contaminant might not have been noticed in the absence of digestion because it was smeared over a broad region of the gel or lurking near the wells.

Partial digestions are more likely if an enzyme has aged beyond its expiration date or been stored at the wrong temperature, or if the researcher is not using the recommended buffer or temperature conditions. If the DNA is not sufficiently pure it may contribute inhibitors to the reaction that interfere with enzymatic activity. If the reaction is not adequately mixed and collected at the bottom of the tube, or if the DNA is not entirely in solution, then the enzyme may not have an opportunity to work on all of the potential cleavage sites.

Problems

The solution to this problem is provided at the end of the chapter.

(vi) Suppose that you want to digest a 500-ng sample of DNA with a restriction enzyme, using 5 units enzyme per µg of DNA. If you have a stock of a restriction enzyme that is 8 units µl^{-1}, how many µl of enzyme are needed in the reaction?

5.4 Restriction mapping

We can learn a few things about a piece of DNA by seeing how it is cut with a specific restriction endonuclease. We found out whether or not the DNA contains the recognition sequence of an enzyme, and if there are multiple sites of digestion then what is the spacing between them? These pieces of information can be brought together to generate a map of the locations of the recognition sequences, which can be used to type and compare DNA fragments. Restriction analysis is a very basic type of DNA sequencing, since the presence of a restriction enzyme site in a DNA fragment implicitly reveals a few nucleotides of sequence. Unfortunately, we don't gather much information about why a restriction enzyme site is lost; the reason might be a substantial change in the DNA or just a single base change at the recognition sequence.

Mapping unknown sequences

In Chapter 4 we considered some simple examples of restriction mapping of a cloned insert in a plasmid, where the map of the vector was already known. One important point about mapping plasmids is that the DNA is supercoiled when it is extracted from *E. coli*, and supercoiled DNA is more compact than nicked, relaxed circular DNA or linear DNA. As explained in Chapter 2, the compact supercoiled DNA usually migrates more rapidly on a gel. An example of this effect is shown in *Figure 5.5A*, where an

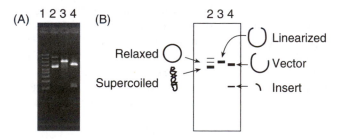

Figure 5.5

Digestion of plasmids with restriction enzymes is detected by gel electrophoresis. (A) An ethidium-bromide-stained agarose gel, in which electrophoresis of DNA was conducted from top to bottom. Lane 1: DNA standards of 1.0, 1.5, 2.0, 2.5, 3.0, 4.0, 5.0, 7.0, and 10.0 kbp (listed bottom to top). Lane 2: the principal band is a supercoiled plasmid. The faint band above the principal band is a nicked, relaxed circle of the same plasmid. The faint band above that is unidentified, but may be a plasmid dimer. Lane 3: the same plasmid after digestion with a restriction enzyme that cuts it one time. This is a linearized form of the plasmid. Lane 4: the same plasmid after digestion with a restriction enzyme that cuts it two times, releasing a small DNA insert. (B) A schematic diagram showing an interpretation of the bands in lanes 2–4 of the gel.

undigested plasmid (lane 2) is run on a gel next to a linearized version in which a restriction enzyme has digested the plasmid at one site (lane 3). Lane 4 shows the plasmid digested with a restriction enzyme that cleaves the DNA at two sites separated by approximately 1.1 kbp, with separation of a DNA insert from the remainder of the vector (see *Figure 5.5B*).

Let's now suppose we have a plasmid of 5.8 kbp with an unknown structure, and for which we want to prepare a restriction map. Suppose we digest it with the restriction endonucleases *Eco*RI and *Sal*I, separately and together. We obtain the following pattern of restriction fragments upon gel electrophoresis (see *Figure 5.6A*):

- *Eco*RI   2.8, 2.0, and 1.0 kbp
- *Sal*I   5.0, and 0.8 kbp
- *Eco*RI + *Sal*I   2.5, 1.5, 1.0, 0.5, and 0.3 kbp

The first thing to check is that we have accounted for all of the DNA fragments. Since each digestion produced a list of fragments that add up to 5.8 kbp, it is not likely that two fragments are co-migrating, or that the digestions were partial. However, we cannot rule out the possibility that a very small fragment has been run off the end of the gel.

We consider the proposition that every fragment in the *Eco*RI-digested DNA is made up of one or more fragments from the *Eco*RI and *Sal*I double-digest, joined together (see *Figure 5.6B*). For example, the 2.8-kbp *Eco*RI fragment is likely to be composed of a 2.5-kbp *Eco*RI to *Sal*I fragment, and a 0.3-kbp *Sal*I to *Eco*RI fragment because the two double-digest pieces (2.5 + 0.3) add up to 2.8 and the *Sal*I site has to be between two *Eco*RI sites. The 2.0-kbp *Eco*RI fragment is likely to be composed of a 1.5-kbp *Eco*RI to *Sal*I fragment plus a 0.5-kbp *Sal*I to *Eco*RI fragment by similar logic. The 1.0-kbp *Eco*RI fragment probably doesn't have a *Sal*I site within it because there aren't any smaller pieces in the double-digest that add up to 1.0 kbp. We have accounted for all of the DNA fragments in the double-digest, and yet used each fragment only once in explaining the composition of the fragments in the *Eco*RI digested DNA (see *Figure 5.6B*).

Every *Sal*I-digested DNA fragment is also made up of some combination of fragments from the double-digest, joined together. The 5.0-kbp *Sal*I fragment is likely to be composed of the 2.5-, 1.5-, and 1.0-kbp fragments in the double-digest; two of these subfragments will have *Sal*I and *Eco*RI ends, and one will have only *Eco*RI ends. We have already established that the 1.0-kbp *Eco*RI fragment probably doesn't have a *Sal*I site in it. Finally, the 0.8-kbp *Sal*I fragment is likely to be composed of a 0.5-kbp *Sal*I to *Eco*RI fragment, and a 0.3-kbp *Eco*RI to *Sal*I fragment (see *Figure 5.6B*).

How do we put all of this information together? We can start at any point and work our way through the evidence logically. The 2.8-kbp *Eco*RI fragment contains a 0.3-kbp *Sal*I to *Eco*RI subfragment, which also appears to be part of a 0.8-kbp *Sal*I fragment when put with another 0.5-kbp *Eco*RI to *Sal*I subfragment. It is therefore likely that the order of fragments in the double-digest is 2.5 kbp (*Eco*RI to *Sal*I) – 0.3 kbp (*Sal*I to *Eco*RI) – 0.5 kbp (*Eco*RI to *Sal*I). The last fragment on the list, the 0.5-kbp fragment, appears to be part of a 2.0-kbp *Eco*RI fragment in conjunction with another 1.5-kbp *Sal*I to *Eco*RI fragment. The 1.5-kbp fragment is part of a much larger 5.0-kbp *Sal*I fragment, by virtue of a 1.0-kbp and a 2.5-kbp subfragment.

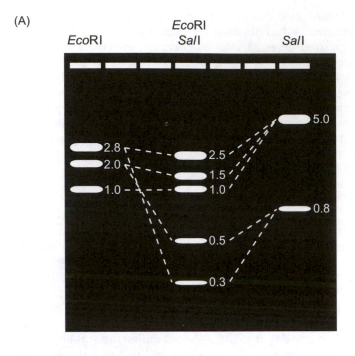

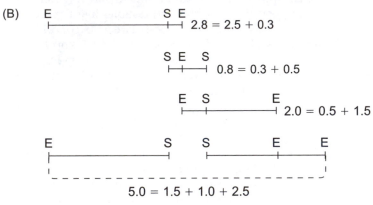

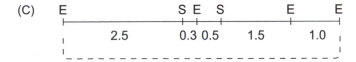

(A)

(B)

(C)

Figure 5.6

Mapping a plasmid with restriction endonucleases. (A) Schematic diagram of an ethidium-bromide-stained gel, showing plasmid DNA digested with *Eco*RI alone (left), *Eco*RI and *Sal*I combined (middle), and *Sal*I alone (right). The sizes of the fragments in kbp is shown next to the bands. The dashed lines indicate how the *Eco*RI and *Sal*I fragments are constituted by one or more fragments from the double-digestion. (B) An intermediate stage in map-making, showing by vertical alignment how the order of fragments can be logically determined from the fragment size data. Abbreviations: E = *Eco*RI and S = *Sal*I. The dashed line at the bottom indicates that the two *Eco*RI sites are the same. (C) A completed map showing the order of restriction fragments. The dashed line indicates that the two *Eco*RI sites are the same, due to circularization of the plasmid.

We have already seen the 2.5-kbp subfragment; we started the analysis with it, and established it as an *Eco*RI to *Sal*I fragment in the double-digestion products. Therefore, the 1.0-kbp *Eco*RI fragment must lie between the 1.5-kbp *Sal*I to *Eco*RI fragment and the 2.5-kbp *Eco*RI to *Sal*I fragment. The spacing of restriction endonuclease recognition sequences in the plasmid is likely to be: *Eco*RI (2.5 kbp), *Sal*I (0.3 kbp), *Eco*RI (0.5 kbp), *Sal*I (1.5 kbp), *Eco*RI (1.0 kbp), *Eco*RI (see *Figure 5.6C*). The map is circular, so the first and last *Eco*RI sites on the list are one and the same.

These types of analyses do not always yield unambiguous maps. For example, if three *Eco*RI sites had occurred in a row, with no intervening *Sal*I sites, it would not have been possible to determine the order of all *Eco*RI fragments in that region. Additional enzymes might be brought in to resolve that particular uncertainty, and to reinforce the information already gathered from the first double-digestion.

Mapping DNA inserts in a known plasmid or bacteriophage vector

Mapping entire plasmids by restriction analysis is usually not necessary if a researcher is using a well-known vector. In that case it may only be the map of the inserted DNA that is in question, with the restriction map or DNA sequence of the surrounding vector being already established. This simplifies the problem because in any restriction endonuclease digestion the vector DNA will generate fragments that are predictable; only the inserted DNA and the junction between the vector and inserted DNA will produce new restriction fragments.

Examples of insert mapping were provided in Chapter 4 and these techniques will usually work well for plasmids with small DNA inserts, but what about mapping bacteriophage λ inserts, cosmids, BACs, and PACs? In the case of bacteriophage λ vectors, a researcher may be faced with a 10-kbp insert in a 40-kbp vector, and in the case of a BAC the researcher may have a 250-kbp insert in a 15-kbp vector. Some information may be gathered about large DNA inserts by using restriction endonucleases that have long recognition sequences and tend to cut the DNA infrequently. For example, the enzymes *Not*I (GC/GGCCGC) and *Pac*I (TTAAT/TAA) have eight nucleotide recognition sequences and produce fragments of DNA that are usually 10 to 20 times larger than the average *Eco*RI or *Bam*HI fragment. Fragments of 30 to 200 kbp are too large to be resolved by conventional gel electrophoresis, but pulsed field gel electrophoresis can be used to help establish a restriction map.

Bacteriophage λ vectors may generate a large number of restriction fragments, most of which map to the bacteriophage arms and will be consistent from clone to clone. The novel fragments are related to the DNA inserts, being entirely inside the insert or lying at the junction between DNA insert and bacteriophage arm. Many bacteriophage λ vectors have a pair of symmetrical multiple cloning sequences separated by a stuffer fragment, as discussed in Chapter 4, and in these cases it is possible to separate the bacteriophage arms from the DNA insert using a single restriction endonuclease that has recognition sequences flanking the cloning site.

Bacteriophage λ DNA is linear, but the arms may be joined by hydrogen bonding at the *cos* sites; in a restriction digestion this produces an alternative fragment consisting of the left arm plus right arm. The matter can be simplified if a researcher heats the restriction endonuclease digestion products to 65°C for 5 min prior to loading them on a gel to separate the *cos* ends.

It may be difficult to identify all of the bands on a gel in a complicated restriction digestion pattern, and some researchers end-label the products of digestion with a [^{32}P]-phosphate so that DNA may be detected by autoradiography instead of by staining. This is particularly helpful in identifying small fragments that do not amount to a significant proportion of the total DNA. For example, a 500-bp DNA fragment is only 1% of the length of the 50 kbp in a bacteriophage λ clone, so if we perform a restriction endonuclease digestion on 1 μg of the clone DNA then we can only expect to have 10 ng in the 500-bp fragment. A band on a gel with only 10 ng of DNA might be easily overlooked. By transferring a [^{32}P]-phosphate to the digestion products and imaging the bands on film, DNA detection is more sensitive and the signal from each band is independent of size. That is, 10 ng of a 500-bp fragment will have the same radioactivity as 100 ng of a 5000-bp fragment. This uniformity of signal makes it easier to assign apparent molecular weights to each fragment.

The researcher can sometimes apply creative approaches to identify the junction between known vector DNA and unknown DNA insert. For example, the vector could be digested using a restriction endonuclease that cuts only in the multiple cloning site, then end-labeled with ^{32}P, and subsequently digested with a second restriction endonuclease. Only fragments in which one or both ends are generated by the first endonuclease will be imaged on film, and this can help to identify the end fragments. Alternatively, the vector alone could be radioactively labeled and used as a probe in a Southern analysis of the digested recombinant plasmid. Of the restriction fragments generated from the unknown DNA insert, only those two that are at the junction with the vector DNA will overlap with the radiolabeled probe.

Some vectors have bacteriophage T3, T7, or SP6 RNA polymerase promoters flanking the multiple cloning site and oriented so that a small sense or antisense RNA probe can be made from the DNA insert. These short end probes can be used in Southern analysis of the recombinant clone, and will point to fragments that are close to the junction between DNA insert and vector DNA.

Alternatively, the sequence at the very ends of a DNA insert may be determined using universal primer sites that lie just within the vector sequence, and these short sequences can be chemically synthesized and used as oligonucleotide probes in a Southern analysis. In either case, determining the ends of a DNA insert is usually the best strategy for initial mapping work. In a similar way, when people work to solve a jigsaw puzzle they will first try to find the edge pieces and build the frame of the puzzle before solving the interior.

The order of restriction endonuclease fragments can sometimes be inferred from the results of a partial digestion. For example, suppose a bacteriophage λ vector has *Bam*HI sites in symmetrical multiple cloning sites that flank the DNA insert. We will imagine that a complete digestion of a recombinant clone DNA generates the left and right arms of the vector, and also five additional *Bam*HI fragments with sizes 6.5, 5.0, 3.0, 1.0, and 0.7 kbp. What is the order of these *Bam*HI fragments in the DNA insert? If we make end probes and test the fragments by Southern analysis, we might discover that the 5.0- and 1.0-kbp fragments lie at the ends of the DNA insert, but that still leaves a question about the order and orientation of the other three fragments. If we performed a partial digestion with *Bam*HI and repeated the Southern analysis, we might learn that one end probe generates one series of bands: 16.2, 11.2, 4.7, 4, and 1 kbp (see *Figure 5.7B*). The other end probe generates a different series of bands: 16.2, 15.2, 12.2, 11.5, and 5 kbp (see *Figure 5.7C*). With the exception of

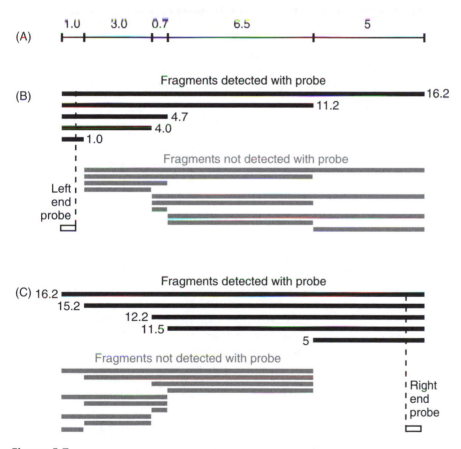

Figure 5.7

Mapping of a phage DNA insert by partial restriction endonuclease digestion and Southern blotting. (A) A completed map showing the order of restriction fragments. (B) Fragments generate by partial DNA digestion. The fragments that are detected by Southern blotting using a 'left end' probe are shown as black lines, and the fragments that do not hybridize to the probe are shown as shaded lines. (C) The same fragments generated by partial DNA digestion, but examined by Southern blotting using a 'right end' probe. The fragments that do not hybridize to the probe are shown as shaded lines.

the 1- and 5-kbp DNA fragments, these sizes represent partially digested DNA fragments; for example, the 4-kbp band is 1 kbp plus 3 kbp, the 4.7-kbp band represents 1 kbp plus 3 kbp plus 0.7 kbp, and the 11.2-kbp band represents 1 kbp plus 3 kbp plus 0.7 kbp plus 6.5 kbp. Only the digestion products that are covered by the end probes are detected; the remaining products are present but not labeled (see shaded fragments in *Figure 5.7B, C*). From the results we see that the order of *Bam*HI fragments in the DNA insert can be inferred, and checked in the other direction using the other end probe. Ultimately, the partial digestion mapping can be used to resolve the restriction map of the DNA insert (see *Figure 5.7A*).

When a fine structure map of a large cloned sequence is desired, it is common for researchers to subdivide or subclone the sequence into a plasmid vector so that the individual parts can be studied in isolation. For example, the aforementioned five *Bam*HI subfragments of a bacteriophage λ clone might be subcloned into five plasmid vectors for more detailed mapping of each subfragment. The overall picture of the bacteriophage λ clone depends on assembly of the five detailed maps into a contiguous 16.2-kbp map. What if we overlooked a very small *Bam*HI subfragment, for example a 50-bp fragment that was too small to visualize on the gel? This is a serious problem, and one solution is to subclone the bacteriophage λ clone in several different ways, for example preparing a second set of overlapping plasmid clones that is based on cloning using a different restriction endonuclease. The 50-bp fragment would be captured in one of these clones and might not be as easily overlooked in a restriction digestion.

Genome restriction mapping

A segment of a genome can be mapped using restriction endonucleases, provided that a probe is available for hybridization to DNA fragments in a Southern blot. A Southern blot of genomic DNA starts with gel electrophoresis of restriction-enzyme-digested products, which usually produces a broad smear of bands of different sizes in the genome (see *Figure 5.8A*). The intensity of DNA staining is proportional to the nanograms of DNA, which is a point that was discussed in Chapter 2. The DNA on the gel is transferred to a nylon or nitrocellulose membrane, which is then probed with a labeled nucleic acid probe (see *Figure 5.8B*). For most applications it is important that the probe be specific to one locus in the genome, and that it not hybridize to repetitive sequences distributed in another part of the genome. For example, suppose a 4.2-kbp fragment of cDNA DNA is used as a probe in a Southern blot of fragments of bacterial genomic DNA, and the following results are obtained (see *Figure 5.8B*):

* *Eco*RI   9.5, and 2.3 kbp
* *Bam*HI   4.0, and 5.0 kbp
* *Eco*RI + *Bam*HI   4.0, 2.3, and 1.0 kbp

Why do the fragments add up to different total lengths? The probe only makes visible the fragments of genomic DNA with which it overlaps. Different restriction endonucleases will produce different collections of

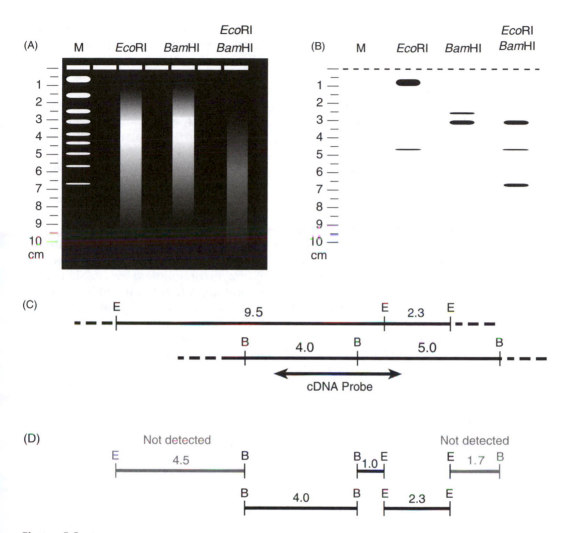

Figure 5.8

Mapping of a genomic DNA by restriction endonuclease digestion and Southern blotting. (A) Schematic diagram of an ethidium-bromide-stained agarose gel. At the far left is a diagram of a cm ruler. The 'M' indicates a lane loaded with DNA standards of 1.0, 1.5, 2.0, 2.5, 3.0, 4.0, 5.0, 7.0, and 10.0 kbp (listed bottom to top). The remaining lanes are loaded with genomic DNA digested with EcoRI, BamHI, or EcoRI plus BamHI, as indicated. Note that no individual bands are discernable in these lanes because the genome produces a wide variety of fragment sizes. (B) Schematic diagram of a Southern blot, following transfer of the gel and hybridization of the blot to a labeled cDNA probe, and autoradiography. (C) Restriction map showing a possible interpretation of the data. Abbreviations: B = BamHI and E = EcoRI. The overlap of the cDNA probe with genomic DNA is indicated by a double-headed arrow. (D) An accounting of the fragments generated by double-digestion with EcoRI and BamHI. Note that the fragments that do not overlap with the cDNA probe are not detected on the Southern blot (shown as shaded lines).

overlapping fragments, so the sum of the sizes of fragments is not expected to be constant as it might be for restriction mapping of a plasmid. What happened to the 9.5- and the 5.0-kbp bands in the double-digestion with *Eco*RI and *Bam*HI? The 9.5- and 5.0-kbp bands were each cut into at least two pieces by the second enzyme, but we only see the fragments that overlap with the probe (see *Figure 5.8B, C*). It is likely that the 9.5-kbp *Eco*RI fragment has at least two *Bam*HI sites within it, and a double-digest of the 9.5-kbp fragment yields a 1.0-kbp *Eco*RI to *Bam*HI subfragment and 4.0-kbp *Bam*HI to *Bam*HI subfragment. The 'missing' 4.5 kbp (from the 9.5-kbp fragment) does not overlap with the probe and is not visualized on the Southern blot (see shaded fragments, *Figure 5.8D*). Similarly, the 5.0-kbp *Bam*HI fragment is digested with *Eco*RI to yield a 1.0-kbp *Bam*HI to *Eco*RI subfragment and a 2.3-kbp *Eco*RI subfragment, and the remaining 2.7 kbp do not overlap with the probe sequence and cannot be visualized.

Using a 4.2-kbp probe we have learned something about the structure of 7.3 kbp of genomic DNA: there is a 4.0-kbp *Bam*HI to *Bam*HI fragment, followed by a 1.0-kbp *Bam*HI to *Eco*RI fragment, followed by a 2.3-kbp *Eco*RI to *Eco*RI fragment (see *Figure 5.8C*). One important point about the signal strength on a Southern blot is that for internally labeled probes, the signal is proportional to the extent of overlap between probe and fragment. For example, we see in *Figure 5.8B* that the 5.0-kbp *Bam*HI fragment is less intense than the 4.0-kbp *Bam*HI fragment. The two fragments are present in equimolar amounts on the gel, but the 5.0-kbp fragment has less sequence overlap with the cDNA probe than the 4.0-kbp fragment (see *Figure 5.8C*).

A more complicated picture might emerge if the probe sequence were a cDNA derived from a eukaryotic mRNA, and the target remained a genomic DNA, because then the probe might overlap with numerous exon sequences distributed over a broad swath of genomic DNA. The cDNA can be subcloned as a series of smaller fragments so that exon-specific probes can be made. Alternatively, the researcher may wish to use chemically synthesized oligonucleotide probes to achieve a high degree of specificity.

Problems

The solutions to these problems are provided at the end of the chapter.

(vii) Suppose that you have a 4400-bp plasmid vector with an *Eco*RI site at nt 1, an *Spe*I site at nt 2800, and a *Bgl*II site at nt 3900. You use the *Eco*RI site to clone an *Eco*RI-digested fragment of 2.6 kbp. Upon treatment of the 7-kbp plasmid with restriction endonucleases, you obtain the following fragment sizes:

- *Eco*RI: 4.4, and 2.6 kbp
- *Eco*RI and *Bgl*II: 3.9, 2.2, 0.5, and 0.4 kbp
- *Bgl*II: 4.3, and 2.7 kbp
- *Bgl*II and *Spe*I: 3.2, 1.5, 1.2, and 1.1 kbp
- *Spe*I: 4.7, and 2.3 kbp

Determine the positions of *Eco*RI, *Bgl*II, and *Spe*I sites in the 7-kbp plasmid.

(viii) Suppose that you use the same 4.4-kbp vector as in the previous problem to clone a 250-bp *Eco*RI-digested fragment at the *Eco*RI site. Upon digestion of the clone with restriction endonucleases, you run the digestion products on a 1.2% agarose gel that allows you to resolve bands between 0.2 and 10 kbp. You find the following:

- *Eco*RI: 4.4, and 0.25 kbp
- *Eco*RI and *Bgl*II: 3.9, 0.5, and 0.15 kbp
- *Bgl*II: 4.0, 0.6, and 0.3 kbp

Try constructing a restriction map. These are strange results! Can you explain what may be wrong with this clone?

(ix) Suppose that you obtain the following restriction digestion data on an unknown plasmid:

- *Xba*I: 4.9, and 3.1 kbp
- *Xba*I and *Sma*I: 4.9, 1.3, 0.8, 0.6, and 0.4 kbp
- *Sma*I: 7.0, 0.6, and 0.4 kbp

Can you construct a restriction map? Is there more than one solution?

(x) Suppose that you perform restriction mapping on an unknown plasmid using the enzymes *Sma*I and *Bam*HI, and examine the fragments by agarose gel electrophoresis. You load on the gel a set of molecular size standards that include fragments ranging in size from 1 kbp to 5 kbp; their migrations are as follows: 1 kbp (6.6 cm), 1.5 kbp (5.6 cm), 2 kbp (4.9 cm), 2.5 kbp (4.3 cm), 3 kbp (3.8 cm), 4 kbp (3.1 cm), and 5 kbp (2.5 cm). For the restriction fragments, you find the following pattern of migration:

- *Sma*I: 3.3 cm, 3.7 cm, and 5.0 cm
- *Sma*I and *Bam*HI: 4.3 cm, 5.0 cm, 5.3 cm, 5.6 cm, and 6.4 cm
- *Bam*HI: 2.7 cm, and 3.1 cm

Determine the restriction map of the plasmid. Hint: the migration of DNA on a gel was discussed in Chapter 2.

Further reading

Ausubel FW, Brent R, Kingston RE, Moore DD, Seidman JG, Smith JA and Struhl K (eds) (1987) *Current Protocols in Molecular Biology*. John Wiley & Sons, New York.

Perbal B (1988) *A Practical Guide to Molecular Cloning*, 2nd Edn. John Wiley & Sons, New York.

Sambrook J, Russell D (2001) *Molecular Cloning: A Laboratory Manual*, 3rd Edn. Cold Spring Harbor Laboratory Press, Cold Spring Harbor, NY.

References

1. Meselson M, Yuan R (1968) DNA restriction enzyme from *E. coli. Nature* **217**: 1110–1114.
2. Linn, S, Arber W (1968) Host specificity of DNA produced by *Escherichia coli*, X. In vitro restriction of phage fd replicative form. *Proc. Natl Acad. Sci. USA* **59**: 1300–1306.

3. Smith HO, Wilcox KW (1970) A restriction enzyme from *Hemophilus influen-zae*. I. Purification and general properties. *J. Mol. Biol.* **51**: 379–391.

4. Wilson GG, Murray NE (1991) Restriction and modification systems. *Annu. Rev. Genet.* **25**: 585–627.

5. Danna K, Nathans D (1971) Specific cleavage of simian virus 40 DNA by restriction endonuclease of *Hemophilus influenzae*. *Proc. Natl Acad. Sci. USA* **68**: 2913–2917.

6. Cohen SN, Chang ACY, Boyer HW, Helling RB (1973) Construction of biologically functional bacterial plasmids in vitro. *Proc. Natl Acad. Sci. USA* **70**: 3240–3244.

7. McClelland M, Nelson M, Raschke E (1994) Effect of site-specific modification on restriction endonucleases and DNA modification methyltransferases. *Nucleic Acids Res.* **22**: 3640–3659.

Solutions to problems

Solutions to problems posed in Section 5.1

(i) *Scr*FI (5′-CCNGG-3′) has two sites (underlined):

```
5′-GAGTACTCCCGGACCGAGTTTACAGACCTGGGTCATGACCGAGCAGG-3′
3′-CTCATGAGGGCCTGGCTCAAATGTCTGGACCCAGTACTGGCTCGTCC-5′
```

There was an embedded trick: note that the sequence 3′-CCTGG-5′ at nucleotide 11 on the bottom strand is not an *Scr*FI site, because it is in the order 3′ to 5′. *Rca*I (5′-TCATGA-3′) has one site (underlined):

```
5′-GAGTACTCCCGGACCGAGTTTACAGACCTGGGTCATGACCGAGCAGG-3′
3′-CTCATGAGGGCCTGGCTCAAATGTCTGGACCCAGTACTGGCTCGTCC-5′
```

Once again, there was a trick: the sequence 3′-TCATGA-5′ at nucleotide 2 on the bottom strand is not an *Rca*I site. *Psp*GI (5′-CCWGG-3″) has one site (underlined):

```
5′-GAGTACTCCCGGACCGAGTTTACAGACCTGGGTCATGACCGAGCAGG-3′
3′-CTCATGAGGGCCTGGCTCAAATGTCTGGACCCAGTACTGGCTCGTCC-5′
```

Note that every *Psp*GI site is also an *Scr*FI site, but not vice-versa.

(ii) *Bmg*B I (5′-CACGTC-3′) has one site (underlined):

```
5′-ATTACAGAGAAGCCACGTCAGTACCGCTTTGCGGCATACAGTACAGT-3′
5′-TAATGTCTCTTCGGTGCAGTCATGGCGAAACGCCGTATGTCATGTCA-5′
```

*Ear*I (5′-CTCTTC-3′) has one site (underlined):

```
5′-ATTACAGAAGAGCCACGTCAGTACCGCTTTGCGGCATACAGTACAGT-3′
5′-TAATGTCTTCTCGGTGCAGTCATGGCGAAACGCCGTATGTCATGTCA-5′
```

Note the trick: the 5′-CTCTTC-3′ reads from right to left on the bottom strand. This is an asymmetric recognition sequence, so if we are only studying the top strand then we need to search for both 5′-CTCTTC-3′ and 5′-GAAGAG-3′ sequences.

*Aci*I (5′-CCGC-3′) has two sites (underlined):

```
5′-ATTACAGAGAAGCCACGTCAGTACCGCTTTGCGGCATACAGTACAGT-3′
5′-TAATGTCTCTTCGGTGCAGTCATGGCGAAACGCCGTATGTCATGTCA-5′
```

Once again, *Aci*I is asymmetric, so the 5′-CCGC-3′ sequence can be read left to right on the top strand, or right to left on the bottom strand. If we are only studying the top strand, we need to search for both 5'-CCGC-3' and 5'-GCGG-3' sequences.

Solutions to problems posed in Section 5.2

(iii) Cleavage of a DNA with *Xba*I (T/CTAGA):

```
5'-GACAAGATTCTAGAGATTACAGT-3'
3'-CTGTTCTAAGATCTCTAATGTCA-5'
```

```
5'-GACAAGATT              CTAGAGATTACAGT-3'
3'-CTGTTCTAAGATC              TCTAATGTCA-5'
```

(Note that this enzyme leaves 5' overhanging ends)
   Cleavage of a DNA with *Sph*I (GCATG/C):

```
5'-GGGACTAGCATGCATCAGGACTA-3'
3'-CCCTGATCGTACGTAGTCCTGAT-5'
```

```
5'-GGGACTAGCATG              CATCAGGACTA-3'
3'-CCCTGATC              GTACGTAGTCCTGAT-5'
```

(Note that this enzyme leaves 3' overhanging ends)
   Cleavage of a DNA with *Eco*RV (GAT/ATC).

```
5'-CCATAGACGATATCAGATCCACG-3'
3'-GGTATCTGCTATAGTCTAGGTGC-5'
```

```
5'-CCATAGACGAT              ATCAGATCCACG-3'
3'-GGTATCTGCTA              TAGTCTAGGTGC-5'
```

(Note that this enzyme leaves blunt ends).

(iv) The enzyme recognition sites for *Bsa*I (GGTCTC(1/5)) and *Bsg*I (GTGCAG(16/14)) are underlined in the starting DNA and the large cleavage product.

```
5'-GGCAAGATCCGAGACCGATGTGCAGAAGAGTACAGATCACGTCTTAG-3'
3'-CCGTTCTAGGCTCTGGCTACACGTCTTCTCATGTCTAGTGCAGAATC-5'
```

```
5'-          GATCCGAGACCGATGTGCAGAAGAGTACAGATCACG          -3'
3'-          GCTCTGGCTACACGTCTTCTCATGTCTAGT          -5'
```

(Note that if these two ends are joined with other cohesive ends during cloning, the *Bsa*I and *Bsg*I sites will be preserved).

(v) *Bam*HI (G/GATCC) and *Bcl*I (T/GATCA) generate 5'-GATC overhanging ends that are compatible; *Spe*I (A/CTAGT), and *Nhe*I (G/CTAGC) generate 5'-CTAG overhanging ends that are compatible; and *Nde*I (CA/TATG), *Bfa*I (C/TAG), and *Ase*I (AT/TAAT) all generate 5'-TA overhanging ends that are compatible; *Ssp*I (AAT/ATT), and *Eco*RV (GAT/ATC) generate blunt ends that are compatible.

Solution to problem posed in Section 5.3

(vi) First, 500 ng of DNA is 0.5 μg, so if we want 5 units μg^{-1} enzyme then the total units needed are 0.5 μg × 5 units μg^{-1} = 2.5 units enzyme. We can then calculate the number of μl, which is 2.5 units ÷ 8 units μl^{-1} = 0.31 μl. If that small volume cannot be dispensed accurately, we might dilute a small amount of the concentrated enzyme into a $1 \times$ buffer so that larger volumes can be handled.

Solutions to problems posed in Section 5.4

(vii) The *Eco*RI site was used for cloning, and upon insertion of the 2.6-kbp fragment the *Eco*RI site at the junction between vector and insert was preserved. The DNA insert appears to not have additional *Eco*RI sites, because digestion of the 7-kbp plasmid with *Eco*RI cleanly separates the insert and vector. Let us say that in the new numbering system for the plasmid, one of the *Eco*RI sites is still at nt 1 and the other is at nt 2600. The positions of the *Bgl*II sites are approximately nt 2200 (internal to the DNA insert) and nt 6500 (formerly 3900 in the vector). The positions of the *Spe*I sites are approximately nt 700 (internal to the DNA insert) and 5400 (formerly 2800 in the vector). If the 7-kbp plasmid were digested with all three enzymes, the order of fragments would be 0.7 kbp (*Eco*RI to *Spe*I), 1.5 kbp (*Spe*I to *Bgl*II), 0.4 kbp (*Bgl*II to *Eco*RI), 2.8 kbp (*Eco*RI to *Spe*I), 1.1 kbp (*Spe*I to *Bgl*II), and 0.5 kbp (*Bgl*II to *Eco*RI).

(viii) The first thing to do is to add up the fragment sizes to see if we have accounted for all of the DNA:

- *Eco*RI: 4.4 + 0.25 = 4.65 kbp
- *Eco*RI and *Bgl*II: 3.9 + 0.5 + 0.15 = 4.55 kbp
- *Bgl*II: 4.0 + 0.6 + 0.3 kbp = 4.9 kbp

The *Eco*RI digestion adds up to the expected 4.65 kbp, but the *Eco*RI plus *Bgl*II digestion is short about 0.1 kbp, and the *Bgl*II digestion has an extra 0.25 kbp. Our original 4.4-kbp vector had a unique *Bgl*II site 500 bp from the *Eco*RI cloning site, so the DNA insert must have two *Bgl*II sites. However, the presence of both 0.6-kbp and 0.3-kbp fragments in the *Bgl*II digestion isn't easy to understand if the DNA insert is only 250 bp.

Suppose that we had a double-insertion during cloning, so that the total amount of inserted DNA was 0.5 kbp. In the case of the *Eco*RI digestion, we would have released two copies of the 250-bp insert for every one copy of 4.4-kbp vector, but perhaps we couldn't tell that on a gel because the two copies of insert co-migrated. The 0.6-kbp *Bgl* II fragment may include 0.5 kbp of DNA from the vector (between the *Bgl*II site and the *Eco*RI cloning site) and 0.1 kbp of DNA from the insert. If two copies of the 250-bp insert were cloned head-to-head, then there will be a second *Bgl*II site located 0.15 + 0.15 = 0.3 kbp from the first one in the insert. That is, the 0.3-kbp *Bgl*II fragment is symmetrical, carrying parts of two inserts. The order of fragments in the 4.9-kbp clone may be as follows: 0.1 kbp (*Eco*RI to *Bgl*II), 0.15 (*Bgl*II to *Eco*RI), 0.15 (*Eco*RI to *Bgl*II), 0.1 kbp (*Bgl*II to *Eco*RI), 3.9 kbp (*Eco*RI to *Bgl*II), 0.5 kbp (*Bgl*II to *Eco*RI). The gel system used in the experiments did not permit us to detect the 0.1-kbp fragments, which explains why the DNA fragments in the double-digestion did not add up to 4.9 kbp.

(ix) The plasmid has two sites for the enzyme *Xba*I and three for the enzyme *Sma*I. All of the *Sma*I sites lie within the 3.1-kbp *Xba*I fragment, and the double-digestion fragments could be ordered in two different ways:

- 1.3 kbp (*Xba*I to *Sma*I), 0.6 kbp (*Sma*I to *Sma*I), 0.4 kbp (*Sma*I to *Sma*I), 0.8 kbp (*Sma*I to *Xba*I), and 4.9 kbp (*Xba*I to *Xba*I);
- 1.3 kbp (*Xba*I to *Sma*I), 0.4 kbp (*Sma*I to *Sma*I), 0.6 kbp (*Sma*I to *Sma*I), 0.8 kbp (*Sma*I to *Xba*I), and 4.9 kbp (*Xba*I to *Xba*I).

These solutions are equally likely, and more analyses with different enzymes would be needed to exclude one of them as possibilities.

(x) The first thing to do is to make a semilogarithmic plot of the sizes of the DNA markers as a function of migration, such as that shown

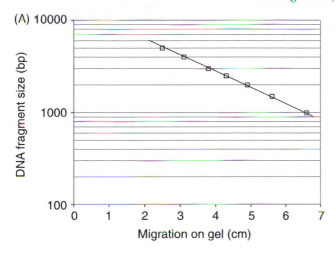

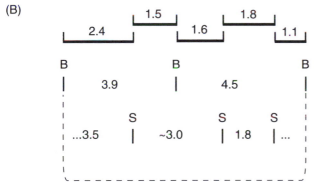

Figure 5.9

Solution to a restriction mapping problem. (A) A semilogarithmic plot of the DNA standards, in which the sizes of the fragments are plotted as a function of migration. (B) An interpretation of the order of double-digestion fragments, with sizes indicated in kbp. The sites for *Bam*HI and *Sma*I are indicated below, along with their spacing in the plasmid. Abbreviations: B = *Bam*HI and S = *Sma*I. The dashed line indicates that the two *Bam*HI sites are the same, due to circularization of the plasmid.

in *Figure 5.9A*. Then, as was explained in Chapter 2, you can look at the migration of an experimental fragment (an x-axis value) and correlate it with an apparent size on the y-axis of the graph. Your numbers may differ slightly from these results:

- *Sma*I: 3.3 cm (3.5 kbp), 3.7 cm (3.0 kbp), and 5.0 cm (1.8 kbp)
- *Sma*I and *Bam*HI: 4.3 cm (2.4 kbp), 5.0 cm (1.8 kbp), 5.3 cm (1.6 kbp), 5.6 cm (1.5 kbp), and 6.4 cm (1.1 kbp)
- *Bam*HI: 2.7 cm (4.5 kbp), and 3.1 cm (3.9 kbp)

The fragments in the three digestions don't quite add up to the same total, but it is close: 8.3 to 8.4 kbp for each. We may make a provisional map based on these data, bearing in mind that the experimental data are likely to be a bit inaccurate and that we might have overlooked a small fragment.

When we consider the 3.5-kbp *Sma*I fragment, it is likely to be composed of the 2.4-kbp and 1.1-kbp fragments from the *Sma*I plus *Bam*HI digestion. The 3.0-kbp *Sma*I fragment is likely to be composed of the 1.6-kbp and 1.5-kbp fragments from the double-digestion (it is at least a close fit, if not exact). The 1.8-kbp *Sma*I fragment appears to be untouched in the double-digestion. When we consider the 3.9-kbp *Bam*HI fragment, it is likely to be composed of the 2.4-kbp and 1.5-kbp fragments from the double-digestion. The 4.5-kbp *Bam*HI fragment may be composed of the three remaining fragments from the double-digestion; the 1.8 kbp, 1.6 kbp, and 1.1 kbp fragments. The fragments in the double-digestion may be ordered as they are shown in *Figure 5.9B*.

Polymerases

6

6.1 Overview

In the late 1970s, an argument might have been made that restriction endonucleases form the most important class of enzymes in the molecular biology laboratory. It would be more difficult to make such a statement today, with Sanger DNA sequencing and the polymerase chain reaction showing what DNA polymerases can accomplish. DNA polymerases can be used in very creative methods of cloning and diagnostic analysis, and this chapter will try to help the reader understand the fundamentals of these methods. We also turn to the subject of bacteriophage RNA polymerases, which are commercially available and allow a researcher to synthesize transcripts *in vitro* using a template with a cognate promoter sequence. These *in vitro* synthesized RNA molecules are useful for many research applications, including probe preparation, *in vitro* translation, and ribozyme research.

What are polymerases?

Polymerases are enzymes that copy nucleic acids. During the copying reaction, a single strand of RNA or DNA serves as a template for synthesis of a second strand of RNA or DNA, following the Watson–Crick base-pairing rules. DNA polymerases make DNA and RNA polymerases make RNA. The well-known functions of DNA polymerases in cells include semiconservative replication of DNA, and repair of damaged or mismatched DNA. The principal role of RNA polymerases is in transcription of RNA. Most RNA polymerases use DNA as a template; however there are many viruses that use RNA-directed RNA polymerases to copy their RNA genomes without ever making DNA. If a DNA polymerase can use an RNA template, it is called an RNA-dependent DNA polymerase, or a reverse transcriptase.

The capabilities of polymerases

By way of introduction, there are a few general statements that can be made about polymerases.

- RNA polymerases may initiate new synthesis, for example at a transcriptional promoter; however DNA polymerases can only extend a pre-existing strand of nucleic acid. In other words, DNA polymerases require a primer, while RNA polymerases do not. Most DNA polymerases will extend an RNA primer, but some require a DNA primer.
- Polymerases add nucleotides to the 3′ end of a nucleic acid, never the 5′ end. That is, polymerases conduct synthesis in a 5′ to 3′ direction.

- Polymerases need to have instructions, in the form of a template to copy. If a polymerase is poised at a 3' end but has no underlying template, then synthesis ceases, unless the polymerase also has a terminal deoxynucleotidyl transferase activity (discussed in Section 6.6).
- Polymerases need to receive the correct nucleoside triphosphate substrates. The substrate must base-pair to the underlying template, and the sugar must match the type of polymerase; for example, DNA polymerases cannot use ribose-containing substrates.

DNA polymerases differ in their Michaelis constants (K_M) for substrate, and this may have a significant effect on their performance in the polymerase chain reaction (as discussed in Chapter 7) and other laboratory procedures such as DNA labeling (as discussed in Sections 6.2 and 6.8). K_M is a substrate concentration at which an enzyme following Michaelis–Menten kinetics achieves half of its maximal velocity. An enzyme with a lower K_M is better able to utilize low concentrations of substrate, and each substrate may have a different K_M value for a given enzyme. For example, E. coli DNA polymerase I has a K_M for nucleoside triphosphate substrates (dNTPs) of 1 to 2 µM (1) whereas Deep VentR® DNA polymerase (New England Biolabs) is reported to have a K_M of 50 µM for dNTPs (2). The relative K_M of these enzymes for DNA primer-template complexes is reversed, with DNA polymerase I having a K_M of 5 nM and Deep VentR® DNA polymerase having a K_M of 0.01 nM. To put these numbers in perspective, 0.01 nM of a 5-kbp plasmid is approximately 1 ng in a volume of 30 µl, and 5 nM of the plasmid is approximately 500 ng in the same volume.

Polymerases also have widely different rates of incorporating errors into a newly synthesized strand of DNA, with T4 DNA polymerase making less than one, Klenow enzyme making approximately 18, and Taq DNA polymerase making nearly 300 errors per million nucleotides incorporated (3,4). DNA polymerases that have proofreading activities tend to have lower error rates, and the reason for this will become clear in the next section, where 3' to 5' exonuclease activities are discussed.

Two problems and two solutions

DNA polymerization reactions can be impeded by two types of problems that require special solutions. The first problem exists when a primer has a mismatch at the 3' end and is not base-paired to the template (see *Figure 6.1A*), while the second occurs when there is an obstruction on the template that blocks forward motion of the polymerase (see *Figure 6.1B*). While some DNA polymerases can solve one or both of these problems on their own, others cannot.

Handling a 3' mismatch problem

Without hydrogen-bonding between a priming nucleotide and template, the polymerase may fail to extend the primer. In other words, a mismatched nucleotide at the 3' end may cause the polymerization reaction to stall. Many, but not all DNA polymerases have a 3' to 5' exonu-

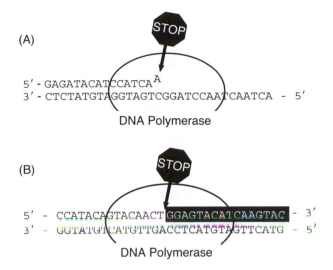

(A)

5′ - GAGATACATCCATCA^A
3′ - CTCTATGTAGGTAGTCGGATCCAATCAATCA - 5′

DNA Polymerase

(B)

5′ - CCATACAGTACAACTGGAGTACATCAAGTAC - 3′
3′ - GGTATGTCATGTTGACCTCATGTAGTTCATG - 5′

DNA Polymerase

Figure 6.1

DNA polymerases face these two challenges. (A) A base mismatch at the 3′ end of a primer may stall forward synthesis. (B) A base-paired nucleic acid (white on black lettering) in the path of synthesis can block synthesis.

cleave activity that can remove a nucleotide at the 3′ end of a DNA, and this is what is needed to correct the mismatch and get the polymerization reaction moving again.

As their name indicates, exonucleases work from the free ends of a nucleic acid, or, to be precise, from a point at which the phosphate backbone is broken and not covalently closed. An exonuclease that removes a nucleotide at a free 3′ end is called a 3′ to 5′ exonuclease, because the direction of enzyme movement is from the 3′ end towards the 5′ end (see *Figure 6.2C*). This direction is the reverse of the polymerization direction, which is 5′ to 3′. When a 3′ to 5′ exonuclease is a part of a polymerase protein it is called a proofreading function because, like the backspace key on a computer, it can erase what was most recently 'written'. DNA polymerases that have a 3′ to 5′ exonuclease activity, such as T4 DNA polymerase, Klenow enzyme, or *Pfu* DNA polymerase, tend to have a lower error rate than polymerases that do not have a 3′ to 5′ exonuclease activity, for example *Taq* DNA polymerase. The proofreading function leads to error correction; a DNA polymerase that lacks a 3′ to 5′ exonuclease activity has no equivalent of a 'backspace key' and has a higher rate of incorporation of errors.

While the 3′ to 5′ exonuclease is always available in an enzyme such as T4 DNA polymerase, it is only revealed when polymerization is not taking place. Let's consider the analogy of a car driving up a hill. Gravity is acting on the car at all times, pulling it down, but as long as the engine has a fuel supply and a road to travel the car keeps moving up the hill. If the engine stalls or the car comes to the end of the road, then the car may slip back down the hill until the engine pushes it forward again.

In comparing an enzyme such as T4 DNA polymerase or Klenow enzyme to a car on a hill, the presence of a 3′ to 5′ exonuclease is analogous to the backward-pulling component of gravity. Having an underlying template is analogous to the car having a road in front (for example, *Figure 6.2A*). Having the correct nucleotide substrate in the enzyme's active site is analogous to the car having fuel in the carburetor. Finally, having a mismatched nucleotide at the 3′ end of a primer is analogous to the car engine stalling (see *Figure 6.2B*). A DNA polymerase tends to conduct 5′ to 3′ synthesis of nucleic acids, so long as there is a template, a supply of the nucleoside triphosphate substrates, and a 3′ primer base-paired to the template. If any of these three conditions is not met, a proofreading DNA polymerase will use its 3′ to 5′ exonuclease activity to subtract nucleotides (see *Figure 6.2C*) until conditions once again favor polymerization (see *Figure 6.2D*).

Handling an obstruction

When double-stranded DNA is being replicated in a cell, the two strands are separated and each may serve as a template for synthesis of a daughter strand. The example of *in vitro* DNA synthesis provided in *Figure 6.2* is

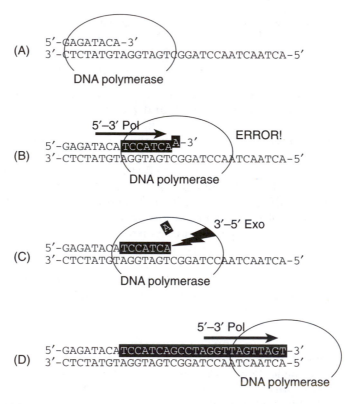

Figure 6.2

A polymerase with a 3′ to 5′ exonuclease activity can proofread mistakes. (A) DNA polymerase at a primer (top strand) bound to a template (bottom strand). (B) Synthesis of new DNA (white on black lettering), ending with a mismatched nucleotide at the 3′ end. (C) The 3′ to 5′ exonuclease activity of the DNA polymerase removes the mismatched nucleotide. (D) DNA synthesis continues (white on black lettering).

a different situation, in that the synthesis is taking place on a template that is already single-stranded and uncovered. No opposing strand of DNA needs to be displaced for the polymerase to proceed. Could a DNA polymerase extend a primer if the template were obstructed by a complementary strand of DNA (for example, the white-on-black lettered nucleotides in *Figure 6.1B*)? There are three possible outcomes for a polymerase that is resting on a free 3′ end of a primer, with an obstructed template lying before it. The polymerase might be unable to proceed with synthesis (see *Figure 6.3A*), it might displace the obstructing strand while performing forward synthesis (see *Figure 6.3B*), or it might use a 5′ to 3′ exonuclease function that can digest the obstructing nucleotides (see *Figure 6.3C*).

Klenow enzyme is a fragment of *E. coli* DNA polymerase I that lacks a 5′ to 3′ exonuclease function. Klenow enzyme has a proofreading activity and can remove a mismatched 3′ nucleotide in a primer; however, it cannot digest nucleotides that stand in the way of forward synthesis on a template (see *Figure 6.3A*). There are many laboratory procedures in which the presence of a 5′ to 3′ exonuclease activity is critical for success,

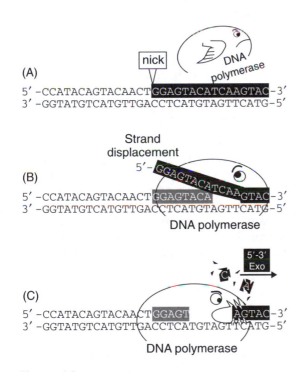

Figure 6.3

When a DNA polymerase reaches a base-paired nucleic acid in the path of synthesis (white on black lettering) there are three possible outcomes. (A) A polymerase may be unable to extend a primer 5′ to 3′ through an occluding DNA strand, and may leave a nick behind. (B) A polymerase may displace the occluding nucleic acid and extend a primer (white on gray lettering). (C) A polymerase may employ a 5′ to 3′ exonuclease activity to digest the occluding nucleic acid and extend a primer (white on gray lettering).

and others that would be ruined by the presence of such an activity. It is important to read the entire label of an enzyme tube before using it in an experiment; Klenow is sometimes sold in tubes marked 'DNA Polymerase I, Large Fragment.'

The DNA polymerase from bacteriophage phi29 is an enzyme with a strand-displacing polymerase function (see *Figure 6.3B*), in addition to a 3' to 5' exonuclease. As the phi29 polymerase synthesizes DNA 5' to 3' on a double-stranded template, it displaces the strand that occludes the template; on a circular template this can lead to a rolling circle mechanism of replication. There are several applications of strand-displacing polymerases in isothermal polymerase chain reaction (as discussed in Chapter 7) and whole genome amplification.

The complete DNA polymerase I enzyme from *E. coli* has a 5' to 3' exonuclease function, in addition to its 5' to 3' polymerase and 3' to 5' exonuclease functions that are associated with Klenow fragment. DNA polymerase I can remove an obstruction on the template, nucleotide by nucleotide, in the 5' to 3' direction (*Figure 6.3C*). *Taq* DNA polymerase is another example of a polymerase with a 5' to 3' exonuclease activity, and this is put to good use in the TaqMan® (Roche Molecular Systems, Inc.) polymerase chain reaction, discussed in Chapter 7.

Making blunt ends

The types of ends found in linear DNA molecules were discussed in Chapter 5, in the context of the 5' and 3' ends left by restriction endonucleases. DNA polymerases are often used to change or fix incompatible ends by the addition or subtraction of nucleotides.

- In an addition reaction, a 3' recessed end can serve as a primer for DNA synthesis using the template provided by the 5' overhanging end of the complementary strand. The polymerase activity of the Klenow enzyme binds to the recessed 3' end and uses it as a primer (see *Figure 6.4*).
- In a subtraction reaction, a 3' overhanging end has no underlying template, and the 3' to 5' exonuclease activity in a DNA polymerase removes nucleotides. The enzyme binds to the overhanging 3' end, and deletes the unpaired nucleotides at the 3' end, one by one (see *Figure 6.5*).

Both addition and subtraction result in a blunt end (see *Figure 6.4B* and *Figure 6.5C*), but not because the enzyme stops working when the overhanging ends are gone. At a blunt end, the Klenow enzyme enters into a futile cycle of removing the 3' nucleotide, and then adding it again, over and over (see *Figure 6.4B, C*, and *Figure 6.5C, D*). As long as the concentration of nucleoside triphosphate substrate is not limiting, the polymerase activity is favored and the net result is a blunt end. However, if substrate is consumed to the point that polymerase activity drops, the exonuclease activity will prevail and the blunt end will not be maintained.

A partial fill-in

A DNA polymerase such as Klenow enzyme or T4 DNA polymerase can be used in a partial fill-in reaction, in which only some of the four dNTP

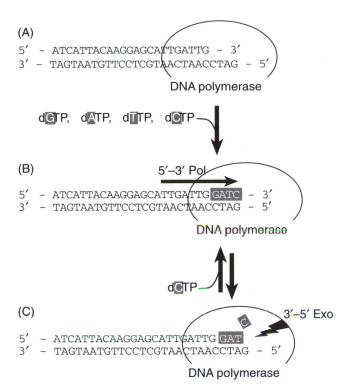

Figure 6.4

A DNA polymerase may fill in a recessed 3′ end, leaving a blunt end. (A) A DNA polymerase at a recessed 3′ end, such as the one left by the enzyme *Bam*HI (G/GATCC). (B) The recessed 3′ end is extended by the polymerase using dNTP substrates, and leaving a blunt end. (C) A DNA polymerase with a 3′ to 5′ exonuclease activity removes the nucleotide at the 3′ end of a blunt-ended molecule. The nucleotide is replaced (up arrow to B) if the nucleoside triphosphate substrate is present, and the B to C to B cycle continues.

substrates are provided to the enzyme. For example, the 4-nt overhanging end left by the enzyme *Xho*I (C/TCGAG) can be converted to a 2-nt overhanging end by treatment with Klenow enzyme, and only the nucleotide substrates dTTP and dCTP. The enzyme *Xho*I leaves this type of end:

```
5′-...GGAGTACATACC     -3′
3′-...CCTCATGTATGGAGCT -5′
```

Upon treatment with Klenow enzyme, and the substrates dTTP and dCTP, the 3′ end (on the right) can be extended so that the 5′ overhanging end is only 2 nt.

```
5′-...GGAGTACATACCTC   -3′
3′-...CCTCATGTATGGAGCT -5′
```

This 5′-TC overhanging end is not compatible with itself for the purposes of ligation, a matter that is discussed in Chapter 4, but would be compatible with a 5′-GA overhanging end.

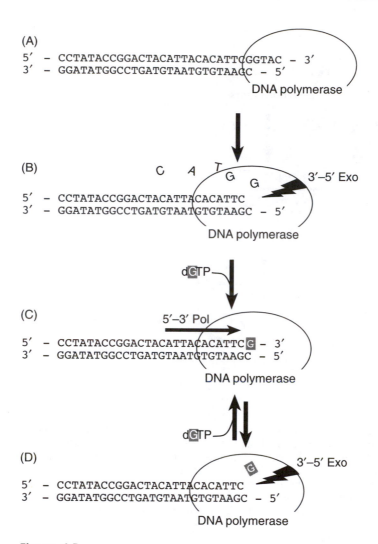

Figure 6.5

A DNA polymerase with a 3′ to 5′ exonuclease activity may digest an overhanging 3′ end, leaving a blunt end. (A) A DNA polymerase at an overhanging 3′ end, such as that left by the enzyme *Kpn*I (GGTAC/C). (B) The overhanging 3′ end is digested by the 3′ to 5′ exonuclease activity, leaving a recessed 3′ end. (C) The recessed 3′ end is extended by the polymerase using dNTP substrate, leaving a blunt end. (D) The 3′ to 5′ exonuclease activity removes the nucleotide at the 3′ end (again). The nucleotide is replaced (up arrow to C) if the nucleoside triphosphate substrate is present, and the C to D to C cycle continues.

We can make 5′-GA overhanging ends by partial fill-in of the 5′ overhanging ends left by restriction endonucleases such as *Bam*HI (G/GATCC), *Bgl*II (A/GATCT), and *Sau*3AI (/GATC). For example, *Bam*HI leaves this type of end:

```
5'-...AATAGTCATAGG        -3'
3'-...TTATCAGTATCCCTAG -5'
```

Upon treatment with Klenow enzyme, and the substrates dGTP and dATP, the 3' end (on the right) can be extended so that the 5' overhanging end is only 2 nt.

```
5'-...AATAGTCATAGGGA    -3'
3'-...TTATCAGTATCCCTAG -5'
```

Ends generated by *Xho*I and partially filled in with dTTP and dCTP are therefore compatible with ends generated by *Bam*HI and partially filled in with dGTP and dATP.

The engineering of DNA ends using a DNA polymerase and dNTP substrates is discussed in more detail in Chapter 7, where applications related to PCR product cloning are presented.

Problems

The solutions to these problems are provided at the end of the chapter.

(i) If the DNA fragments below were treated with Klenow enzyme or T4 DNA polymerase, both of which have a 3' to 5' exonuclease activity but lack a 5' to 3' exonuclease activity, and the four dNTP substrates (dGTP, dATP, dTTP, and dCTP), what would be the final products?

```
(A) 5'- AATTGTACAGGCCCACCTGGGAATACATGATCAG -3'
    3'-     CATGTCCGGGTGGACCCTTATGTACT        -5'
```

```
(B) 5'- GATCTAAGAGCTACAGTTACTTACTACCCT    -3'
    3'-     ATTCTCGATGTCAATGAATGATGGGAGGCC -5'
```

```
(C) 5'-     TATTACATACTTCGGAGAGTCAGGAT    -3'
    3'- AATTATAATGTATGAAGCCTCTCAGTCCTATCGA -5'
```

```
(D) 5'-      GAGAGATCAGATCATGGGATCAGGATAGCT -3'
    3'- GCGCCTCTCTAGTCTAGTACCCTAGTCCTA    -5'
```

(ii) Suppose that you have a DNA polymerase that is like Klenow enzyme, but lacks a 3' to 5' exonuclease activity (this Exo-minus Klenow is a commercially available enzyme). If you used that enzyme and the four substrates, would your answers to the previous problem be the same or different?

(iii) Suppose that you digest a plasmid vector with the enzyme *Eco*RI (G/AATTC) to linearize it at a cloning site, and digest a DNA fragment with the enzyme *Kpn*I (GGTAC/C), in preparation for cloning. You treat both the vector and fragment with Klenow enzyme and the four dNTP substrates (dGTP, dATP, dTTP, and dCTP), to make blunt ends.
   A. Draw a diagram showing what the *Eco*RI and *Kpn*I digested ends look like before and after treatment with Klenow enzyme.
   B. What would be the consequence of not adding dGTP to the reaction with Klenow enzyme?
   C. If the blunted *Eco*RI ends were joined to each other by ligation (after Klenow treatment), would an *Eco*RI site be regenerated in the DNA sequence?

D. If the blunted *Eco*RI and *Kpn*I ends were joined by ligation (after Klenow treatment), would an *Eco*RI site be regenerated in the DNA sequence?

6.2 DNA labeling

Extending a primer on single-stranded DNA

Imagine a DNA polymerase such as Klenow enzyme is provided with a long, unobstructed DNA template, a primer, and an adequate supply of dATP, dGTP, dCTP, and dTTP substrates. The enzyme would be able to speed down the template, extending the primer and building a second DNA strand complementary to the DNA template. This 5' to 3' synthesis may be called primer extension because it is a straightforward example of DNA polymerization. Because the template is unobstructed, a 5' to 3' exonuclease is not involved, and a 3' to 5' exonuclease is only involved for proofreading occasional errors.

There are many applications for this *in vitro* DNA synthesis, several of which will be discussed in this section. For example, if the DNA polymerase is provided with a primer that is specially labeled, perhaps with a radioactive phosphate or fluorescent dye at its 5' end, the long DNA strand prepared by primer extension will carry the special marker. There is a related application in which a primer carries a mutation or mismatched region that does not base-pair to the template (see Section 6.3). If the mismatch is not at the 3' end the primer can be extended on the template so that the mutation is incorporated into the longer DNA. This is an important step in site-specific mutagenesis.

Setting aside the idea of using an altered DNA primer, it is also possible in primer extension to use nucleotide substrates that are different from the usual ones, dGTP, dATP, dCTP, dTTP, to make a DNA molecule that is internally labeled with a marker. For example, if the alpha phosphate of dCTP (proximal to the deoxyribose) is constituted with a ^{32}P isotope, then when the [^{32}P]-dCTP is incorporated into a growing DNA strand the radioactive phosphate becomes part of the phosphodiester backbone. That is, the long DNA strand is made radioactive during synthesis. Nonradioactive internal markings can be used as well. For example, if dUTP is substituted for dTTP, many polymerases will incorporate it into the growing DNA so that the synthesized DNA is made up of the nucleotide bases G, A, C, and U. This has important applications in site-directed mutagenesis, when it is useful to be able to mark a template DNA strand and its complementary strand differently.

Sometimes, a primer extension method is used as a measuring device to see how far a template extends beyond a primer, or to see how far synthesis can proceed before it is interrupted. For example, a specific 5' cleavage site in an RNA can be located by using the RNA as a template and employing a reverse transcriptase enzyme to synthesize a cDNA (as discussed in Section 6.5). Primer extension from a specific sequence in the RNA will lead to a cDNA product that extends from the primer to the cleavage site, the point at which the template stops. The length of that cDNA, which can be determined by gel electrophoresis, gives information about the distance to the cleavage site; it is almost as if a little measuring tape had been pulled out between the primer site and the 5′ end of the RNA. DNA sequencing by the Sanger method is also a type of primer extension, in which the different lengths of DNA products represent the distances between a common primer and many sites of interruption by dideoxyribonucleotide chain termination.

DNA labeling by random hexamer primer extension

In a primer extension reaction, the DNA or RNA that is copied is determined in part by the primer chosen, because the primer is the starting point for new synthesis. So, for example, if a random collection of small primers is used to initiate synthesis, then a large single-stranded template can be randomly covered with small double-stranded segments. This is called random hexamer priming because the collection of primers consists of all possible six-nucleotide sequences (5). During synthesis, chemically or radioactively labeled substrates may be incorporated into the product and this is a common way to make labeled probes from an unlabeled template, or make a cDNA from an RNA template.

How is it possible for random primers to find binding sites on a template to start synthesis? First, the primers are small and so can be added to a reaction at a high molar concentration compared to the template. That high concentration helps to reduce the amount of time it takes a given primer to anneal. Second, a six-nucleotide sequence binds to a template with a melting temperature very near to room temperature, so while these random primers may seem small they are still stable enough on the template to serve as a starting point for synthesis. Finally, any given six-nucleotide sequence has a reasonable chance of being represented in a sequence of several kbp. Many protocols for random hexamer priming follow this general pattern.

i.   Make the DNA template single-stranded, perhaps by incubation in a boiling water bath.
ii. Allow the random primers to anneal to the single-stranded template, by incubation on ice.
iii. Add the deoxyribonucleotide substrates and DNA polymerase and raise the temperature to improve the performance of the enzyme; to 37°C for Klenow enzyme, or 42°C for an AMV (avian myeloblastosis virus) reverse transcriptase. Some protocols advocate a pre-treatment at 25°C for 10 min to allow brief extension of the primer, increasing the stability of the pairing at higher temperatures.

In step 1, the high temperature causes breakage of the hydrogen bonds in the double-stranded template. The DNA polymerase is left out during step 1 because it would be denatured at the high temperature, unless it is a thermostable enzyme of the type used in polymerase chain reactions. In step 2, the low temperature is allowing hydrogen bonds to form between the single-stranded template and the short primers. In step 3, the polymerase begins to extend the primer, adding nucleotides to its 3′ end, and the stability of the duplex DNA increases because more hydrogen bonds are holding the duplex together. There are many commercially available kits that use slightly different protocols, and it is important to follow the manufacturer's recommendations. However, every protocol must start with a step that allows for initial separation of the strands of a double-stranded template, then a condition that favors binding of the short primers. The polymerase employed might be Klenow enzyme or T4 DNA polymerase, because no 5′ to 3′ exonuclease is needed or desired. If the template is RNA, then the primer extension would be conducted using a reverse transcriptase.

The products of *in vitro* synthesis from random hexamers may be in the order of hundreds of nucleotides in length, but they may not make a full-length copy of the single-stranded template. One reason is that the synthesis is starting from multiple primers distributed on the template. Depending on the polymerase used, a bound primer may block the 5′ to 3′ progress of a downstream polymerase, unless the polymerase is strand displacing (see *Figure 6.3B*) or has a 5′ to 3′ exonuclease activity (see *Figure 6.3A*).

A second reason for short products may have to do with the specific conditions of the reaction, particularly if a radiolabeled nucleotide is being incorporated into the growing DNA chains at a high specific activity. Radioactivity is measured in units of Curies (Ci), with one Ci being 2.22×10^{12} disintegrations per minute. It is usually desirable to make a probe that has a very high specific activity, which is to say that it has a very high content of radioactivity per nanogram of product, because the limit of detection of hybridization experiments depends on the specific activity. For example if α-[^{32}P]-dCTP is incorporated into the growing DNA strand, then ideally every cytidylate nucleotide in the product would have a [^{32}P]-phosphate because that would maximize the radioactivity per nanogram. The dCTP substrate is simply being used for the purposes of illustration; any of the other three nucleotides could have been labeled instead. In setting up such a reaction with radioactive dCTP, the substrates would consist only of dATP, dGTP, dTTP, and α-[^{32}P]-dCTP. No unlabeled dCTP would be included because adding nonradioactive dCTP would only serve to reduce the specific activity of the product. This causes a problem that will be explained in more detail later (see Section 6.8), but suffice it to say that the enzyme may fail to make forward progress if one or more of the deoxyribonucleotide substrates are in short supply, with the consequence that the products are shortened.

Nick translation

DNA polymerases that possess a 5′ to 3′ exonuclease function may start at a nick in a DNA (see *Figure 6.6A*), digest the sequence that occludes the template, and re-synthesize the complementary sequence that was just removed (see *Figure 6.6B*). As a result, a nick in the phosphodiester backbone is shifted each time a nucleotide is excised and then re-added. The excision and re-synthesis of a sequence in the laboratory, the shifting of a nick in a 5′ to 3′ direction (see *Figure 6.6C*), allows the replacement of regular DNA with radioactive or chemically labeled DNA. This is called nick translation, where the word 'translation' has its more mathematical meaning of movement in space rather than conversion between languages. The nick translation procedure involves two separate events:

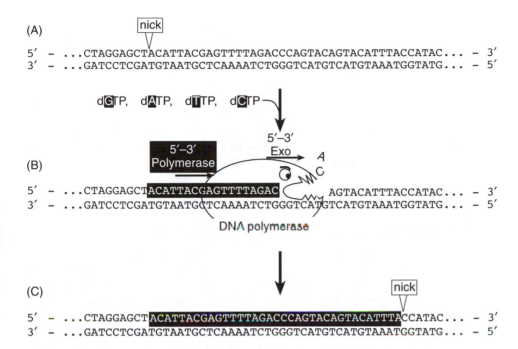

Figure 6.6

Nick translation by DNA polymerase I *in vitro*. (A) A double-stranded DNA with a nick in the top strand. (B) *E. coli* DNA polymerase I enters at the nick and digests the DNA using a 5′ to 3′ exonuclease, and synthesizes new DNA (white on black lettering) using the free 3′ end as a primer. In many applications of this method, the newly synthesized DNA is radioactively or chemically labeled. (C) Upon departure of the DNA polymerase, a nick is left between the newly synthesized DNA and the undigested old DNA.

- limited, random nicking of a double-stranded DNA by DNase I;
- translation of the introduced nicks with DNA polymerase I, using a mixture of labeled and unlabeled nucleoside triphosphate substrates.

These reactions can be performed sequentially or concurrently. However, if the DNase I remains active when the DNA polymerase I is added, then it will not only nick the template but also steadily shorten the products of synthesis.

The relative amounts of template DNA, DNase I, and DNA polymerase I, as well as the timing of the reaction are all critical parameters. If the activity of DNase I is too low then labeling may be inefficient because there are too few nicks created, and hence too few sites of priming. Some view the nick translation as being a 'fussy' reaction because these conditions need to be optimized. There are commercially available kits, or mixtures of DNA polymerase I and DNase I, that may ease the difficulty of getting the reaction to work. It is important to reiterate that many DNA polymerases, including Klenow enzyme and T4 DNA polymerase, do not have a 5′ to 3′ exonuclease and so will not conduct a nick translation. DNA polymerase I from *E. coli* is the enzyme typically used, because it has a 5′ to 3′ exonuclease.

Nick translation is used to label DNA, and some of the important issues related to specific activity are discussed in Section 6.8. With radioactively labeled nucleoside triphosphates, the problems of the K_M of the enzyme *versus* the concentration of limiting substrate, and the overall yield of synthesis are important considerations. With nonradioactive labeling methods, the concentration of labeled substrate is usually not problematic, but determining the overall yield of synthesis from a given amount of template is just as important. If a large amount of template produces only a very small amount of new synthesis, then the limit of detection of the probe may be inadequate for the experiment.

6.3 Site-specific mutagenesis

Primer match

Site-specific mutagenesis is typically based on the extension of a mismatched primer on a template. The primer, which is designed by the researcher and constructed by *in vitro* chemical synthesis, bears a mutation that is to be fixed in the DNA sequence. This cannot be accomplished with very short primers, such as the random hexamer primers used in labeling, because a mutation would dislodge the primer from the template. With longer primers, the number of base pairs between primer and template increases and the 3′ end can be stably bound to the template even if the 5′ end is not.

Temperature also plays a role because the binding of a primer to a template is based on hydrogen-bonding between the nucleotide bases. Higher temperatures lead to breakage of hydrogen bonds. That is why a researcher will often use a boiling water bath to denature double-stranded DNA into single strands, and an ice bath to bring together single DNA strands and short primers.

If a specific DNA primer has a mismatch with the template, how many additional nucleotides are needed at the 3′ end to ensure that it is anchored to the template, and can be extended? This is a key question in methods of site-specific mutagenesis because the primer is not random, but designed to base-pair to a specific template sequence. Suppose that a 15-nt primer has a mismatch with the DNA template at or near its 5′ end. If incubated with the template on ice, the primer might be bound to the template at both the 5′ and 3′ ends and only a short unpaired region of single-stranded DNA would betray the mismatch. An unpaired region surrounded by double-stranded DNA would look like a small 'bubble' of single-stranded DNA. If the sample is taken off ice and allowed to rise in temperature, the mismatch 'bubble' would gradually extend as hydrogen bonds break, until the 5′ end was no longer base-paired. The molecule might be called a 'Y' structure at this point, because of its shape: a double-stranded base with two single-stranded arms. Provided that the double-stranded base is of sufficient length, a DNA polymerase can still extend the 3′ end of the primer. As the temperature rises further the arms lengthen and the double-stranded base shortens, until finally the entire primer is dislodged from the template.

The destabilization of the primer is like a zipper traveling down a piece of clothing. The rate at which the primer unzips depends on the temperature and the base composition, for each A-T base pair has two hydrogen bonds and each G-C base pair has three. A GC-rich DNA sequence will stay base-paired to a higher temperature, because the additional hydrogen bonds increase the strength of the base-pairing. Researchers use formulae to calculate melting temperatures (T_M) of short oligonucleotides, such as:

$$T_M = 2(A+T) + 4(G+C)$$

In this formula A+T is the number of A and T base pairs and G+C is the number of G and C base pairs. These calculations can help to predict the minimum number of nucleotides of perfect complementary sequence needed at the 3′ end of a primer.

For example, if Klenow enzyme is being employed in the primer extension at a temperature of 37°C then 12 nucleotides of perfect match at the 3′ end of the primer is sufficient if there are seven G-C base pairs and five A-T base pairs. On the other hand, if the primer extension were conducted with a thermostable DNA polymerase at an annealing temperature of 55°C, then about nine G-C base pairs and nine A-T base pairs of perfect match, or 18 nucleotides total, would provide stability at the 3′ end. This is not to say that fewer nucleotides will uniformly lead to failure in an experiment, or that every experiment would succeed with just the number recommended by the calculation. Even a primer that is too short to work most of the time may still work some of the time, and perhaps often enough be captured once by cloning.

In one way of conducting site-specific mutagenesis, a 'parental' DNA plasmid template is annealed to a mismatched oligonucleotide primer, which is extended to make a double-stranded DNA 'daughter' strand. The parental DNA is a single-stranded circle, and might have been made either by denaturation of a double-stranded DNA, or by synthesis of a single-stranded product from a filamentous bacteriophage origin of replication (discussed in Chapter 4).

Adding a restriction site

Let us suppose that a researcher wants to add a restriction enzyme recognition site, such as *Bam*HI (GGATCC) to a sequence, between nucleotides 143 and 144 of the following sequence, which is shown as it would look on a computer printout in GenBank® (US Department of Health and Human Services) format:

```
121 gcctaacgtc acagtctctg aaaatgcaca gggatgcctg gctacctcgc
```

By convention, the 5′ to 3′ direction of the sequence is from left to right. The other strand of the DNA is not represented, but could be imagined running antiparallel to the first. The researcher wants to change this sequence to the following, where the inserted restriction endonuclease sequence is in capital letters, and both strands are shown:

```
121 gcctaacgtc acagtctctg aaaGGATCCa tgcacaggga tgcctggcta cctcgc
121 cggattgcag tgtcagagac tttCCTAGGt acgtgtccct acggaccgat ggagcg
```

Before oligonucleotide primers can be planned, it must be decided which of these two DNA strands will be the template for mutagenesis. We'll make the assumption that the sequence is cloned in a circular vector, for example a plasmid, as discussed in Chapter 4. Some of the methods of template generation, for example production in filamentous bacteriophage capsids, causes only one of the two strands to be available. Other methods, such as simple denaturation of a double-stranded plasmid, might allow either strand to be the template.

If the bottom DNA strand is selected as a template, this being the strand that has a 5′ to 3′ direction running from right to left on the page, then an oligonucleotide primer that binds to it will have a sequence that is antiparallel and resembles the top DNA strand. The oligonucleotide sequence 5′-cagtctctgaaaGGATCCatgcacagggat has the restriction endonuclease recognition sequence embedded into the top DNA strand, and also includes a dozen nucleotides upstream and downstream to allow annealing at 37°C, the temperature at which Klenow enzyme would be used. Only the right side of the oligonucleotide sequence will be extended, that being the 3′ end, but annealing of the 5′ end serves as a 'bumper' for a Klenow enzyme that went around the vector circle and is completing synthesis from the left. If this oligonucleotide primer is annealed to the parent DNA sequence, it will have two regions of perfect match, separated by a looped out region of six nucleotides. The protocol for extension would be just two simple steps.

1. Combine the oligonucleotide primer and template in a tube and place in a boiling water bath for 5 min. This denatures the template.
2. Cool the mixture to 37°C and add buffer, dGTP, dATP, dCTP, dTTP, and Klenow enzyme. This extends the primers.

We have assumed that the bottom strand is the template, but suppose instead that the top strand is the template. Then the oligonucleotide designed would run antiparallel to the first one: gtcagagactttCCTAGGtacgtgtccccta-5′. There are two important statements to make about this approach. First, the 5′ to 3′ direction of the sequence just written is from right to left, and so it needs to be rewritten in the conventional left to right format before it is ordered from an oligonucleotide supplier: 5′-atccctgtgcatGGATCCtttcagagctg. Second, this second oligonucleotide primer would never be used in the same experiment with the first oligonucleotide, because they would anneal to each other instead of to the cloned DNA sequence in the vector.

Making a point mutation

Oligonucleotide design

We have considered how to insert a six-nucleotide restriction endonuclease recognition sequence by mutagenesis; let us now go back to the original sequence and consider how to change a single nucleotide without increasing the length of the sequence.

```
121 gcctaacgtc acagtctctg aaaatgcGca gggatgcctg gctacctcgc
121 cggattgcag tgtcagagac ttttacgCgt ccctacggac cgatggagcg
```

Once again, the change is shown in capital letters, and the researcher must select which of the two strands will serve as a template. If it is the bottom strand then the oligonucleotide primer will have a 5′ to 3′ direction running from left to right, just as was the case with the top strand. A suitable primer might be 5′-ctctgaaaatgcGcagggatgcctg because this will anneal to the bottom strand of the original sequence with only a single mismatch in the middle of the sequence. On the other hand if the top strand is to serve as template then the suitable primer would be a version that is antiparallel to the first oligonucleotide; but again, these would never be used in the same experiment. Rewritten in the conventional format, with the 5′ to 3′ direction running from left to right, this alternative oligonucleotide primer could be ordered from a supplier as the sequence 5′-caggcatccctgCgcattttcagag.

Denaturation of a template from a double-stranded DNA is straightforward, just a matter of heating the sample in a boiling water bath, but some discussion of the effects of the heat may be in order here. It has already been mentioned that heating the DNA disrupts the hydrogen bonds, but plasmids are not a uniform sample. Heating a supercoiled plasmid DNA in a boiling water bath will cause it to collapse into an unusable clump of DNA. The relaxed circular plasmid DNAs that have a single nick in one of their strands will separate upon heating into one single-stranded circle and one linear single-stranded DNA, corresponding to the covalently closed strand and the nicked strand. If the relaxed circular plasmid DNA has nicks in both strands, perhaps widely separated so that the double-stranded DNA maintains its circular shape, heating will separate the DNA into two linear single-stranded molecules. It is the circular single-stranded plasmid DNA that is wanted as a template in site-directed mutagenesis, so the best starting material for denaturation would be a plasmid that was nicked an average of once per molecule.

Phagemids bear an origin of replication from a filamentous bacteriophage, such as f1, and these can be used to generate single-stranded DNA molecules, as explained in Chapter 4. The cells are infected with an M13 helper phage that activates the origin of replication, which replicates only one of the two strands. The helper phage also provides the necessary trans-acting proteins to allow the single-stranded plasmid sequence to be packaged into a phage capsid, and released from the cell into the growth medium. Preparing a template from a phagemid has some advantage over the previously described method of denaturing a double-stranded molecule with heat, because the oligonucleotide primer does not have to compete with a complementary strand for binding to the template.

What's so bad about having extra parental strands? The problem is that the parental strands bear the unmodified DNA sequence, the sequence that has not been changed by site-specific mutagenesis. These parental strands are a constant reminder of the starting point of the experiment, not the end point; if they persist at too high a level in the DNA sample they will also be recovered later at a high level among the clones, making the modified clones that much harder to locate in the collection.

One concern about the daughter strand is that it may be corrected back to the parental version in the E. coli cell by a methylation-directed mismatch repair system, and for that reason it is common for researchers to level the playing field by using a mutS−E. coli strain that is deficient in this mismatch repair. A second concern is that after completion of synthesis the daughter strand has a nick at the 5′ end of the oligonucleotide primer, and this nick may reduce the success of the daughter strand relative to the parental strand that is not nicked. Some researchers suggest using oligonucleotides that have a 5′ phosphate group, either purchased in this form from a supplier or added using the enzyme polynucleotide kinase, so that the *E. coli* may readily rebuild a phosphate backbone between the 5′ phosphate of the oligonucleotide primer sequence and the 3′ end of the extended sequence. Some researchers may even add a DNA ligase to the reaction *in vitro*, so that the *E. coli* is given a covalently closed daughter strand during transformation. This is not required for success, but it may increase the efficiency of the process slightly.

Altered Sites®-based mutagenesis

There are a number of methods of positive selection for the daughter strand in mutagenesis. One approach is to link the desired mutation in a sequence to a selectable mutation elsewhere on the same strand. For example, the Altered Sites® plasmid pALTER® (Promega Corporation) carries a mutated version of the selectable marker gene β-lactamase that normally destroys the antibiotic ampicillin. The plasmid would confer ampicillin resistance to a host cell, except that the mutation in the β-lactamase gene renders the selectable marker gene nonfunctional. The researcher may use a different selectable marker such as tetracycline resistance while in the process of cloning a gene of interest. The basic principle of the Altered Sites® approach is that two changes in a plasmid may be linked on a daughter strand, with one of the modifications being a convenient selectable marker.

During mutagenesis of the Altered Sites® plasmid, the gene of interest and the nonfunctional β-lactamase (ampicillin resistance) gene are changed simultaneously, by extending one primer that anneals to the region of the gene of interest and generates a mutation, and another primer that anneals to the β-lactamase gene and restores it to working order (see *Figure 6.7A*). A third primer is used to inactivate the tetracycline resistance gene, for reasons that will be discussed later. Synthesis is complete when each of the

Figure 6.7

Site-directed mutagenesis using the Altered Sites® system. (A) Annealing of oligonucleotides to a single-stranded plasmid circle. A mismatch between oligonucleotide and plasmid is indicated by a '^' in the line segment. The plasmid includes a cloned gene of interest (top), a defective Amp^R gene (left, with a dot indicating the mutation), and a Tet^R gene (right). (B) Extension of oligonucleotide primers *in vitro*, using a DNA polymerase such as Klenow enzyme. The dashed lines indicate DNA synthesis of the daughter strand, using the parental strand as the template. (C) Following transformation into *E. coli*, the products of replication of the parental strand (on the left) lead to ampicillin-sensitive cells that lack the mutation in the gene of interest, while the products of replication of the daughter strand (on the right) lead to ampicillin-resistant, tetracycline-sensitive cells that are likely to carry a mutation in the gene of interest.

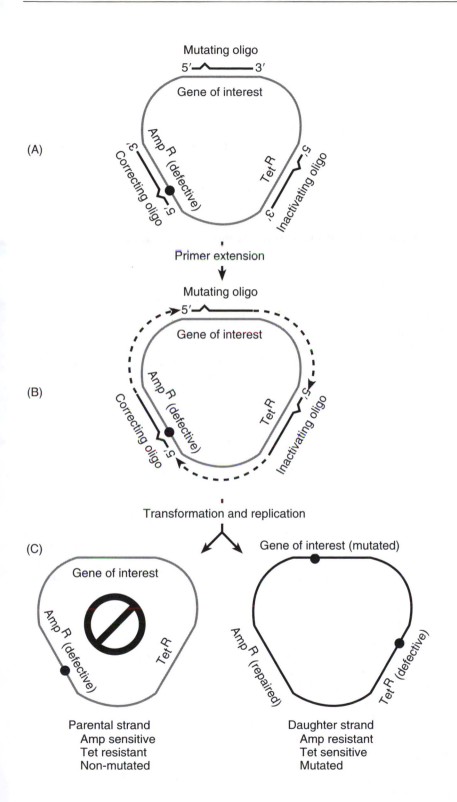

primers is extended so the daughter strand is made in three parts, although they are not covalently connected (see *Figure 6.7B*).

The daughter strand carrying the mutagenized gene of interest and the restored β-lactamase gene is base-paired to the parent strand at the point of transformation, and the *E. coli* host cell separates them during replication (see *Figure 6.7C*). The plasmid is passed on to the *E. coli* daughter cells in two different forms, with about half the cells inheriting a double-stranded plasmid with the daughter strand sequence and half the cells inheriting a plasmid with the parental strand sequence. The cells that inherit the daughter strand sequence are very likely to have the ampicillin resistance gene that has been restored to working order, and those cells can survive in medium containing ampicillin. On the other hand, the cells that inherit the parental DNA sequence will not have a corrected ampicillin resistance gene, and cannot survive in medium containing ampicillin.

The ability of the plasmid to confer ampicillin resistance is only interesting to the researcher because it indicates inheritance of a daughter strand, one that was synthesized *in vitro*. That same strand is likely to bear the mutation that the researcher really cares about, the mutation in the gene of interest. A daughter strand that is made solely from the primer covering the gene of interest would not have a corrected ampicillin resistance gene and would not appear among the colonies on a bacteriological plate with ampicillin.

There is only one slight elaboration of the method, which can be appreciated if the reader thinks about the possibilities of doing a second round of mutagenesis on the gene of interest. Since the ampicillin resistance gene has already been corrected during the first round of mutagenesis, what leverage can be used to add a second mutation to the first? The original Altered Sites® plasmid had two selectable marker genes: the aforementioned ampicillin resistance gene that was in some need of repair, and a tetracycline resistance gene that was in good repair.

The prescient researcher performs the mutagenesis in conjunction with a third primer that has the role of inactivating the tetracycline resistance marker in the daughter strand (see *Figure 6.7A*). As a consequence, the first mutation in the gene of interest is carried on a plasmid that confers ampicillin resistance but is not capable of conveying tetracycline resistance (see *Figure 6.7C*). There is always more to do in an experiment, so the next mutation can be added to the gene of interest and selected by co-repair of the tetracycline gene; and again, the wise researcher would simultaneously inactivate the ampicillin resistance gene during this second round of mutagenesis so that a third round could be conveniently performed.

Besides the Altered Sites® approach, there are other ways of disadvantaging the parental strand and favoring the daughter strand in a site-specific mutagenesis. These methods rely on the fact that the parental strand and the daughter strand are made under different conditions, with the parental strand probably being made in a cell and the daughter strand being made *in vitro*.

Methylated templates

If the parental DNA strand is made in an *E. coli* cell that has a Dam+ phenotype, then the adenylate nucleotides in the sequences GATC will be methylated. The restriction endonuclease *Dpn*I cleaves the recognition sequence GATC only if the adenylate nucleotide is methylated, as discussed in Chapter 5. GATC is a common sequence in DNA, and so a plasmid grown in a *dam+ E. coli* would tend to be digested into small fragments by *Dpn*I.

If a single strand from such a plasmid was used as a template for primer extension *in vitro*, the polymerase would make the daughter strand with unmethylated adenylate nucleotides. A mutation could be directed to a gene of interest by an oligonucleotide primer, and the daughter strand carrying the mutation could be distinguished from the unmutated parent strand by its lack of methyl adenosine. What happens if such a parent–daughter hybrid is treated with *Dpn*I? The enzyme nicks the parent strand at every GATC sequence, because those sites have methylated adenylate nucleotides, but leaves the corresponding GATC sequences in the daughter strand untouched. Upon transformation of the duplex DNA into *E. coli*, the parent strand is disadvantaged because of the additional nicks in the phosphodiester backbone. While these can be repaired, the daughter strand is favored because it is more intact.

Uridylate-containing templates

A related approach involves growing the parental plasmid in a strain of *E. coli* that lacks dUTPase and uracil *N*-glycosylase enzymes, called a Dut– Ung– phenotype. In wild-type *E. coli*, the dUTPase keeps the pool of dUTP at low levels so that uridylate nucleotides are infrequently incorporated into DNA. If uridylate nucleotides do occur in the DNA, either by chance incorporation or deamination of cytosine bases, the enzyme uracil *N*-glycosylase excises the uracil by hydrolysis of the bond between the uracil and the deoxyribose. If the parental plasmid is grown in a *dut– ung–* strain of *E. coli*, then it will carry a mixture of thymidylate and uridylate nucleotides (see *Figure 6.8A*). If used as a template for primer extension in a site-specific mutagenesis, the daughter strand will be synthesized *in vitro* and can be distinguished from the parent because it will carry only thymidylate nucleotides (see *Figure 6.8B*). The duplex can then be treated with the enzyme uracil *N*-glycosylase *in vitro*, or simply transformed into a wild-type *E. coli* strain that has an Ung+ phenotype (see *Figure 6.8C*). In either case the uracil bases in the parental DNA will be excised, creating damage in that specific strand, and favoring the undamaged daughter strand.

Problems

The solution to this problem is provided at the end of the chapter.

(iv) Suppose that you wish to make a mutation in the following sequence, changing the nucleotide indicated by a lowercase 'g' to an 'A' nucleotide. The sequence is cloned in a phagemid, and the single strand shown below is the one that will be extracted from virions. Design an oligonucleotide that will be suitable for annealing and extension, and will carry the mutation in the *in vitro*-synthesized daughter strand.

5'–...AAGTCAGACGGAGCCGTTAGGAGTgAGTTACCCATCCGATCAGATCACT...–3'

6.4 DNA sequencing

In DNA sequencing by the Sanger method, a DNA polymerase is used to measure a series of distances; in this case, between a specific primer binding site on a template and all of the G, A, T, and C nucleotides that lie in the path of primer extension (6). DNA synthesis is conducted by

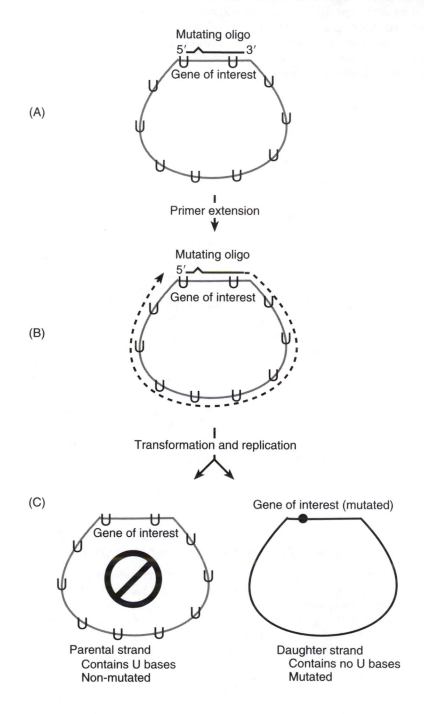

connecting the 5′ phosphate of a nucleotide substrate to a 3′ hydroxyl group of a primer. An important fact is that a DNA molecule cannot serve as a primer if its free 3′ end has no hydroxyl group. A DNA polymerase will incorporate a nucleotide that has a 3′-H instead of a 3′-OH, a type of sugar that is called dideoxyribose instead of deoxyribose. However, once incorporated the dideoxyribose cannot serve as a primer or be extended further. The incorporation of the dideoxyribose nucleotide terminates the growth of the DNA chain; we say that the dideoxyribonucleotide is a chain terminator.

Suppose we conduct a primer extension reaction with a DNA polymerase and four nucleotide substrates, dGTP, dCTP, dTTP, and ddATP, where ddATP stands for dideoxyadenosine triphosphate. That is, the first three triphosphates carry the usual deoxyribose sugar and the last bears a dideoxyribose sugar. Suppose that the primer is 5′ GGATACGATCAGGAT-GCT and it is bound to a template (top strand) as follows, and extended from right to left (bottom strand):

```
5′-GACTACCGAGTACAGACAAAGCATCCTGATCGTATCC
                        TCGTAGGACTAGCATAGG-5′
```

What will be the product of synthesis? The first eight nucleotides added will have a deoxyribose sugar, with a 3′-OH that serves as a primer for continued extension of the DNA, then a ddATP will be used as a substrate for the ninth addition and the chain will be terminated. The sole product of synthesis on the bottom strand will be

```
5′-GGATACGATCAGGATGCTTTTGTCTGTA
```

where the newly synthesized DNA is underlined and the strand has been turned around so that it reads 5′ to 3′ from left to right. This is an interesting way of finding out the identity of a single nucleotide. If the template has a T nucleotide at the given position nine nucleotides from the 3′ end of the primer, then the product will be extended by nine nucleotides to the dideoxyadenosine chain terminator, and be 27 nt in length overall. If the T is mutated to a different nucleotide, for example C, the dideoxyadenosine will be incorporated at a later stage in the extension and the product will be longer by seven nucleotides:

```
5′-GGATACGATCAGGATGCTTTTGTCTGTGCTCGGTA
```

Figure 6.8

Site-directed mutagenesis using a plasmid carrying deoxyuridylate nucleotides. (A) Annealing of an oligonucleotide to a single-stranded plasmid circle carrying some deoxyuridylate nucleotides (U) in place of deoxythymidylate nucleotides. A mismatch between oligonucleotide and plasmid is indicated by a '^' in the line segment. (B) Extension of the oligonucleotide primer *in vitro*, using a DNA polymerase such as Klenow enzyme. The dashed line indicates DNA synthesis of the deoxythymidylate-carrying daughter strand, using the parental strand as template. (C) Following transformation into *E. coli*, the parental strand (on the left) is disrupted by the host's uracil *N*-glycosylase, while the daughter strand (on the right), carrying the mutation in the gene of interest, is not sensitive to the uracil *N*-glycosylase.

This could be easily measured by gel electrophoresis and used as a diagnostic test for a single-base mutation.

Suppose now that the reaction contains not only ddATP, but also dATP, and that one out of every 100 adenylate nucleotides incorporated will have a dideoxyribose sugar on average. As a result there will be three products of synthesis:

5′-GGATACGATCAGGATGCT<u>TTGTCTGTA</u>

and

5′-GGATACGATCAGGATGCT<u>TTGTCTGTACTCGGTA</u>

will be present in small amounts, each terminating with a dideoxyribonucleotide, and the predominant product will represent primer extension to the end of the template

5′-GGATACGATCAGGATGCT<u>TTGTCTGTACTCGGTAGTC</u>

Since the probability of incorporation of a dideoxyribose sugar is low, the probability of chain termination is also low.

Nonetheless, we have gained some important information, since both the 27- and 34-nt products were seen among the results. We now know that there are T nucleotides in the template, lying exactly 9 and 16 nt downstream of the 3′ end of the primer. More generally, with ddATP added to the primer extension reaction it is possible to make a collection of all of the products that end with an adenylate nucleotide. If extension is performed on a single-stranded template that is a kilobase in length, the collection might have about 250 primer extension products ending with an A base. Suppose that there were four separate reactions being conducted, each based on the addition of a different dideoxyribonucleotide triphosphate substrate:

- reaction 'G': (<u>ddGTP</u> + dGTP), dATP, dTTP, and dCTP;
- reaction 'A': dGTP, (<u>ddATP</u> + dATP), dTTP, and dCTP;
- reaction 'T': dGTP, dATP, (<u>ddTTP</u> + dTTP), and dCTP;
- reaction 'C': dGTP, dATP, dTTP, and (<u>ddCTP</u> + dCTP).

The reaction labeled 'G' would generate all of the primer extension products ending with a guanylate nucleotide, and the 'A', 'T', and 'C' reactions would collect all products ending with adenylate, thymidylate, and cytidylate nucleotides, respectively (see *Figure 6.9A*).

An interesting point about these four collections of products is that no products in any two different reactions should have the same length.

Figure 6.9

DNA sequencing using chain-terminating nucleotides. (A) A DNA oligonucleotide (open rectangle) is annealed to a template (gray line) and extended in four reactions, each with a different ddNTP chain-terminating nucleotide. The reaction products are of various lengths representing different chain termination sites (octagon shapes). (B) If the products of the four reactions are ordered according to size, by gel electrophoresis, the sequence of chain-termination sites and hence the nucleotide sequence can be inferred (bottom). For illustration, the 'T' chain termination sites are marked with vertical dashed lines.

Suppose that the 'G' and the 'A' reactions could both generate a fragment of length 384 nt. Then there must be a single point on the template, exactly 384 nt from the 5' end of the primer, to which these two

(A)

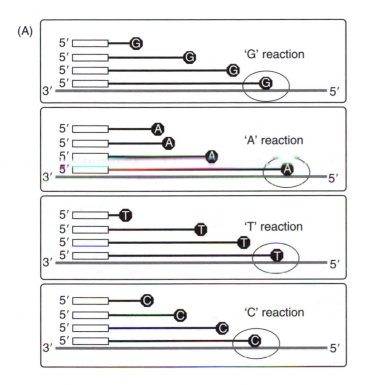

(B)

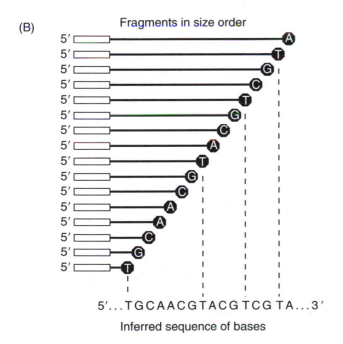

Fragments in size order

5'...T G C A A C G T A C G T C G T A...3'

Inferred sequence of bases

nucleotides can both base-pair. That isn't possible if the template is a single sequence, so the four reactions 'G', 'A', 'T', and 'C' should not have any reaction products that are the same size. Furthermore, if the four collections of DNA fragments were mixed together, then every possible extension length would be represented (see *Figure 6.9B*). Each reaction product is like a little tape measure that has been pulled out from the 5' end of the primer to the 3' chain termination event. When all four reactions are put together and the fragments separated by size, we can infer the order of nucleotides in the template (see bottom of *Figure 6.9B*).

In a DNA sequencing reaction, a primer of approximately 20 nt is annealed to a template DNA and extended with a DNA polymerase such as Klenow enzyme or T7 DNA polymerase. The steps in one simple protocol for sequencing are as follows.

i.   The template and primer are heated to 95–100°C so that they become denatured.
ii. The template and primer are cooled to a temperature at which the primer can anneal to the template. Depending on the length and composition of the primer, this temperature will probably be in the range of 55 to 65°C.
iii. The annealed template and primer are divided into four aliquots, with the respective nucleoside triphosphate mixtures described earlier as a 'G reaction', 'A reaction', 'T reaction', and 'C reaction', and the temperature is dropped to 37°C.
iv. A DNA polymerase is added, and the polymerization commences. In this protocol it is assumed that the DNA polymerase is not thermostable, so it cannot be added during the high temperature steps.
v.   A 'chase' solution is added that consists of the four regular deoxyribonucleoside triphosphates. This step lengthens DNA fragments that were stalled during synthesis but not terminated by incorporation of a dideoxyribonucleotide, and reduces the background of bands.
vi. The products of the reaction are purified and separated by polyacrylamide gel electrophoresis, and the lengths of the fragments are assessed. Each nucleotide in the sequence is represented by the unique length of a terminated fragment, so a size-ordered collection of these fragments allows the sequence to be inferred (*Figure 6.9B*).

Sequencing primer

The oligonucleotide primer selected for DNA sequencing must be very specific because if the primer anneals to more than one site in the template DNA, the extension products from the two sites will not tell a consistent story about the DNA sequence. Furthermore, the primer must be of high quality, and not be a mixture of lengths. This is a problem related to the coupling efficiency during synthesis of the primer.

Oligonucleotides are synthesized chemically, built up from the 3' end to the 5' end by addition of phosphoramidites. Incomplete coupling means that the primer is shortened at its 5' end, which can lead to shadow bands if the primer is used for DNA sequencing. Remembering that the 5' end of the primer must serve as the common starting point for the little 'tape measure', we see how an inconsistent starting point could become a problem. If a primer is a mixture of 21-nt and 20-nt oligonucleotides, with the variability being at the 5′ end, and it is extended to a common chain termination event that is 30 nt beyond the 3′ end of the primer, then the products that were terminated at that specific nucleotide will be a mixture of 51-nt and 50-nt lengths. This problem would be perpetuated at every point during primer extension, like a bothersome echo, and the true sequence would be more difficult to discern.

A common strategy for DNA sequencing is to clone a sequence into a plasmid or bacteriophage vector, and to use primers that anneal to the vector sequences immediately adjacent to the cloned DNA. As discussed in Chapter 4, vectors often have 'universal' M13 forward and reverse primers, the uses of which date back to the filamentous bacteriophage vectors once commonly used for preparation of template. There are also common primers that correspond to bacteriophage RNA polymerase promoter sequences, for example T7, T3, and SP6, as discussed in Chapter 4 and Section 6.7. In this case the promoter sequence is not being used as a transcriptional promoter, but as a target for oligonucleotide binding. When an unknown DNA sequence is cloned into a plasmid vector, we can start pulling out the little 'tape measure' from these known sites and extend them into sequences that we do not yet know. There is nothing special about universal primer sequences, other than the fact that we know what they are; many other sequences would do just as nicely in the part, and it is common for a researcher to design and purchase custom-made oligonucleotide primers for a sequencing project.

The most important design principle is that the 3′ end of the oligonucleotide is the end that is extended by the DNA polymerase, and 5′ to 3′ is therefore the direction of DNA sequencing. A double-stranded DNA may be sequenced in either direction, but these different directions involve using primers that anneal to different template strands. Sequencing in either direction, while keeping synthesis 5′ to 3′, may remind us that we can drive a car in either direction between two cities, but in doing so we must drive on the correct side of the road. The sequencing primer is like an entrance ramp to the highway, and the researcher designing the experiment needs to check to make sure it points in the right direction.

A sequencing primer must anneal to a single place in a template. The tolerance for mismatches during annealing of a primer to a template has already been discussed (in Section 6.3), and the same principles hold for DNA sequencing primers. The sequence composition and complexity of a prospective sequencing primer must also be considered. For example, a 21-nt sequence such as 5′ ATTATGACAATAATACATGAT may be unique in a template but inadvisable as a sequencing primer because it is very AT-rich and will have a low annealing temperature. On the other hand, the primer 5′ CAAAAACCCCCGTGTGTGTGT has a balanced single-nucleotide composition, but may also perform poorly because it has low complexity at the dinucleotide level; the GT repeat may allow initiation of DNA synthesis from multiple points in a template. A primer sequence such as 5′ TGCCGAATACAGACTAGTCTG has a well-balanced nucleotide composition and looks reasonably complex; however, the last 12 nt at the 3′ end form a dyad symmetry that could cause the primer to base-pair to itself, interfering with annealing to the template:

```
5' -TGCCGAATACAGACTAGTCTG
         GTCTGATCAGACATAAGCCGT-5'
```

Finally, it should be remembered that the first few nucleotides added to the 3' end of a primer during sequencing may not be easily resolved, and the ability of a researcher to read sequence close to the primer depends on the method of separation of fragments and the purification of sequencing products prior to gel loading. The very first nucleotides that are desired to be readable in a sequence should lie 20 to 30 nucleotides downstream of the 3' end of the primer, and the most readable sequence may lie further away still. Just as a photographer may back-up with a camera to frame a picture well, so too must a researcher sometimes back-up with a primer to get the best information from a template.

Sequencing strategies

There are several strategies for sequencing an unknown fragment of DNA. One is to make successive deletions in a fragment, eliminating DNA that has already been sequenced and bringing new DNA closer to a universal primer binding site. A more cost-effective approach may be to design new primers that match sequence that has been collected, so that each successive sequencing experiment confirms some of the previous sequence and reaches a little further into the unknown regions. Provided that the new primer binding site is unique in the template, this can be a rapid method of sequencing a cloned DNA.

In a very large sequencing project, an unknown DNA may be initially partitioned into artificial chromosomes of several hundred kbp, and each of these may be broken randomly into smaller overlapping pieces for sequencing. Such random collections are often called shotgun libraries, because like the broad firing pattern of a real shotgun, the libraries are not taking close aim at any one sequence. The sequencing of random elements in such a collection may not seem efficient, but progress is made from many points simultaneously using universal primers, and information from multiple overlapping clones may be assembled into contiguous sequence files called contigs.

In any sequencing strategy, it is necessary to repeat the sequencing several times, and to sequence both strands of DNA for comparison. This redundancy helps to reduce the number of errors in a sequence, because manual and machine calling of nucleotides is imperfect.

Electrophoresis of sequencing reaction products

It is challenging to separate the products of sequencing reactions by electrophoresis because the migration of DNA fragments in a gel depends on both charge and conformation. The charge on a single-stranded DNA molecule is a straightforward matter; charge is directly proportional to DNA length, because there is one negatively charged phosphate group per nucleotide in the phosphodiester backbone. On the other hand, a single-stranded DNA molecule can bend and form hydrogen-bonded stems and loops with itself, causing a mixture of odd shapes that will migrate differently through a gel matrix.

This is a problem, because in DNA sequencing analysis we need to be able to determine the size order of a collection of DNA fragments (see *Figure 6.9B*). For example, we must be able to resolve a 700-nt DNA from a 699-nt DNA on a gel if we want to sequence 700 nucleotides. Raising the temperature of the running gel and adding denaturing substances such as urea to the gel will disrupt hydrogen bonding between nucleotides and allow the gel separation to be based on DNA size (length) instead of shape. DNA sequencing slab gels are made with cross-linked polyacrylamide containing 6 to 8 M urea, and are run at a temperature of 50°C to 55°C. Capillary gel electrophoresis systems use a different separation matrix, usually one that is not cross-linked, but they are also run in the presence of urea and at high temperature.

Differences in migration are more pronounced between a 69- and 70-nt fragment than they are between a 699- and 700-nt fragment, and as a consequence the vertical spacing between bands is greater at the bottom of the gel than the top (the migration of DNA on a gel was discussed in Chapters 4 and 5). One strategy for making the spacing more even is to pour a slab gel with a gradient of polyacrylamide. The gradient gel has a higher percentage of acrylamide at the bottom than at the top, and faster moving DNA fragments are slowed down as they run further into the gel. This gradient allows the gel to be run for a longer time and permits more data to be collected from a single sequencing reaction.

The heating of a gel to the temperatures used in electrophoresis may depend on a combination of Joule heating and external temperature control. Joule heating depends on electrical power dissipation, the current and resistance. For example, an electrophoresis buffer with a higher ionic strength will provide for more Joule heating, as will an increase in the electrophoresis voltage. Many gel systems are pre-run so that the gel reaches an appropriate running temperature prior to sample loading. External temperature controls may include applying a uniformly heated plate to a slab gel to even-out temperature differences across the face of the gel, or blowing air around a capillary electrophoresis system to provide a uniform running temperature.

Detection using radioisotopes

Large sequencing projects are usually performed by automated methods that involve the use of fluorescently tagged products; however, radioactive labeling of sequencing products is still performed by many laboratories. For example, RNA can be sequenced directly during primer extension using a reverse transcriptase and radiolabeled nucleotide substrates. In isotopic sequencing reactions, a small amount of radioactive label is usually added as a tracer to make the products visible by autoradiography. This can be accomplished in several ways, for example 5′ end-labeling of the primer with polynucleotide kinase, or direct incorporation of a radiolabel during polymerization. End-labeling involves the transfer of the terminal (γ) phosphate from a [^{32}P]- or [^{33}P]-ATP molecule to the 5′ end of a primer.

If the primer is not end-labeled, radioactivity can be directly incorporated into the growing DNA chain by adding a trace amount of α-[^{32}P]-dCTP, or another one of the deoxyribonucleoside triphosphate substrates labeled on the alpha phosphate. In this case the specific activity does not need to be as high as for random hexamer primer labeling, where a probe is being made for hybridization (as discussed in Section 6.8). The specific activity of the labeled deoxyribonucleoside triphosphate can be reduced to the range of 100 to 800 Ci mmol^{-1} by mixing labeled substrate with unlabeled substrate, or by purchasing substrate at a reduced specific activity from the radioisotope supplier. One advantage of end-labeling compared to direct incorporation methods is that the radioactive products represent extensions of a specific primer. When the radioactivity is incorporated as a nucleoside triphosphate during extension, then there are opportunities for nonspecific incorporation resulting from accidental priming events.

It is common in isotopic DNA sequencing to use a ^{33}P- or ^{35}S-labeled nucleoside triphosphate, instead of ^{32}P, because these two isotopes produce a less energetic β emission during decay than ^{32}P and a sharper band in autoradiography, albeit with a longer film exposure time. When ^{35}S is used in nucleoside triphosphates, it substitutes for one of the oxygen atoms to make a phosphorothioate. ^{33}P and ^{35}S isotopes have longer half-lives than ^{32}P, which is an advantage if the labeled substrate is going to be used over a period of several weeks. The respective half lives of these isotopes are: ^{32}P, 14.3 days; ^{33}P, 25.4 days; and ^{35}S, 87.4 days. The radioactive decay of these isotopes over time is shown in *Figure 6.10*, and can be predicted by the formula:

$$A = A_0 e^{-kt}$$

where A_0 is the starting amount of a radioactive substance, A is the amount of radioactive material after an elapsed time t, and k is a constant that depends on the units of t and the radioisotope (see inset in *Figure 6.10*). For example, k = 0.04847 if the radioisotope is ^{32}P and time is measured in days. If we have a sample of 100 µCi of ^{32}P on 30 April, how much remains on 14 May? We can set A_0 = 100 µCi and t = 14 days and calculate that:

$$A = 100e^{-(0.04847)(14)} = 50.7 \text{ µCi}$$

The passage of 14 days is almost one half-life, so a little more than half of the original material remains.

Film autoradiography of sequencing gels

When DNA sequencing reaction products are to be detected by radioactivity, they are first separated on a thin polyacrylamide slab gel poured between two glass plates. Following electrophoresis, the plates are separated and the gel is transferred to a piece of adsorbent paper. The gel is then dried in a heated, vacuum gel dryer so that it is flat and will not stick to a piece of X-ray or autoradiography film. Depending on the isotope being used and the method of labeling, the exposure of the gel to film may be conducted in a film cassette for a period of 12 hours to 1 week. Several differently timed exposures may be necessary to extract all of the information from the experiment. The gel and film are usually placed in a −70°C freezer during exposure, to increase the longevity of

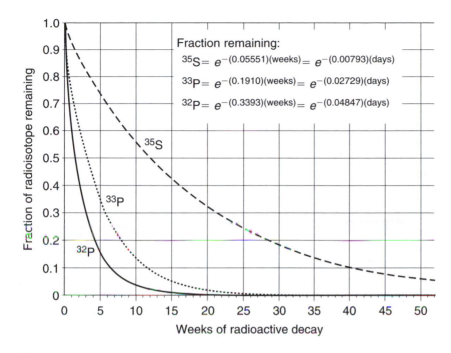

Figure 6.10

The radioactive decay of ^{35}S, ^{33}P, and, ^{32}P as a function of time. The fraction of remaining radioisotope after a time period t is predicted by the expression e^{-kt}, where k is a constant related to the isotope and the units of time. The inset formulae provide constants k for the three isotopes, with time measured in weeks and days.

the activated silver granules in the film and increase the intensity of the exposure. The low temperature causes a chemical effect in the film but does not alter the rate of radioactive decay.

The smaller products of the reaction, the DNA fragments that terminated closer to the primer, will migrate more quickly in the gel and will be located at the bottom of the gel when electrophoresis is concluded. The larger products, the ones extended further from the primer, are closer to the top of the gel. The lengthening of the DNA primer in the 5′ to 3′ direction is represented in the order of DNA fragments, from the bottom to the top of the gel. The products of the four separate sequencing reactions are run in adjacent lanes so that fragments ending in G, A, T, and C dideoxyribonucleotides can be ordered on the gel from bottom to top (see *Figure 6.11*). The sequencing ladder is the pattern of bands in these four lanes, and to read the ladder the researcher starts at a band at the bottom of the gel and looks for the next higher band in any of the four lanes, representing an extension of one nucleotide (see *Figure 6.11B*). The search for the next higher band is repeated again and again, climbing the ladder until the band order can no longer be discerned. At each 'step' in the ladder, we learn whether that fragment is terminated with a ddG, ddA, ddT, or ddC nucleotide, and the sequence of stop points tells us the 5′ to 3′ sequence of bases.

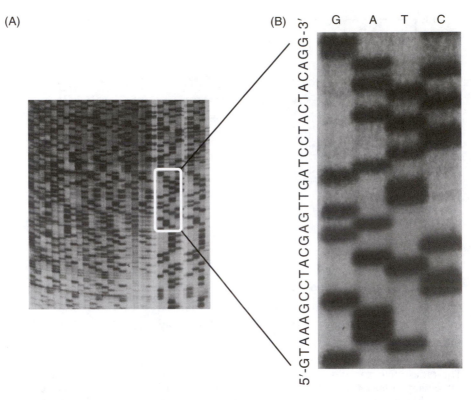

Figure 6.11

Autoradiogram of a DNA sequencing gel. The sequencing reaction products were labeled with ^{35}S. (A) The entire gel is shown, with multiple sequencing experiments. (B) The white rectangular segment of the entire gel is enlarged, with the 'G', 'A', 'T', and 'C' reaction lanes labeled. The DNA sequence that was inferred from the order of bands is shown on the left. Autoradiogram data: Courtesy of Dr. Aida Metzenberg.

There is a common artifact on a sequencing gel that is called compression, and it is due to secondary structure formation and neighbor effects between the nucleotide bases at the 3' end of the single-stranded DNA fragments. Before DNA samples are loaded on a sequencing gel, they are denatured in formamide at a high temperature to make the DNA single-stranded. Running the gel at 50°C, and with 7 M urea also tends to prevent most secondary structure; however, the problem may not be entirely resolved by these measures. For example, a researcher may find that stretches of cytidylate nucleotides, for example CC, CCC, and CCCC are compressed vertically on the gel. Four cytidylate nucleotides in a row may not look like four evenly spaced bands, and could be easily misread as three. If the other strand is also sequenced for comparison, it may clearly show four complementary guanylate nucleotides at that position, and that is one reason that redundancy in sequencing is important.

Compression artifacts can sometimes be reduced by the addition of formamide to the gel to increase the extent of denaturation of the single-stranded DNA. Another approach is to use analogues of dGTP in the sequencing reaction, such as dITP or 7-deaza-dGTP. These base analogues form fewer hydrogen bonds with cytidylate nucleotides, and when they are incorporated into a sequencing fragment, there is less opportunity for secondary structure formation.

Detection using fluorescent dyes

Most DNA sequencing, and certainly all high-throughput sequencing projects, are now performed using chemically labeled primers or substrates. One very common approach is to label DNA primers or dideoxyribonucleotides with four different fluorescent dyes to identify terminations at ddG, ddA, ddT, and ddC. The fluorescent dyes are identified by their emission patterns, after excitation with a laser. For example, the BigDye™-labeling system (Applied Biosystems Inc.) uses the dyes FAM-dR110, FAM-dR6G, FAM-dTAMRA, and FAM-dROX to represent ddG, ddA, ddT, and ddC terminations, respectively.

If the primer is dye-labeled then four separate reactions are run; each reaction contains a single labeled primer and only one corresponding type of dideoxyribonucleotide triphosphate. For example, the ddGTP-containing reaction would contain a primer labeled with FAM-dR110, and the ddATP-containing reaction would contain a primer labeled with FAM-dR6G. Upon completion of the reactions the contents of the four separate tubes may be combined before loading on a gel, because the specific dye-label at the 5′ end of the sequencing product uniquely identifies the dideoxyribonucleotide at the 3′ end.

When it is the dideoxyribonucleotide triphosphate that is dye-labeled, then the sequencing reaction can be conducted in a single tube. There is no longer any need to conduct four separate reactions because the termination with ddG, ddA, ddT, or ddC and the attachment of a specific dye to the sequence are simultaneous; the dideoxyribonucleotides carry their own 'colors' with them, and mark the type of chain termination that has occurred.

The products of the sequencing reaction are run in a single lane of a slab gel, or in a single glass capillary, and a laser beam is used to excite the dyes as they migrate through a detection window. The sequence of the bands as they emerge from the end of the gel is used to determine the order of steps in the sequencing ladder; again, the shorter fragments migrate more quickly than the longer fragments, and emerge from the gel first. The light emission from each dye-labeled fragment is separated by wavelength, either by a filter set or a prism, and mapped onto a set of photodiodes or a charge coupled device (CCD) chip. The fluorescent dyes have spectral overlap, and a single sweep of the laser may excite multiple dyes simultaneously, and consequently the raw emission data must be manipulated mathematically to permit estimation of the amount of each of the four dyes that are present. When slab gels are being used, a fifth nonspecific dye is sometimes added to the sample to help identify the track of a lane as the gel is running.

The 'raw' emission data vary considerably in background and signal strength along the sequencing track, and background subtraction and signal amplification are necessary components of analysis. There are also

(A)

(B)

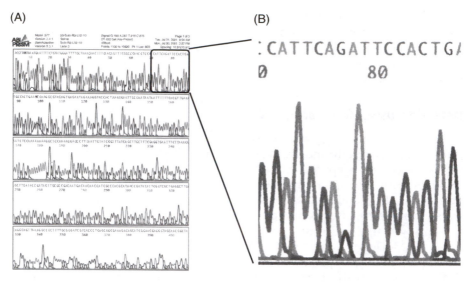

Figure 6.12

Chromatogram generated by an automated DNA sequencer. (A) The entire first page of the chromatogram is shown. (B) The black rectangular segment of the chromatogram is enlarged, with the machine-read nucleotide sequence over the chromatographic peaks. The original chromatogram is printed in four colors, with each nucleotide base represented by a different peak color. Chromatogram data: Courtesy of Dr. Saima Zubair.

corrections that must be made for differences in fragment mobility, depending on which of the four dyes is affixed. After these mathematical manipulations are completed, the software of an automated DNA sequencer generates a chromatogram or 'electropherogram', showing the relative levels of each dideoxyribonucleotide at any point in the sequencing run (see *Figure 6.12*). The software also searches for peaks that are spaced appropriately in the data, and 'calls' the base order by printing a 'G', 'A', 'T', or 'C' over the peaks. If a base cannot be called because the results are ambiguous, the software may call it as an 'N' nucleotide, where 'N' represents any of the four bases. Reading one of these electropherograms from left to right shows the sequence of chain terminations during the electrophoresis run, which in turn represents the 5' to 3' order of bases in the DNA sequence (see *Figure 6.12B*). A high-quality sequencing reaction will have very few 'N' markings or miscalled bases on the electropherogram, whereas lower quality results may need to be reviewed carefully and corrected by eye.

Loading sequencing gels

DNA sequencing products are denatured with heat and denaturing chemicals such as formamide before they are loaded on a gel. If the sequencing system is based on gel-filled capillaries, the sample may be injected as a 'plug' by an automated loader. If a polyacrylamide slab gel is being loaded by hand, then a comb is used to separate the lanes of the gel. The urea in the gel may gradually diffuse into the well and form a dense pad that prevents loading of a sample; this problem requires that the well be flushed with running buffer immediately before loading.

There are combs that can be used to form wells in the polyacrylamide for sample injection, but it is difficult to get even polymerization and uniform wells by this approach. A very common method of separating lanes is through use of a sharkstooth comb that is applied to the flat top of the gel after polymerization. The sharkstooth comb is essentially a spacer that has room for sample injection, and it is gently pressed into the gel surface to isolate each lane. The gel surface is flat and recessed by a few millimeters below the glass plate, and the gaps between the teeth are used for sample injection. If the sharkstooth comb is not pressed firmly enough into the gel, the samples in different lanes may run together and cross-contaminate each other. If the sharkstooth comb is pressed too firmly into the gel it may cause the gel surface to pucker and form a 'pillow' shape in each lane, or it may damage the surface of the gel. The sharkstooth comb is inserted after polymerization of the gel and before loading, and may be removed only after the sample has completely entered the gel.

Even with these precautions it is possible for samples to gradually move between wells as they settle into the spaces that are left for them. Researchers often load a few samples at a time, turning on the power briefly to run the samples into the gel between loading steps. This method limits the amount of time for sample movement between wells, but obviously leads to lanes that are not synchronized with each other. If this staggered loading is done with a nonautomated sequencing gel, the kind that is used with radioactively labeled materials, then it is important that each set of G, A, T, and C lanes be loaded simultaneously so that the migration of bands in the four lanes can be compared. In nonautomated sequencing methods it is common for a researcher to load two sets of the same four G, A, T, and C lanes at different times. The first set of four reactions is run for a longer period of time than the second, and shows better separation of the larger DNA fragments. In order to reveal this separation the shorter fragments are run off the end of the gel and lost, but they can be seen in the second set of lanes that were loaded later and run for a shorter period of time. In reading such a gel, a researcher would start by collecting data from the second set of loaded lanes and then switch to the overlapping information from the first set of loaded lanes to extend the readable data.

If a staggered loading method is used with an automated sequencer, there may be some interference between adjacent lanes if the tracks are run very close together. For example, if there is a substantial amount of background 'garbage' fluorescence emerging from the gel at the beginning of data collection, it may obliterate data signals in adjacent lanes that have been run for a longer time.

Problems

The solutions to these problems are provided at the end of the chapter.

(v) The sequence below is part of a cDNA for the *Mus musculus* IgG Fc receptor IIb (GenBank® locus NM_010187):

```
  1  tgtgcctgca  gctgactcgc  tccagagctg  atgggaatcc  tgccgttcct
 51  actgatcccc  atggagagca  actggactgt  ccatgtgttc  tcacggactt
101  tgtgccatat  gctactgtgg  acagccgtgc  taaatcttgc  tgctgggact
151  catgatcttc  caaaggctgt  ggtcaaactc  gagcccccgt  ggatccaggt
```

Suppose that you obtained the 20-nt sequence

5'-GCTGACTCGCTCCAGAGCTG-3'

(from nt 11–30 of the cDNA) as an oligonucleotide sequencing primer, and prepared a set of 'G', 'A', 'T', and 'C' reactions carrying the respective dideoxyribonucleotide chain terminators. What would be the lengths of the first 10 chain-terminated extension products in the 'G' reaction, including the primer?

(vi) Suppose that you have obtained two additional sequencing primers that you could use in the experiment described in the previous problem.
A. You have a 21-nt oligonucleotide that is extended by one nt at the 3' end:

5'-GCTGACTCGCTCCAGAGCTGA-3'

If you used this oligonucleotide as a sequencing primer, would the lengths of the first 10 chain-terminated extension products in the 'G' reaction be different than they were in the previous problem?
B. You have a 21-nt oligonucleotide that is extended by one nt at the 5' end:

5'-AGCTGACTCGCTCCAGAGCTG-3'

If you used this oligonucleotide as a sequencing primer, would the lengths of the first 10 chain-terminated extension products in the 'G' reaction be different than they were in the previous problem?

(vii) Design a 20-nt oligonucleotide sequencing primer that you could use to obtain sequence information from the complementary strand of the cDNA described in Problem (v).
A. What is the sequence of this new primer, written 5' to 3'?
B. Suppose that you used this primer in DNA sequencing and that the template DNA was heterogeneous. Half of the DNA template molecules have an insertion of one extra 'A' nucleotide at nt 175-177, to make a string of four instead of three 'A' nucleotides. Ignoring the fact that it is difficult to obtain sequence information near to the end of a primer, draw a diagram of the sequencing ladder, as it would appear for the first 10 extension products.

(viii) Suppose that you perform a sequencing reaction using ^{33}P-labeled nucleotides. The half-life of ^{33}P is 25.4 days. You expose the sequencing gel to X-ray film to obtain an autoradiogram, starting at noon on April 30 and ending at 4:00 PM on May 4. You immediately start exposing the autoradiogram to a second fresh sheet of film at 4:00 PM, while you develop the first sheet of film. After consideration of the data on the first sheet of film, you decide that you want the second film to be exposed to exactly five times more radioactive disintegrations than the first. On what date and at what time should you stop the second exposure?

6.5 Reverse transcriptases

A polymerase with reverse transcriptase activity can act as an RNA-directed DNA polymerase, synthesizing DNA while using an RNA sequence as a template (7). Several important applications of reverse transcriptases include primer extension mapping of RNA structure, and synthesis of complementary DNA (cDNA) for cloning or PCR amplification. The components of a reverse transcriptase reaction include an RNA template, a primer that can be extended at its 3′ end, and the deoxyribonucleotide substrates dGTP, dATP, dTTP, and dCTP.

There are many commercially available reverse transcriptases with slightly different properties and optimal conditions. There are even some enzymes that are widely known as DNA polymerases but have an ability to also act as a reverse transcriptase. For example, *Taq* DNA polymerase is a thermostable enzyme used in PCR, but it also has a reverse transcriptase activity.

cDNA synthesis

In a cDNA synthesis reaction, an RNA is annealed to a complementary primer, usually a DNA oligonucleotide, which is then extended at its 3′ end by a reverse transcriptase. One common approach for preparing cDNA from polyadenylated RNA is to use an oligo(dT) primer of 12 to 15 nt in length, because the deoxythymidylate nucleotides in the DNA primer base-pair to the poly(A) at the 3′ end of the RNA. Other methods involve using a nonspecific primer, such as random hexamers, or a specific primer that may match only a single species of RNA. If the reverse transcriptase extends a primer from the 3′ end of the RNA to the natural 5′ end of the RNA, then we say the cDNA is 'full length'.

If the RNA template is partially degraded, then the reverse transcriptase will not be able to extend the cDNA beyond the break in the template and the cDNA will not be full length. cDNA synthesis is usually performed in the presence of RNase inhibitors such as RNasin® (Promega Corp.), and other reagents that are kept RNase-free (a lab practice that is discussed in Chapter 2). Disposable plasticware is used in all manipulations of RNA, and a researcher will wear disposable gloves to minimize exposure of the sample to environmental RNases.

RNA secondary structure is a serious impediment to reverse transcriptases, and a typical method for cDNA synthesis will involve a step in which the RNA template is heat denatured at 68°C to 70°C for 5 min prior to using it in a reaction. The high temperature destabilizes hydrogen-bonding in stem structures and other intramolecular interactions that might interfere with the polymerase. The usual temperature for using reverse transcriptases in a reaction is 37°C to 42°C, but when a researcher is attempting to make a cDNA of an RNA that may have interfering secondary structures, a temperature slightly higher than 42°C may be selected. AMV and MMLV (Moloney murine leukemia virus) reverse transcriptases have slightly different temperature optima, and a researcher should carefully consult the manufacturer's instructions for use. Thermostable enzymes such as *Tth* reverse transcriptase are available, and these might be useful for cDNA synthesis on templates that form extensive secondary structures.

DNA can serve as a template for reverse transcriptases as readily as can RNA, so a researcher trying to make a definitive statement about a putative RNA in a sample must be concerned about the possibility that a product of synthesis came from a DNA contaminant. The RNA sample can be treated with an RNase-free DNase enzyme prior to use as a template in reverse transcription; if the template is truly an RNA then the cDNA product should still be generated after this enzymatic treatment.

Conversely, if the RNA in a template sample is specifically destroyed, then a cDNA should not be generated unless it is being made from a contaminating DNA. RNA can be specifically hydrolyzed by treatment of the sample with 0.2 N NaOH at room temperature for an hour, followed by neutralization with a 0.1 M Tris-Cl, pH 7.5 buffer and an equivalent amount of 0.1 N HCl, and then purification to remove the excess salt. The contaminating DNA will be denatured into single strands by the alkali treatment, but will be otherwise undamaged. Alternatively, RNase A could be added to the RNA sample to destroy the RNA template, but the researcher would be wise to use heat-inactivated RNase, in which the DNase had been destroyed, and perform the RNase reaction in the presence of EDTA to inhibit any residual DNase contaminant.

When cDNA has been prepared from a eukaryotic mRNA, the absence of intervening sequences in the product might be taken as encouraging evidence that the template was a spliced RNA and not a genomic DNA contaminant. However, a pseudogene DNA sequence may lack intron sequences, and a nuclear precursor mRNA may still contain introns, so the presence or absence of introns in a cDNA is not a conclusive proof of the origin of the template.

A final word of caution is in order, because it is not widely appreciated that some DNA polymerases such as *Taq* can act as reverse transcriptases. When a researcher performs a PCR reaction with *Taq* DNA polymerase, it is possible that the product originates from an initial RNA template. The finding of an amplified DNA product in PCR cannot be taken as evidence that the original template sequence was DNA; it could have been due to a contaminating RNA in the DNA sample.

RT-PCR

Reverse transcription, followed by a polymerase chain reaction (RT-PCR) can be used to generate a specific cDNA from a collection of RNA molecules (see *Figure 6.13*). With some knowledge of the sequence of a transcript, a specific oligonucleotide can be used that will base-pair at the 3′ end of the RNA and be extended with the enzyme reverse transcriptase towards the 5′ end of the RNA. An RNA/DNA heteroduplex, or 'first-strand' is the product of the reverse transcription reaction and this can be introduced as a template into a polymerase chain reaction, discussed in Chapter 7. If the reactions are being run sequentially in two different buffers, it is usually helpful to transfer only 1–10% of the reverse transcription reaction to the second reaction, to minimize inhibition of the polymerase chain reaction. However, there are commercially available systems that are

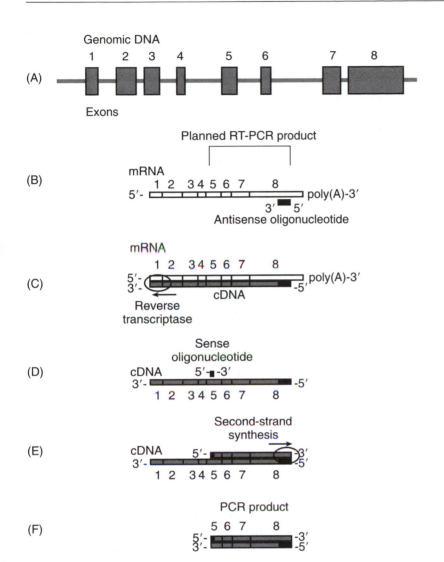

Figure 6.13

Reverse transcription followed by PCR (RT-PCR). (A) The genomic structure of an imaginary gene. Exons 1–8 are indicated by rectangular areas. (B) The spliced mRNA, with exons 1–8 forming a contiguous sequence. The planned region for RT-PCR is indicated by the bracket (top) and a 3′ 'antisense' oligonucleotide primer (black rectangle) is shown annealed in the exon 8 region. (C) Extension of the antisense oligonucleotide primer using reverse transcriptase. The first-strand cDNA product is indicated by the gray rectangles. (D) A 5′ 'sense' oligonucleotide primer is shown annealed to the cDNA in exon 5. (E) Extension of the sense oligonucleotide primer using a DNA polymerase to generate a second-strand cDNA product. This may be conducted during the first PCR cycle. (F) Continuation of the PCR using both the exon 5 and exon 8 oligonucleotide primers leads to an exon 5–8 product.

optimized so that RT-PCR can be completed in a single tube and reaction mixture. The polymerase chain reaction requires two oligonucleotide primers, one of which may be the oligonucleotide that was originally used to prime the cDNA synthesis, and the second of which will anneal to the first strand of the cDNA.

One application for RT-PCR is to amplify a specific sequence from a collection of transcripts, for a specific cloning project. For example, a researcher might decide to clone a sequence encoding a specific domain from a human protein, and consult the known genomic sequence of the gene (see *Figure 6.13A*). Let us suppose that the desired sequence starts in exon 5 and ends in exon 8 of the mRNA (see *Figure 6.13B*). The researcher might plan to capture the cDNA by RT-PCR, and would design a first oligonucleotide that anneals to the mRNA in exon 8 (see *Figure 6.13B*). In a reverse transcription reaction using mRNA from an appropriate human tissue or cell source, this oligonucleotide would be extended towards the beginning of the transcript (see *Figure 6.13C*). If the first-strand cDNA were full length, its 3′ end would lie in exon 1 and its 5′ end would be the same as the 5′ end of the oligonucleotide. For the PCR part of the reaction, the researcher might use this oligonucleotide in conjunction with a second oligonucleotide that is designed to anneal to the exon 5 sequence in the first-strand cDNA (see *Figure 6.13D*). The first PCR cycle would generate the second strand of the cDNA, which would have its 5′ end in exon 5 and its 3′ end in exon 8 (see *Figure 6.13E*). Continued cycling would amplify the exon 5 to exon 8 fragment as a double-stranded DNA, suitable for cloning (see *Figure 6.13F*).

If only an internal sequence is known for a gene, a researcher might use a specific primer in that region to direct the first-strand cDNA synthesis and add a homopolymer tail such as oligo(dC) to the 3′ end of the cDNA using the enzyme terminal deoxynucleotidyl transferase (discussed in Section 6.6). This oligo(dC) tail serves as an annealing site for an oligo(dG) primer that initiates the second strand of cDNA synthesis, and serves as a primer in continued PCR cycling. This technique is termed rapid amplification of cDNA ends (RACE), and is useful for capturing the 5′ end of a gene when only a partial sequence is known. A related technique involves the addition of a specific oligonucleotide sequence to the 3′ end of the first strand of cDNA, using the enzyme T4 RNA ligase. This specific added sequence may serve as a target for the annealing of a complementary oligonucleotide that will prime the second-strand cDNA.

A second application for RT-PCR is the relative quantification of a specific RNA in a population of transcripts. A researcher using this technique may be asking a question about gene regulation; for example, what is the difference in the steady state level of a specific mRNA when a cell line is grown under two different conditions? If the RNA samples are too small for detection by Northern blot analysis, then RT-PCR might be considered as an alternative. The RNA samples would be run in parallel RT-PCR reactions, and the rate of generation of product would be taken as an indirect measure of the original amount of RNA template. If one sample began to show a product of cDNA synthesis in cycle 22 of PCR, and a second sample consistently showed the same amount of product in cycle 25 of PCR, then it would be inferred that the first sample had significantly more mRNA template than the second.

There are some significant challenges to making a quantitative argument about RNA levels by RT-PCR, not the least of which is that the amplified products of synthesis are many steps removed from the original templates. If the reverse transcription reaction is more efficiently completed with one RNA sample than another, then the relative amounts of first-strand cDNA synthesized may not adequately represent the original amounts of RNA template. The PCR would also have to be of similar efficiency between the two cDNA templates, if quantitative conclusions are to be drawn. Many researchers use a 'real time PCR' method, discussed in Chapter 7, to keep track of the reaction efficiency and improve their confidence in the quantitative data.

cDNA cloning

A cDNA can be preserved in a cloning vector, and this is a common technique used when preparing a cDNA library, which is a representation of expressed transcripts from a cell sample, as discussed in Chapter 4. For example, cDNA might be prepared from polyadenylated RNA of a eukaryotic cell that had been first purified by binding to an oligo(dT) cellulose column. The first-strand cDNA could be extended from either an oligo(dT) primer or a random hexamer primer. Alternatively, a researcher might use a sequence-specific primer so that only one species of RNA in the population is primed for cDNA synthesis.

There are several strategies for priming synthesis of the second strand of the cDNA, the sense strand that replaces the RNA in the heteroduplex. One common method is to use the enzyme RNase H to generate random nicks in the RNA of the RNA/DNA heteroduplex, and extend these nicks using an polymerase with a 5' to 3' exonuclease activity, such as *E. coli* DNA polymerase I. The nicks are all translated toward the 3' end of the sense strand. Many reverse transcriptases carry an inherent RNase H activity, and a researcher may supplement this by adding additional RNase H at the start of second-strand synthesis. The nucleotides at the 5' end of the sense strand cannot be replaced with DNA using this method, because there is no primer that can generate them, so several nucleotides may be lost as the 5' end is made blunt by the polymerase.

An alternative method involves the addition of a homopolymer tail or a specific oligonucleotide sequence, as discussed earlier for the RACE method, and the use of a complementary homopolymer or specific primer for initiation of second-strand synthesis. An advantage of this approach is that the sequence at the very 5' end of the RNA may be recovered in the cDNA. If a specific oligonucleotide is added to the 3' end of a first-strand cDNA using T4 RNA ligase, the sequence may carry a convenient restriction endonuclease recognition sequence for cloning (as discussed in Chapter 5), or generate a cleverly designed blunt end that can later be converted to a 5' overhanging end using Klenow enzyme and a limited number of dNTP substrates (as discussed in Chapter 7).

A second strand may also be self-primed by formation of a hairpin loop at the 3' end of the cDNA first-strand. This short hairpin can form naturally as a result of the action of reverse transcriptase, and if the RNA strand is destroyed by alkali or separated from the DNA by heat denaturation, the hairpin-loop primer may be extended to generate a second strand. It may be of interest to note that the first- and second-strand cDNA are covalently connected by a hairpin structure, but it does present an immediate problem for cloning: how can we clone a linear DNA that has only one end? The answer is that the hairpin end has to be cleaved with an endonuclease such as S1 nuclease that digests single-stranded DNA. The hairpin has a small single-stranded loop, and cleavage with S1 will create a blunt end that can be used in ligation.

Once a double-stranded cDNA is prepared, a researcher may use a variety of techniques to ligate it into a cloning vector. It is difficult to obtain a high efficiency of ligation with blunt ends, and if the purpose is to generate a large cDNA library then it may be useful to add linkers or adaptors to the cDNA ends to generate a cohesive end.

For example, the pair of oligonucleotides 5'-AATTCGCC-3' and 5'-GGCG-3' can anneal to form an adaptor; a partially double-stranded sequence with a four nucleotide 5'-AATT overhanging end:

```
5'-AATTCGCC-3'
3'-    GCGG-5'
```

The blunt end of this adaptor can be ligated to the blunt ends of cDNA, transferring a 5'-AATT overhanging end to each cDNA end:

```
5'-AATTCGCCNNNNNNNNNNNNNNNNNNNNNNNNNNNNNNNNNNNNNNGGCG    -3'
3'-    GCGGNNNNNNNNNNNNNNNNNNNNNNNNNNNNNNNNNNNNNNNCCGCTTAA-5'
```

In this scenario, only the 5'-GGCG-3' oligonucleotide is phosphorylated at its 5' end, and the 5'-AATTCGCC-3' oligonucleotide has a 5' hydroxyl group. This organization of phosphate groups allows the blunt end of the cDNA to be ligated to the blunt end of the adaptor, but prevents the 5'-AATT ends from being ligated. After ligation to the cDNA, the new 5'-AATT end is compatible with ends generated by the restriction endonuclease EcoRI (G/AATTC) and no digestion of the fragment is required. As explained in Chapter 5, the restriction endonuclease leaves a 5' phosphate group and this is sufficient to allow ligation of one of the two 5'-AATT overhanging ends at each junction.

An alternative strategy is to add double-stranded linkers to the ends of a cDNA, carrying a restriction endonuclease recognition sequence, and then generate the overhanging ends by treatment of the concatenated linker-cDNA-linker sequence with the restriction endonuclease. For example, we could add EcoRI sites to the ends of a blunt double-stranded cDNA by annealing the symmetric oligonucleotide 5'-GCGAATTCGC-3' to itself to make a 10-bp blunt fragment, and ligating that fragment to the blunt cDNA ends:

```
5'-GCGAATTCGCNNNNNNNNNNNNNNNNNNNNNNNNNNNNNNNNNNNNNGCGAATTCGC-3'
3'-CGCTTAAGCGNNNNNNNNNNNNNNNNNNNNNNNNNNNNNNNNNNNNNCGCTTAAGCG-5'
```

This reaction can be driven to completion because the 10-bp fragment is small and can be added to the ligation reaction at a very high molar concentration. The 5'-GCGAATTCGC-3' sequence includes the EcoRI recognition sequence (G/AATTC), plus two nucleotides 5' to the sequence to improve the efficiency of digestion near the end of the fragment.

If the 5'-GCGAATTCGC-3'oligonucleotide is not phosphorylated and bears a 5′ hydroxyl group, only a single linker can be ligated to the ends of each cDNA; the cDNA has 5′ phosphates which will allow the first linker to be ligated. However, if the oligonucleotide is phosphorylated then a long concatenated sequence of linkers may be added to each cDNA end. Upon treatment of the ligated cDNA and linker with the specific restriction endonuclease, in this example *Eco*RI, excess linker will be digested away and only a single linker sequence will remain, yielding a 5′-AATT cohesive end:

```
5′-    AATTCGCNNNNNNNNNNNNNNNNNNNNNNNNNNNNNNNNNNNGCG        -3′
3′-        GCGNNNNNNNNNNNNNNNNNNNNNNNNNNNNNNNNNNNCGCTTAA    -5′
```

One problem with adding linkers to the end of a blunt cDNA fragment, and then digesting the products with a specific restriction endonuclease, is that the cDNA may have internal recognition sequences for the same restriction endonuclease:

```
5′-GCGAATTCGCNNNNNNNGAATTCNNNNNNNNNNNNNNNNNNNNNGCGAATTCGC-3′
3′-CGCTTAAGCGNNNNNNNCTTAAGNNNNNNNNNNNNNNNNNNNNNCGCTTAAGCG-5′
```

Digestion of the cDNA to expose cohesive ends in the linkers may break the cDNA into pieces, which though individually suitable for cloning would not be full length:

```
5′-AATTCGCNNNNNNNG        AATTCNNNNNNNNNNNNNNNNNNNNNGCG        -3′
3′-    GCGNNNNNNNCTTAA        GNNNNNNNNNNNNNNNNNNNNNCGCTTAA-5′
```

How likely is that? As discussed in Chapter 5, in a random sequence that is composed of 25% of each nucleotide G, A, T, and C, we would expect the average separation between 5′-GAATTC-3′ sequences to be 4096 bp. Most cDNA fragments will not be that long, but we may estimate from a Poisson distribution that the chance of a 1-kbp cDNA having an internal *Eco*RI site is a little over 20%.

A solution is to treat the cDNA with a specific restriction methylase prior to addition of linkers, so that the internal recognition sequences are protected from digestion. For example, a cDNA might be treated with the enzyme *Eco*RI methylase, which methylates the second adenosine in the sequence GAATTC, and renders it insensitive to the endonuclease *Eco*RI (G/AATTC). Then, upon addition of the unmethylated linker sequence 5′-GCGAATTCGC-3′, the cDNA is only susceptible to digestion with *Eco*RI at the linker sites. After cloning and replication of the DNA in *E. coli*, the methylation pattern on the internal *Eco*RI sites is not preserved and the GAATTC sequences within the cDNA will be susceptible to *Eco*RI again. That is, the methylation protection is only temporary and will not interfere with restriction mapping work done at a later time.

Restriction endonucleases are not the only way to generate cohesive ends, and if specific oligonucleotides are being used to prime synthesis of the first and second strand of the cDNA, then a bit of clever design can be employed. For example, a researcher might use the oligonucleotide 5′-AATTC-CTTTTTTTTTTTTTTT to prime the first-strand synthesis instead of oligo(dT). The 3′ end of the first strand of cDNA synthesis might be extended with oligo(dC) using the enzyme terminal deoxynucleotidyl transferase and the substrate dCTP, and the second-strand synthesis primed with the specific oligonucleotide 5′-AATTCCGGGGGGGGGGGGGGGGG instead of oligo(dG).

Upon completion of synthesis of the second strand of the cDNA, both 5′ ends will have the sequence 5′-AATTCC, but more importantly the 3′ ends will have the complementary sequence 3′-TTAAGG. A researcher can take this blunt cDNA and add Klenow enzyme, plus only the substrate dGTP, and remove four nucleotides from each 3′ end. These four dA and dT nucleotides will be removed because the Klenow enzyme has a 3′ to 5′ exonuclease activity; once removed these dA and dT nucleotides cannot be replaced. The dG nucleotide can be replaced when it is removed, so the enzyme only manages to remove the first four nucleotides at the 3′ ends, exposing a 5′-AATT overhanging end that is compatible with an *Eco*RI site. This technique of controlling the structure of ends by use of Klenow enzyme is based on the capabilities of the enzyme, as discussed in Section 6.1, and explained in more detail in Chapter 7.

Primer extension

Reverse transcriptase enzymes are useful in a technique termed 'primer extension', in which an oligonucleotide primer is extended on an RNA template until it encounters an obstruction or the 5′ end of the transcript. The length of the cDNA product tells us the distance between the 5′ end of the primer and the end of synthesis, and that can be a useful piece of information (see *Figure 6.14*). For example, suppose a researcher wants to map the promoter for a gene and know where transcription is initiated. With a bit of knowledge of the internal sequence of the transcript, the researcher might design an oligonucleotide that anneals to the RNA in exon 1 and can be extended by reverse transcriptase. If the cDNA is labeled and run on a denaturing polyacrylamide gel, the kind that might be used for DNA sequencing (see *Figure 6.14A*), its size can be accurately determined by comparison to known standards. If the length of the cDNA product is 145 nt, for example, then the 5′ end of the transcript is likely to be 144 nt upstream of the 5′ nucleotide of the primer.

It is common for researchers to use a sequencing ladder as an adjacent size marker for a gel, when determining the size of a primer extension reaction product (see *Figure 6.14B*). For example, a researcher might perform a primer extension reaction to determine the 5′ end of an RNA, and also perform DNA sequencing on a cloned DNA template using the same primer, to generate a concordant ladder for size comparison. If the sequence is well known, a researcher might prepare a size marker by conducting an abbreviated DNA sequencing reaction using only a single ddNTP, for example ddATP. An 'A-track' lane would show the position of all of the uridylate or thymidylate nucleotides in the template, and could be used to map the sizes of primer extension products in adjacent lanes of a gel.

One caveat is that the primer extension product only acts as a little 'measuring tape' to show the distance between the 5′ end of the primer and the point at which reverse transcription ceases. If the reverse transcriptase encounters a secondary structure in the RNA, an obstruction that it cannot pass, then synthesis will cease before the reverse transcriptase reaches the true 5′ end of the template. These false stops may be seen in any primer extension reaction as background bands,

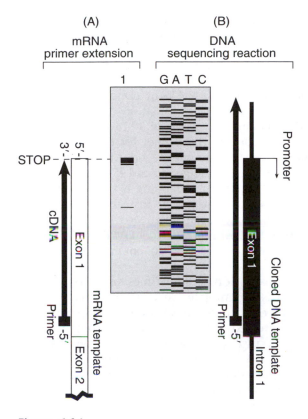

Figure 6.14

Primer extension of mRNA to map the 5' end of a transcript. A schematic diagram of an autoradiogram of a polyacrylamide gel is shown. (A) Lane 1 of the gel shows radioactively labeled extension products from a primer annealed to exon 1, using reverse transcriptase. The diagram to the left of the gel shows the structure of the 5' end of the mRNA, and the upward arrow indicates the direction of primer extension (cDNA synthesis) corresponding to the increase in sizes of products on the gel. The major band in lane 1 represents run-off reverse transcription, and corresponds to the 5' end of exon 1 (vertical dashed line). (B) Lanes 'G', 'A', 'T', and 'C' of the gel represent radioactively labeled DNA sequencing products from a cloned genomic DNA of the exon 1 region, using the same oligonucleotide primer. The diagram to the right of the gel shows the structure of the genomic DNA, and the upward arrow indicates the direction of sequencing. The sequencing ladder is used as a size marker to indicate the length of the run-off reverse transcript in lane 1. From these data, the location of the 5' end of the mRNA can be inferred.

termed 'pausing' or 'strong stop' sites (see minor bands in lane 1 of *Figure 6.14A*). The RNA template is heat-denatured prior to use in a primer extension reaction; however, snapback loops and internal secondary structures can readily reform during a reaction at 37°C. Use of a thermostable enzyme such as *Tth* may allow primer extension reactions to be conducted at a higher temperature that inhibits secondary structures in the template.

A second concern is that a primer extension reaction on an RNA may produce a cDNA that contains portions of multiple exons. If the 5′-most exon of a transcript has not been previously identified, then it may come as a shock to a researcher to have discovered it during the course of a primer extension experiment. In the previous example, a researcher extended a primer on an RNA to generate a 145-nt cDNA, and planned to use that information to discover the site of initiation of transcription. Suppose that the 145-nt cDNA had an undiscovered 30-nt exon 1 at its 5′ end, and unbeknownst to the researcher the primer for cDNA synthesis was located in exon 2. The researcher looking at the 145-nt product might mistakenly assume that the promoter for the gene lies 145 nt upstream of the primer site in the genomic DNA; that is, mapping it to a point 30 nt upstream of the 3′ splice site in the first intron instead of to its true location at the 5′ end of exon 1. One way to discover the 5′-most exon of a cDNA would be to perform a technique such as RACE, and another would be to use the primer extension reaction to obtain a sequencing result.

On the other hand, primer extension is sometimes used to discover obstructions on an RNA (or DNA) template, such as proteins that are bound to it. If a protein factor or complex is suspected of binding to a specific location on an RNA, primer extension might be used to help locate the point at which the reverse transcriptase bumps into the obstruction. This does not precisely define the boundary of the binding site because the bound protein factor might sterically interfere with the reverse transcriptase while it was still many nucleotides away; however, it can be taken as a single piece of evidence that is consistent with other observations of a probable binding site.

Primer extension sequencing

Reverse transcriptases can be used in conjunction with dideoxyribonucleotide substrates to perform indirect sequencing of an RNA. The oligonucleotide primer may be radioactively labeled at its 5′ end by use of the enzyme polynucleotide kinase and the substrate γ-[^{32}P]-ATP, and then annealed to an RNA template. Following annealing, the reaction may be split into four separate tubes and each aliquot extended in a reaction mixture containing a different dideoxyribonucleotide substrate: ddGTP, ddATP, ddTTP, and ddCTP. Just as with Sanger DNA sequencing (discussed in Section 6.4), the cDNA reaction products in the four separate tubes represent the potential 3′ chain termination sites for each nucleotide base. Put together as a sequencing ladder, the RNA sequence can be inferred from the cDNA chain termination results. It is not possible to infer the identity of the 5′-most nucleotide of the RNA by primer extension sequencing, because all four reactions will show a 'strong stop' at that fragment size. That is, we cannot distinguish between chain termination by dideoxyribonucleotide incorporation and run-off reverse transcription.

Sometimes a very specific question needs to be asked about an RNA sequence, such as whether it is 5′-AUACGuGCCGAUACUAGAUGGCGUAU-3′, or 5′-AUACGaGCCGAUACUAGAUGGCGUAU-3′, the lowercase letter indicating the polymorphic nucleotide U or A. The 3' end of this sequence could be annealed to a DNA oligonucleotide primer 5'-ATACGCCATCTAG-TATCGGC-3' that had been 5' end-labeled with polynucleotide kinase and γ-[^{32}P]-ATP:

```
5'-...NNNNNNNNNNNAUACGuGCCGAUACUAGAUGGCGUAUNNNNNNNNNN...-3'
            3'-CGGCTATGATCTACCGCATA-5'

5'-...NNNNNNNNNNNAUACGaGCCGAUACUAGAUGGCGUAUNNNNNNNNNN...-3'
            3'-CGGCTATGATCTACCGCATA-5'
```

The primer could be extended using reverse transcriptase in two separate reactions: the first reaction would contain dGTP, ddATP, dTTP, and dCTP, and the second would contain dGTP, dATP, ddTTP, and dCTP. If the RNA template were 5'-AUACGuGCCGAUACUAGAUGGCGUAU-3' then the first reaction would extend the primer by only one nucleotide and the second reaction would extend it by four nucleotides:

```
5'-...NNNNNNNNNNNAUACGuGCCGAUACUAGAUGGCGUAUNNNNNNNNNN...-3'
            3'-ACGGCTATGATCTACCGCATA-5'

5'-...NNNNNNNNNNNAUACGuGCCGAUACUAGAUGGCGUAUNNNNNNNNNN...-3'
            3'-TGCACGGCTATGATCTACCGCATA-5'
```

If the RNA template were 5'-AUACGaGCCGAUACUAGAUGGCGUAU-3', then the first reaction would extend the primer by five nucleotides and the second reaction would extend the primer by only one nucleotide:

```
5'-...NNNNNNNNNNNAUACGaGCCGAUACUAGAUGGCGUAUNNNNNNNNNN...-3'
            3'-ATGCTCGGCTATGATCTACCGCATA-5'

5'-...NNNNNNNNNNNAUACGaGCCGAUACUAGAUGGCGUAUNNNNNNNNNN...-3'
            3'-TCGGCTATGATCTACCGCATA-5'
```

These extensions of one to five nucleotides could be detected on a high-percentage polyacrylamide gel containing urea, and taken as indirect evidence for the underlying RNA sequence.

Mapping the 3' end

Primer extension of a DNA oligonucleotide can be used to map the 5' end of an RNA transcript, but what about the 3' end? Polymerases extend primers at their 3' end, so there is no way to anneal and extend an oligonucleotide towards the 3' end of an RNA. One approach is to ligate the 3' end of an RNA to a specific oligonucleotide using the enzyme T4 RNA ligase, and then use a specific complementary oligonucleotide primer and the enzyme reverse transcriptase, to determine the sequence of the 3' end of the RNA indirectly.

Alternatively, the 3'-most nucleotide of an RNA can be inferred by using a technique called splint labeling, in which the 3' end of the RNA is used as a primer on a matching DNA oligonucleotide template. If the 3' end anneals perfectly to a specific oligonucleotide, it can be extended and potentially radiolabeled with an isotope such as α-[^{32}P]-dCTP. If the 3' end is mismatched with the DNA oligonucleotide then it cannot be extended, and therefore cannot be radiolabeled. The addition of ^{32}P to the RNA end, actually due to the addition of one or more labeled deoxyribonucleotides, can be taken as evidence of an annealed 3' end.

6.6 Terminal deoxynucleotidyl transferases

Tailing

Terminal deoxynucleotidyl transferases (TdT) are template-independent DNA polymerases that extend the 3′ end of a DNA, often using whatever dNTP substrates happen to arrive at the catalytic site of the enzyme. When a DNA is extended using a TdT, we say that it has been 'tailed' by the enzyme, or that we have performed a 'tailing' reaction. TdT can tail the 3′ end of a single-stranded DNA, or a double-stranded DNA with a blunt, recessed, or overhanging 3′ end. For example, if bovine TdT is provided with a DNA primer and the substrate dCTP, it will catalyze the synthesis of an oligo(dC) tail at the 3′ end of the DNA. If the enzyme is provided with both dCTP and dGTP in equal proportion, it will catalyze the synthesis of random heteropolymeric extensions of varying lengths, such as 5′-GCGCCG or 5′-CGGGCGCCGG.

Homopolymer addition

An important application of TdT is the homopolymeric tailing of DNA fragments, for annealing to a complementary sequence. Several examples were given in Section 6.5 in which a cDNA might be extended with an oligo(dC) tail so that an oligo(dG) primer might be used to prime synthesis of a complementary strand. A related and very simple method can be used for bringing together a DNA fragment and cloning vector. If a DNA fragment is tailed at both ends with oligo(dC) and a DNA cloning vector is linearized with a restriction endonuclease leaving a blunt end, and then tailed with oligo(dG) at both ends, then the fragment and vector DNA will have complementary overhanging 3′ ends. The oligo(dC) and oligo(dG) tails might be slightly different lengths, when taken pair-wise, but if the two are annealed and the single-stranded gaps in sequence filled with Klenow enzyme, then the 5′ and 3′ ends of the DNA can be ligated using T4 DNA ligase.

3′ end labeling

Another application of TdT is the 3′ end-labeling of a fragment of DNA through transfer of a radiolabeled or dye-labeled nucleotide. For example, a DNA fragment could be tailed with α-[^{32}P]-dCTP, which would add a radiolabeled phosphate during the course of nucleotide addition. This technique is useful when conducting experiments such as DNA footprinting reactions, in which a DNA is specifically labeled at the 5′ or 3′ end, exposed to proteins that may potentially bind, and then chemically or enzymatically degraded. Where the DNA is protected from degradation, a blank area or 'footprint' appears in the ladder of degraded fragments on a denaturing polyacrylamide gel, the type one might use for DNA sequencing. The 3′ end-labeling technique is also useful for labeling DNA molecules with fluorescent dyes; a cDNA could be 3′ end-labeled with Cy™3-dCTP (Amersham Biosciences, Ltd.) to generate a fluorescent target for use in microarray studies, against a panel of fixed probe sequences.

Taq polymerase as a terminal transferase

Some DNA polymerases have a limited terminal transferase activity; for example, *Taq* DNA polymerase adds an untemplated adenylate nucleotide to the 3' ends of many of its products. Some *Taq* enzyme products have a blunt end and others have a 3' overhanging end of one 'A' nucleotide. This 3' overhanging end makes it difficult to clone the reaction products into a restriction site in a plasmid that leaves a blunt end after digestion, such as the enzyme *Sma*I (CCC/GGG). However, a researcher may take advantage of the overhanging 'A' residue by ligating the fragment into a linearized plasmid that has a single 3' overhanging 'T' residue. For example, the pCR® and TOPO TA Cloning® vectors (Invitrogen Corp.) are distributed in a linear form, with a 3'-T overhanging nucleotide included for this specific application.

Not all DNA polymerases used in PCR have a terminal transferase activity. For example, the thermostable polymerase *Pfu* leaves a blunt-ended PCR product that would not be suitable for use with a plasmid incorporating a TA cloning system. In this situation, the PCR product might still be tailed with a single adenylate nucleotide by addition of *Taq* DNA polymerase and dATP, to make the product compatible with a TA cloning system.

Chain termination during tailing

TdT will tend to add multiple deoxyribonucleotide residues to a 3' end, but chain termination will occur if it incorporates a dideoxyribonucleotide. The reason that a ddNTP substrate terminates the addition of nucleotides by TdT is that dideoxyribonucleotides lack a 3' hydroxyl group that is necessary for continued chain elongation (as discussed in Section 6.4). Nucleotide addition by TdT can therefore be limited to a single nucleotide, if only dideoxyribonucleotides are offered to TdT as a substrate. Such a method is useful in preparing a vector so that it will have a 3'-T overhanging nucleotide for TA cloning methods; the vector can be linearized at a multiple cloning sequence using a restriction endonuclease leaving a blunt end, then treated with TdT and ddTTP.

A word about ligation between this tailed vector and a PCR product may be in order because the 3'-T nucleotide on the vector has a dideoxyribose sugar that lacks a 3' hydroxyl, and the oligonucleotides used to prepare the PCR product usually lack 5' phosphates. One DNA strand at each end of the PCR product is not joined to the vector, but the other strand has a 5' phosphate from the vector and a 3' hydroxyl from the overhanging adenylate nucleotide so the ligation can be half-completed, leaving only a nick at each end of the PCR product. However, it would be unwise to use TdT and ddATP to add a single dideoxyadenylate nucleotide to a PCR product made with *Pfu* polymerase. In that scenario, the ends of the PCR product would very likely have neither 5' phosphates or 3' hydroxyl groups, and a ligation reaction with a TA vector will probably not succeed using the enzyme T4 DNA ligase.

Detection of fragmented DNA by TUNEL

One important application of terminal transferases is in labeling DNA ends generated during the process of apoptosis. A fragmented chromosome in a cell undergoing programmed cell death will have many free 3′ ends, which are susceptible to extension by TdT. A histological section of a tissue may be fixed and exposed to the enzyme TdT, along with a deoxyribonucleotide substrate labeled with a chemical marker such as biotin or a fluorescent dye such as Cy™3 (Amersham Biosciences, Ltd.). The enzyme will extend the 3′ ends of the broken chromosome, and the extent of labeling of the chromosome overall will be directly related to the number of breaks.

A method based on this approach is referred to as TdT-mediated dUTP nick end labeling (TUNEL), in which dUTP (deoxyuridine triphosphate) is the carrier of the chemical prosthetic group or dye-label. We infer that cells that are highly fluorescent have fragmented chromosomes because the dye-labeled dUTP was extensively incorporated. The TUNEL assay can be performed *in situ* on a slide or tissue section, or in solution using a fluorometer. If the dUTP label is biotin, digoxigenin, or some other type of tag, the chromosome breaks might be detected by an alternate approach, perhaps one involving an enzyme-linked intermediate such as avidinylated horseradish peroxidase, and a colorimetric or fluorogenic substrate.

6.7 RNA polymerases

RNA polymerases are useful in the laboratory because they can be used *in vitro*, in the generation of an RNA transcript from a DNA template (8). RNA prepared *in vitro* may be used as a probe in hybridization, or in experimental procedures such as *in vitro* splicing extracts, or as an mRNA in *in vitro* translation extracts. RNA can also be introduced into cells by transfection, where it may have biological activities, or expressed from a bacteriophage promoter in a cell expressing the cognate bacteriophage RNA polymerase gene (as discussed in Chapter 4).

Bacteriophage T3, T7, and Sp6 promoters

Transcriptional promoters from bacteriophage T3, T7, and Sp6 are sometimes included in cloning vectors and used to direct transcription of a cloned sequence. The template is usually linearized with a restriction endonuclease so that a run-off transcript of defined size is generated. Promoter sequences may also be included in a PCR product, allowing direct transcription of RNA from the linear DNA fragment. For example, the sequence 5′-TAATACGACTCACTATA^GGGAGG-3′ is a consensus promoter for T7 RNA polymerase, where the caret symbol indicates the start of transcription. When the 23-nt sequence is incorporated into a DNA sequence, an RNA transcript can be generated with T7 RNA polymerase from the last 6 nt of the promoter sequence: 5′-pppGGGAGG...-3′.

The bacteriophage RNA polymerase promoters are highly specific to the species of RNA polymerase; for example, a T3 or Sp6 RNA polymerase will not initiate transcription from a T7 RNA polymerase promoter. The consensus promoters for T3 and Sp6 RNA polymerase are 5′-AATTAACC-CTCACTAAA^GGGAGA-3′ and 5′-ATTTAGGTGACACTATA^GAAGNG-3′, respectively. A plasmid could have two different bacteriophage RNA polymerase promoters flanking the multiple cloning site on complementary DNA strands, and the use of one or the other polymerase would yield sense or antisense RNA for the cloned sequence.

Reaction conditions

The DNA template used in an *in vitro* transcription system needs to be free of contaminating RNA and protein. In particular the DNA needs to be free of RNase that may have purposefully been added during the preparation of a plasmid from bacteria, or may have been a contaminant in the restriction enzyme used to linearize the DNA. To remove RNase activities from a template, the DNA can be incubated at 50°C for 30 min, with 100 µg ml^{-1} of the enzyme proteinase K, in a solution of 50 mM Tris-Cl, pH 8.0, and 0.5% sodium dodecyl sulfate (SDS). Following this proteolytic digestion, the proteins can be extracted with 1 volume of a 1:1 mixture of phenol and chloroform, and the residual phenol extracted with 1 volume of chloroform. The DNA in solution is precipitated with 300 mM sodium acetate and 2 volumes of ethanol, washed once with 70% ethanol, dried briefly, and re-suspended to a final concentration of approximately 0.5 mg ml^{-1}.

The reaction is assembled using nuclease-free water, a mixture of rNTP substrates (ribonucleoside triphosphates, not deoxyribonucleotide triphosphates), a $10 \times$ stock reaction buffer, the purified DNA template, and a bacteriophage RNA polymerase. Because the reaction buffer contains spermidine, the reaction solution is assembled at room temperature to prevent aggregation and precipitation of the template. The conditions recommended by the manufacturer should be followed closely; however, here is an example of the reaction conditions recommended by Stratagene® for their T7 and T3 RNA polymerase enzymes:

- 40 mM Tris-Cl, pH 8.0
- 8 mM MgCl$_2$
- 50 mM NaCl
- 2 mM spermidine
- 30 mM dithiothreitol (DTT)
- 400 µM of each rNTP
- 0.04 µg µl^{-1} DNA template
- 0.4 units µl^{-1} bacteriophage RNA polymerase.

One unit of enzyme (from that supplier) is the amount that will catalyze the incorporation of 1 nmol of rNTP into trichloroacetic acid-insoluble material, over a period of 1 hour at 37°C. The 1 nmol of incorporated rNTP would have a mass of approximately 0.33 µg, as incorporated nucleotides have an average formula weight of 330 Da. Therefore, if 10 units of enzyme were used in a 25-µl reaction, then over a 30-min incubation the researcher might expect a yield of approximately 1.6 µg of RNA (10 units × 0.33 µg × 0.5 hours). If the transcription reaction is efficient, multiple copies of an RNA transcript can be made from each DNA template.

Following completion of the transcription reaction, the DNA template may be degraded by digestion with RNase-free DNase, and the RNA collected and purified by LiCl or ethanol precipitation. A small amount of radioactive ribonucleotide may be added to the reaction to keep track of the yield, through a calculation of specific activity (discussed in more detail in Section 6.8). For example, if 1 µCi of α-[^{32}P]-UTP is added to a 25-µl reaction containing 400 µM UTP, that is a reaction containing 10 nmol of UTP, the specific activity of the UTP will be 1 µCi per 10 nmol UTP. If the reaction product includes equimolar amounts of each of the four bases, then the 10 nmol UTP would be one quarter of all of the GTP, ATP, UTP, and CTP incorporated. We calculate that 1 µCi is 2.22×10^6 disintegrations per minute (dpm), because 1 Ci is 2.22×10^{12} dpm. Furthermore, 40 nmol of rNTP is approximately 13 µg because nucleotides have an average formula weight of about 330 Da, so each µg of RNA product would have a radioactivity of 168 000 dpm when freshly made. If 1% of the aforementioned *in vitro* transcription reaction were set aside for TCA precipitation, and yielded 2700 dpm, then we would infer an overall yield of 1.6 µg (100×2700 dpm $\div$ 168 000 dpm µg^{-1} = 1.6 µg).

RNA probes

RNA probes are readily prepared using bacteriophage RNA polymerases, and they can be made to very high specific activity. One advantage of using RNA probes, instead of nick-translated or random oligo-primed DNA probes, is that the DNA probe strands may renature in solution instead of annealing to the nucleic acids cross-linked on the Southern or Northern blot membrane. Once an RNA probe is cleared of residual DNA template by treatment with RNase-free DNase, it is specific for a single target strand (sense or anti-sense) and does not hybridize to other probe molecules. Researchers may find that the added specificity of a single-stranded RNA probe reduces nonspecific background levels in a blotting experiment.

When RNA probes are made with a very high specific activity, for example in a reaction containing 400 µM of each of the three ribonucleoside triphosphates GTP, ATP, and CTP, and 100 µCi of α-[^{32}P]-UTP (3000 Ci mmol^{-1}), the products may not achieve full length during synthesis. This problem is related to the low concentration of the one radiolabeled substrate compared to the K_M of the RNA polymerase, and is discussed in more detail in Section 6.8. This foreshortening of product may only be a concern if the researcher wants the RNA probe to be a uniformly labeled RNA version of the full length of the cloned sequence in a plasmid. If the cloned sequence is long, most of the ^{32}P probe may be representative of sequences near to the promoter.

End-specific probes

One application of RNA polymerase promoters is the generation of probes from the ends of long cloned sequences for use in genomic mapping. For example, the vector Lambda FIX® II (Stratagene) has T7 and T3 RNA polymerase promoters flanking the cloning site, which might hold up to 23 kbp of genomic DNA. A Lambda FIX® II clone DNA can be used as a template in two separate labeling reactions, one with T7 RNA polymerase and another with T3 RNA polymerase, and end-specific probes will be generated. These probes are radiolabeled with high specific activity; they are not long transcripts, but they represent the sequences at the junction between the cloned fragment and the vector and can be used to define the ends of the clone by Southern blot. Furthermore, they can be used as probes to screen the library of clones for overlapping candidates. For example, a hypothetical clone containing genes in the order 'H-I-J-K' can be used to generate end probes 'H' and 'K', which may then identify the overlapping clones 'F-G-H-I' and 'K-L-M-N':

```
        H-I-J-K
  F-G-H-I
              K-L-M-N
```

We then have all the genes between F and N represented, and can continue our chromosome walking in both directions. Provided that none of the end probes contain repetitive sequences that might be represented elsewhere in the genome, causing us to accidentally skip to a different part of the genome when we intended to take only a small step, our ordering of the clone candidates by end probe should present an accurate picture of the genome structure.

Transcripts for translation

RNA molecules prepared by *in vitro* synthesis using RNA polymerase may be introduced into cell-free translation extracts, for the purposes of directing protein synthesis. The benefit of this method to the researcher is that it allows the preparation of small amounts of a specific protein corresponding to an open reading frame in a cloned gene or PCR product. Translation extracts from rabbit reticulocytes or wheat germ can be readily prepared in a laboratory, or obtained from commercial sources. Typically, the extracts are pre-treated with the enzyme micrococcal nuclease to destroy endogenous mRNA, so that the RNA added by the researcher is a primary target for the ribosomes. The translated protein may be folded correctly in the *in vitro* extract, and may be recognizable by antibodies specific to the protein, for example in indirect immunoprecipitation or Western blotting methods. Some *in vitro* translated proteins may also have the enzymatic activity of the native gene product, which may allow a researcher to investigate the structure and function of the enzyme by studying the products of mutated versions of the gene.

Eukaryotic mRNA molecules do not have the equivalent of a ribosome binding site, or Shine–Dalgarno sequence that would be required for efficient ribosome binding. However, the immediate neighborhood of the AUG start codon does have an effect on the efficiency of initiation, and the most favorable consensus sequence is often referred to as Kozak's rules. The more critical aspects of Kozak's rules are that there be an adenylate nucleotide at the -3 position, and a guanylate nucleotide at the $+1$ position, for example the 10 nt sequence 5'-GCCACCAUGG-3' containing an AUG start codon at nucleotides 7–9.

It is most often the first AUG triplet in an mRNA, the AUG closest to the 5′ end, that is used for initiation of translation. A researcher engineering a DNA template to make a transcript for *in vitro* translation will want to plan carefully the 5′ untranslated sequence, to make certain that it carries no extraneous AUG triplets. Furthermore, the mRNA should be generally free of sequences that might interfere with translation, such as mRNA secondary structures. The engineering of the template may either be done through cloning sequences into a DNA vector that has a T3, T7, or Sp6 RNA polymerase promoter flanking a multiple cloning site, or through generation of a polymerase chain reaction product that carries the specific promoter sequence. The construction will also have a 5′ untranslated region, an AUG start codon and protein coding sequence, a stop codon, and optionally a 3′ untranslated sequence that will be included in the RNA transcript. A plasmid-cloned template would need to be linearized at the 3′ end to allow for run-off transcription, while a PCR product template is already in a linear form by design.

Capping and polyadenylation

Native eukaryotic mRNA molecules are usually polyadenylated, and bear a 5′ cap structure of 7-methyl-GTP. The 7-methyl-GTP cap can be included during synthesis of the RNA by RNA polymerases *in vitro*, by adding the cap analog $^{m7}G(5′)ppp(5′)G$ in place of some of the GTP in the reaction mixture. A poly(A) tail may also be added enzymatically after the transcription reaction is complete, by incubation of the RNA with the enzyme poly(A) polymerase and ATP substrate. A poly(A) tail may also be incorporated during transcription of the *in vitro* RNA product, if the template is engineered so that a poly(dT) sequence is included in the DNA. These modifications to the mRNA increase its stability, and improve its readability by ribosomes; however, they are not usually an essential modification for mRNA added to cell-free translation extracts. Including a 5′ cap may decrease the efficiency of RNA production *in vitro*, at the same time that it increases the efficiency of translation *in vitro*.

Transcription/translation extracts

Transcription *in vitro* and translation *in vitro* are often performed in separate reactions, because the optimal conditions for the RNA polymerase are different to those for the ribosome. However, there are commercially available extracts that have been developed so that both transcription and translation can be performed in a single tube; for example the PROTEINscript™ II (Ambion, Inc.) and the TNT® (Promega Corp.) systems. A DNA template is added to the reaction, and the sequential reactions of transcription and translation are conducted in a coupled system. This is convenient because it allows for a wide variety of templates to be tested rapidly; however, the overall yield of protein product may be less that might be obtained in separate reactions because of the compromises that are made in finding a suitable reaction buffer.

6.8 The special problem of specific activity

Many of the techniques presented in this chapter involve a nucleic acid labeling step, and a small calculation may help to explain a problem. Suppose a dCTP substrate could be purchased in which every single alpha phosphate had a ^{32}P isotope; that would be an ideal substrate for making a radiolabeled probe. Given a mmole of this special carrier-free α-[^{32}P]-dCTP, how radioactive would it be? In 1 min, approximately 1.9×10^{16} ^{32}P atoms in the sample would be lost to radioactive decay, or approximately 33 nmol of the original 1 mmol. The calculation is based on equation $A = A_0\, e^{-kt}$ presented in Section 6.4, and *Figure 6.10*. In this calculation t is the time elapsed in minutes and k is a constant equal to approximately 3.27×10^{-5}.

Given 1 mmole of carrier-free α-[^{32}P]-dCTP, which at its freshest produces 1.9×10^{16} disintegrations per minute, the corresponding measure of radioactivity would be approximately 8800 Ci. In other words, α-[^{32}P]-dCTP cannot have a specific activity that is higher than about 8800 Ci mmol^{-1}.

Suppliers of ^{32}P isotopes often provide a product that is 3000 to 6000 Ci mmol^{-1}, which is very close to the maximum specific activity. A single radioactive labeling reaction might consist of as much as 100 µCi. The amount of dCTP in 1.0×10^{-4} Ci of α-[^{32}P]-dCTP can be determined by dividing by the specific activity, with appropriate adjustment for the units:

$$1.0 \times 10^{-4}\ \text{Ci} \div 6000\ \text{Ci mmol}^{-1} = 1.66 \times 10^{-8}\ \text{mmol} = 16.6\ \text{pmol}$$

If a random hexamer primer labeling reaction contains these 16.6 pmol of radioactive dCTP in a reaction volume of 50 microliters, then the dCTP concentration is:

$$1.66 \times 10^{-11}\ \text{mol} \div 5 \times 10^{-5}\ \text{l} = 3.3 \times 10^{-7}\ \text{M} = 0.33\ \text{µM dCTP}$$

That's not very much, considering that nonradioactive dATP, dTTP, and dGTP might be present in the reaction at concentrations of about 100 µM. When the enzyme is adding an A, T, or G nucleotide to a growing DNA it should do it quite well, but what happens when C is the next nucleotide to add and there isn't much of that substrate in the reaction?

Enzyme performance

One measure of enzyme performance is the Michaelis–Menten parameters, and in the case of Klenow enzyme the K_M for nucleotide substrates is approximately 2 µM. If the actual concentration of dCTP in a reaction is only 0.33 µM, one sixth of the K_M of the enzyme, then the velocity of the enzyme would be about 14% of its maximum velocity (see *Figure 6.15*). The velocity of an enzyme, V, under Michaelis–Menten kinetics is described by:

$$V = V_{max}\ [S/(S + K_M)]$$

where V_{max} is the maximum velocity, S is the substrate concentration, and K_M is an enzyme-substrate Michaelis constant.

As bleak as this sounds, Klenow enzyme may still be a good choice for doing the work because some DNA polymerases have an even higher K_M

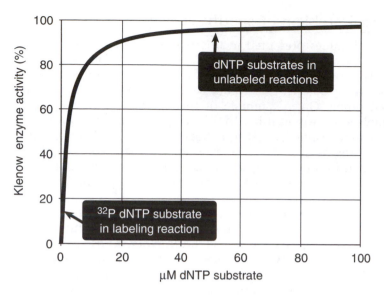

Figure 6.15

A Michaelis–Menten model of Klenow enzyme activity (percentage of V_{max}) as a function of dNTP substrate concentration. The K_M of Klenow enzyme for dNTP substrate is 2 µM. The amount of [^{32}P]-dNTP in a labeling reaction is approximately 0.33 µM (assuming 100 µCi of 6000 Ci mmol^{-1} dNTP substrate is placed in a reaction of 100 µl) (see arrow, lower left). The amount of dNTP used in unlabeled synthesis, for example filling in a recessed 3′ end to make it blunt, is often 50 µM or more (see arrow, upper right).

for nucleotide substrates, as discussed in Section 6.1, and the reduction in velocity would be even more severe if they were employed. For example, a polymerase with a K_M of 50 µM for nucleotide substrate would operate at only 0.6% of its maximum velocity when trying to make use of 0.33 µM dCTP.

Waiting for a scarce substrate

We can imagine, then, that a DNA polymerase might need to be very patient in waiting for radioactively labeled nucleotide substrates because these substrates are used at low molar concentrations. Would it help to add more of the limiting nucleotide? Yes, very much so, but then in most laboratories, 100 µCi of ^{32}P is considered to be a substantial amount by radiation safety standards. Adding more of the radioactive substrate to the reaction would be both expensive and dangerous. The carrier-free ^{32}P substrates are typically sold at a 'fresh' concentration of 10 µCi µl^{-1}, which is 1.66 µM (if the specific activity is 6000 Ci mmol^{-1}), so it would be difficult to increase the concentration of radioactive substrate in a labeling reaction without first concentrating the substrate by vacuum desiccation (which results in yet another piece of equipment to decontaminate).

Would it help to conduct a labeling reaction in a volume smaller than 50 µl so that the concentration of the substrate could be increased? Yes, but working with smaller volumes may be challenging unless the pipetting of small volumes can be performed with accuracy and precision. Would it help to make both the dCTP and the dATP substrates radioactive? Yes and no; while the product would be more heavily labeled with radioactivity the K_M problem would simply be compounded with a second substrate. Would it help to supplement the 16.6 pmol of α-[^{32}P]-dCTP with nonradioactive dCTP, to increase the overall concentration? Yes and no, because while the enzyme would operate with greater velocity, it would add proportionately fewer radioactive phosphates, and the specific activity of the product would be proportionately reduced (see *Figure 6.16*). If the radiolabeled DNA is to be used as a probe in a hybridization reaction, this reduced specific activity will make the probe less sensitive in detecting small quantities of target.

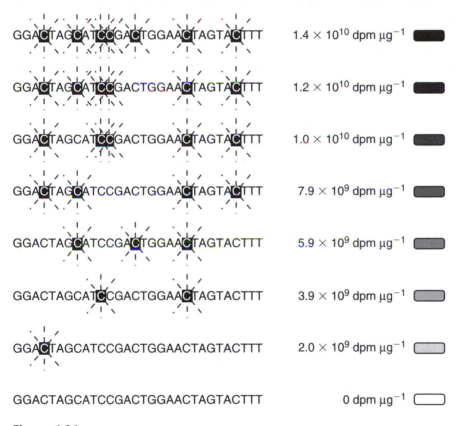

GGACTAGCATCCGACTGGAACTAGTACTTT	1.4×10^{10} dpm μg^{-1}
GGACTAGCATCCGACTGGAACTAGTACTTT	1.2×10^{10} dpm μg^{-1}
GGACTAGCATCCGACTGGAACTAGTACTTT	1.0×10^{10} dpm μg^{-1}
GGACTAGCATCCGACTGGAACTAGTACTTT	7.9×10^{9} dpm μg^{-1}
GGACTAGCATCCGACTGGAACTAGTACTTT	5.9×10^{9} dpm μg^{-1}
GGACTAGCATCCGACTGGAACTAGTACTTT	3.9×10^{9} dpm μg^{-1}
GGACTAGCATCCGACTGGAACTAGTACTTT	2.0×10^{9} dpm μg^{-1}
GGACTAGCATCCGACTGGAACTAGTACTTT	0 dpm μg^{-1}

Figure 6.16

A schematic representation of specific activity in a ^{32}P-radiolabeled molecule. The 30-nt sequence has seven 'C' nucleotides, and versions of the sequence with seven to zero ^{32}P-labeled 'C' nucleotides (white on black lettering), are shown from top to bottom. The specific activity of each 30 nt DNA, in units of dpm μg^{-1}, is indicated on the right (assuming a formula weight of 9900 Da for the 30-mer, and that carrier-free ^{32}P is 8800 Ci mmol^{-1}). A schematic diagram of an autoradiogram gel band, with increasing intensity corresponding to specific activity of the radiolabeled molecule, is shown at the far right.

Table 6.1 Specific activity of internally labeled DNA probes

Substrate DNA template	Product Specific activity[a]	Radioactivity in product sample		
		0.1 pg	1 pg[b]	10 pg
20 ng	2.80×10^9 dpm μg^{-1}	280 dpm	2800 dpm	28 000 dpm
50 ng	1.60×10^9 dpm μg^{-1}	160 dpm	1600 dpm	16 000 dpm
100 ng	0.92×10^9 dpm μg^{-1}	92 dpm	920 dpm	9200 dpm
200 ng	0.50×10^9 dpm μg^{-1}	50 dpm	500 dpm	5000 dpm
500 ng	0.21×10^9 dpm μg^{-1}	21 dpm	210 dpm	2100 dpm
1000 ng	0.11×10^9 dpm μg^{-1}	11 dpm	110 dpm	1100 dpm

[a] Assuming that 50 µCi of radioactive substrate is incorporated into the DNA template, and that there is a total of 20 ng of *in vitro* DNA synthesis.
[b] As a point of reference, 9×10^5 copies of a 1-kbp DNA is 1 pg.

There is a 3′ to 5′ exonuclease-deficient version of Klenow enzyme that is commercially available and this enzyme reportedly shows very fast kinetics in random primer labeling reactions. Regular Klenow enzyme may engage its exonuclease activity and chew away nucleotides at the 3′ end, if the necessary dNTP substrate does not arrive at the enzyme's active site, and this is a matter discussed in Section 6.1. The 'Exo-minus' Klenow enzyme may have the virtue of patience in waiting for a low concentration substrate in a state that is nondestructive.

The problem of limiting concentration of a nucleotide, such as α-[^{32}P]-dCTP, has been discussed in terms of its potential effects on the rate of enzymatic synthesis, but there is a second problem related to yield of product. In the aforementioned example of 100 µCi or 16.6 pmol of carrier-free α-[^{32}P]-dCTP, the mass represented by the 16.6 pmol is 7.4 ng; however with the loss of the β and γ phosphates during incorporation only 5.2 ng of the mass can be included in the DNA polymer. If the substrates dGTP, dATP, and dTTP are similarly incorporated, then the total theoretical yield of newly synthesized DNA polymer is limited to about 20–25 ng. This is a critical point, because the specific activity of the product is diluted if excess template is added (see *Table 6.1*).

Estimating specific activity

The specific activity of a radiolabeled nucleic acid is an indication of how radioactive it is per µg or pmol (see *Figure 6.16*). A pure sample of ^{32}P, for example, is approximately 8800 Ci mmol^{-1} or 1.9×10^{16} dpm mmol^{-1}. In the research setting it is common to work with µCi to mCi amounts of ^{32}P, and fmol to pmol amounts of DNA fragments. For example, if we have 1 ng of a 1-kbp DNA fragment, that amount corresponds to 1.5 fmol. If every molecule in the DNA sample were labeled with a single ^{32}P, then 1.5 fmol of ^{32}P would be associated with the 1.5 fmol of DNA, and the sample would have 1.3×10^{-8} Ci of radioactivity (1.5×10^{-15} mol $\times 8.8 \times 10^6$ Ci mol^{-1}) or 24 000 dpm (1.5×10^{-15} mol $\times 1.9 \times 10^{19}$ dpm mol^{-1}). On a µg basis, the specific activity of such a DNA sample would be 2.4×10^7 dpm per µg, which is too low for it to be used as a Southern blot probe but is fine for tracer studies on a gel. If there were 10 ^{32}P associated with the 1.5 fmol of DNA, then the sample would be 10 times more radioactive and the specific activity would be 10 times greater. In a random hexamer-primed labeling, we might hope to get 100–200 ^{32}P atoms associated with a 1-kbp DNA, or a specific activity of 2.4–4.8×10^9 dpm per μg. We need this level of radioactivity in a probe to be able to detect small amounts of target.

Specific activities of products can be estimated by measuring the amount of radioactivity incorporated into a large DNA polymer, for example by measuring the radioactive disintegrations per minute in a sample precipitated with trichloroacetic acid, and then dividing that measure of radioactivity by the total mass or moles of DNA. A high specific activity of product would typically be in the range of about 1.5×10^9 dpm μg^{-1}. Forty ng of such a product would then have a level of radioactivity that is 6×10^7 dpm, or about 27 μCi:

$$1.5 \times 10^9 \text{ dpm μg}^{-1} \times 4 \times 10^{-2} \text{ μg} = 6 \times 10^7 \text{ dpm}$$
$$6 \times 10^7 \text{ dpm} \div 2.22 \times 10^{12} \text{ dpm Ci}^{-1} = 2.7 \times 10^{-5} \text{ Ci} = 27 \text{ μCi}$$

If the reaction were initiated with 100 μCi, that 27 μCi would correspond to a little less than a third of the labeled nucleotides being incorporated into a product; about 5 pmol or 1.6 ng of incorporated deoxycytidylate nucleotides (assuming the specific activity of the [^{32}P]-dCTP is 6000 Ci mmol^{-1}). If the DNA were composed of the four nucleotide bases in equal proportion, then the 1.6 ng of radiolabeled nucleotides would be a part of approximately 7 ng of labeled DNA, which happens to be base-paired to an additional 7 ng of unlabeled template. So, of the 40 ng of product that is labeled, only 14 ng was involved in the reaction, being either template or new synthesis. How do we explain the remaining 26 ng that was uninvolved? The most charitable statement that can be made is to say that it was a 'spectator' to the reaction and never played the role of template.

However, there is an important lesson hidden in this. Excess DNA template dilutes out the specific activity of the product. The specific activity of the product, which must be high for obtaining sensitive probes, can be lowered by having too many DNA 'spectators' in the reaction and not enough DNA 'players'. For example, if a researcher adds 200 ng of a template to a labeling reaction limited by 100 μCi of α-[^{32}P]-dCTP, in which only 20 ng of new DNA synthesis could possibly occur, then the 220 ng of total DNA has less than one-tenth of its maximal specific activity. It could not have a specific activity greater than 100 μCi per 220 ng, which is:

$$2.2 \times 10^8 \text{ dpm} \div 2.2 \times 10^{-1} \text{ μg} = 10^9 \text{ dpm μg}^{-1}$$

If only half of the 100 μCi is incorporated, which is more realistic, the specific activity would be half of that level, or 5×10^8 dpm μg^{-1} (see *Table 6.1*). If the researcher had started the reaction with a full 1 μg of template and achieved the incorporation of 50 μCi, the specific activity of the product would be about 1.1×10^8 dpm μg^{-1} (see *Table 6.1*).

Why is specific activity so important?

The specific activity, or level of radioactivity in each gram or mole of a nucleic acid, determines how easily we can detect it in a scintillation counter, on a piece of X-ray film, or on a phosphorimager plate. If a probe molecule has less radioactivity per mole, then it is more difficult to detect after binding to a target. If a target molecule is short, it is also more difficult to detect because it will have little probe bound to it.

Using human DNA as an example, 10 µg is a size of sample that might be digested and used in Southern blotting, and based on a molecular mass of about 4×10^{12} daltons this corresponds to 1.5 million copies of each unique sequence. However, in nucleic acid blotting experiments the efficiency of hybridization is not 100%. If we assume 10% hybridization efficiency then 150 000 copies of probe would hybridize to the 1.5 million targets. How much radioactivity would that represent if the probe had a specific activity of 1.5×10^9 dpm µg^{-1}? Let us assume that the target and probe overlap by 1 kbp, and use the figure that 1 kb of probe has a formula weight of 3.3×10^5 Da. The 150 000 copies of probe sequence would provide a signal of approximately 120 dpm, which is detectable on X-ray film:

$$1.5 \times 10^5 \text{ copies} \div 6.02 \times 10^{23} \text{ copies mol}^{-1} = 2.5 \times 10^{-19} \text{ mol}$$
$$2.5 \times 10^{-19} \text{ mol} \times 3.3 \times 10^5 \text{ g mol}^{-1} = 8.2 \times 10^{-14} \text{ g} = 8.2 \times 10^{-8} \text{ µg}$$
$$8.2 \times 10^{-8} \text{ µg} \times 1.5 \times 10^9 \text{ dpm µg}^{-1} = 125 \text{ dpm}$$

Assuming this specific activity and efficiency of hybridization, a 1000-bp restriction fragment would be readily detectable, but a 100-bp fragment would be more challenging to detect as it would have one-tenth the amount of overlapping probe. Furthermore, if the probe had a specific activity 10 times lower, for example if the researcher performed a labeling reaction with too much template, then the signal would drop linearly.

This relationship between specific activity and signal is outlined in *Table 6.1*, and some sample calculations are provided. For example, 0.1 pg of ^{32}P-labeled DNA with a specific activity of 1.6×10^9 dpm µg^{-1} is about 160 dpm, but if the specific activity is dropped to 1.1×10^8 dpm µg^{-1}, as in the last row of *Table 6.1*, then that 0.1 pg of ^{32}P-labeled DNA is about 11 dpm. The calculation of specific activity in the last row of *Table 6.1* is detailed below:

$$50 \text{ µCi} \times 2.2 \times 10^6 \text{ dpm µCi}^{-1} = 1.1 \times 10^8 \text{ dpm}$$
$$1.1 \times 10^8 \text{ dpm} \div 1.02 \text{ µg} = 1.1 \times 10^8 \text{ dpm µg}^{-1}$$

And we calculate the radioactivity in a sample of 0.1 pg:

$$0.1 \text{ pg} \times 1 \times 10^{-6} \text{ µg pg}^{-1} \times 1.1 \times 10^8 \text{ dpm µg}^{-1} = 11 \text{ dpm}$$

That 11 dpm is below the level of background radiation, and would be difficult to detect experimentally.

When labeled substrate is not limiting

There are many commercially available kits that allow the production of nonradioactive-labeled DNA by random hexamer priming. A DNA polymerase may accept a modified nucleotide, such as a dUTP carrying a fluorescent group (e.g., fluorescein-12-dUTP), and incorporate it into a growing chain during synthesis. The fluorescent DNA product may be used as a probe and detected directly, for example in fluorescent *in situ* hybridization (FISH), or detected indirectly using an anti-fluorescein antibody conjugated to a reporter enzyme. Other common approaches are to incorporate biotinylated or digoxigenin-labeled nucleotides and detect them, respectively, with a reporter molecule linked to avidin or to an anti-digoxigenin antibody.

A significant difference between the methods of radioactive labeling and nonradioactive labeling is in the amounts of labeled substrate that can be added to the primer extension reaction. As discussed earlier, a carrier-free radioactive substrate would typically be added in very limiting amounts, perhaps on the order of 0.3 μM, with concomitant negative effects on the kinetic behavior of the enzyme and the overall yield of labeled product. With a nonradioactive-labeled substrate, which could be added to a primer extension reaction at levels approaching mM concentrations, the kinetic barriers can be overcome and the overall yield of labeled product in a reaction can be substantially higher. A μg of template could be added to such a reaction, and given time with the enzyme, nearly a μg of newly synthesized product obtained.

Reaction yield is still a concern

The instructions accompanying commercially available kits often warn the user to determine the yield of synthesis, because a single nonradioactive labeling reaction can be so productive that researchers accidentally use too much of the product in a hybridization reaction. This may lead to higher background signals on a Southern blot, for example.

A typical way to determine the yield of synthesis in a nonradioactive labeling reaction is to compare the signal generated from a measured amount of the product to a known standard. A label such as fluorescein can be detected directly using an ultraviolet light, but labels such as biotin or digoxigenin can only be revealed after binding and enzymatic steps. In either case, the reaction products are first separated from unincorporated labeled nucleotides that might otherwise contribute a background signal. The comparison between labeled product and a standard allows some interpretation of the results.

Suppose a researcher determines that 1% of the labeling reaction produces approximately the same colorimetric 'spot' as 1 ng of a labeled control DNA provided by the company. What does that imply? The entire yield of newly synthesized DNA in the researcher's sample might be judged to be 100 ng, assuming that the density of incorporated label in the control DNA and sample DNA are the same. However, the yield obtained by the researcher might not be as high as the yield obtained by the company. If the company produced 800 ng of labeled DNA for every μg of template, and the researcher only managed to produce 100 ng of labeled DNA for every μg of template, then the results of the spot test are misleading. The 1 ng of DNA from the company has an average label density that is nearly five times greater than that obtained by the researcher, so 1ng of labeled control DNA from the company produces the same intensity of colorimetric spot as 5 ng from the researcher.

How did this happen? The sample and the control are very likely synthesized using identical mixtures of nucleotide substrates, so newly synthesized DNA has the same density of label. However, less of the template DNA is copied in the researcher's sample than in the company's control, and so the researcher's reaction product has too many DNA 'spectators' and not enough 'players'. This is effectively the same problem related to specific activity that was discussed earlier, in the context of radioactive probes. If a researcher adds 1 µg of a template to a nonradioactive-labeling reaction and only achieves a yield of synthesis of 50 ng, then the labeled DNA may not provide enough sensitivity as a probe molecule.

Problems

The solutions to these problems are provided at the end of the chapter.

(ix) Suppose that you want to end-label a DNA oligonucleotide that is 30 nt in length, using the substrate γ-[^{32}P]-ATP and the enzyme polynucleotide kinase.

A. If it is May 7 today and the vial of γ-[^{32}P]-ATP is marked '1.00 mCi in 0.100 ml; 6000 Ci mmol^{-1}, as of April 30', then how many µl should you use if you want to have exactly 50 µCi (as of May 7)?

B. If you have a 5-µM solution of the DNA oligonucleotide, how many µl should you use if you want the end-labeling reaction to be conducted with equimolar amounts of γ-[^{32}P]-ATP and DNA oligonucleotide substrates?

C. Suppose that you complete the labeling reaction with polynucleotide kinase, purify the oligonucleotide so that it is free of unincorporated γ-[^{32}P]-ATP, and determine that 0.1% of the oligonucleotide sample is 40 000 dpm. Assume that the oligonucleotide formula weight is 9900 Da, and that 1 Ci is 2.2×10^{12} dpm. What percentage of the oligonucleotide was labeled with ^{32}P, and what is the specific activity of your oligonucleotide in units of dpm µg^{-1}, or Ci mmol^{-1}?

(x) Suppose that you prepare a ^{32}P-labeled probe for use in a Southern blot by random hexamer-primed synthesis of a 4.5-kbp plasmid carrying a cloned sequence of 1 kbp. The synthesis is conducted using α-[^{32}P]-dCTP, dTTP, dGTP, dATP, 50 ng of plasmid, and Klenow enzyme. After purifying the product to remove unincorporated α-[^{32}P]-dCTP, you test 0.1% of the sample and determine that it has 5000 dpm. This is a bit of a lower level of incorporation than you are used to seeing.

A. What is the specific activity of the probe, in units of dpm µg^{-1}?

B. You had planned to use this probe in a Southern analysis to detect a single-copy 5-kbp restriction fragment, from an organism that has a genome of 5×10^8 bp (the 5-kbp restriction fragment only overlaps 1 kbp with the probe). Let us assume that the efficiency of probe hybridization to the blot is 5%, and that you need 100 dpm hybridized to a band on your Southern blot to have a detectable signal. How many µg of restriction enzyme-digested DNA from the organism would need to be loaded on the gel, to achieve a detectable signal on the Southern blot?

Further reading

Ausubel FW, Brent R, Kingston RE, Moore DD, Seidman JG, Smith JA and Struhl K (eds) (1987) *Current Protocols in Molecular Biology*. John Wiley & Sons, New York.

Perbal B (1988) *A Practical Guide to Molecular Cloning*, 2nd Edn. John Wiley & Sons, New York.

Sambrook J, Russell D (2001) *Molecular Cloning: A Laboratory Manual*, 3rd Edn. Cold Spring Harbor Laboratory Press, Cold Spring Harbor, NY.

References

1. McClure WR, Jovin TM (1975) The steady-state kinetic parameters and non-processivity of *Escherichia coli* deoxyribonucleic acid polymerase I. *J. Biol. Chem.* **250**: 4073–4080.

2. Kong HM, Kucera RB, Jack WE (1993) Characterization of a DNA polymerase from the hyperthermophile archaea *Thermococcus litoralis*. Vent DNA polymerase, steady-state kinetics, thermal stability, processivity, strand displacement, and exonuclease activities. *J. Biol. Chem.* **268**: 1965–1975.

3. Kunkel TA, Loeb LA, Goodman MF (1984) On the fidelity of DNA replication. The accuracy of T4 DNA polymerases in copying phi X174 DNA in vitro. *J. Biol. Chem.* **259**: 1539–1545.

4. Tindall KR, Kunkel TA (1988) Fidelity of DNA synthesis by the Thermus aquaticus DNA polymerase. *Biochemistry* **27**: 6008–6013.

5. Feinberg AP, Vogelstein B, (1984) A technique for radiolabeling DNA restriction endonuclease fragments to high specific activity. *Anal. Biochem.* **137**: 266–267.

6. Sanger F, Nicklen S, Coulson AR (1977) DNA sequencing with chain-terminating inhibitors. *Proc. Natl Acad. Sci. USA* **74**: 5463–5467.

7. Temin H, Baltimore D (1972) RNA-directed DNA synthesis and RNA tumor viruses. In: *Advances in Virus Research*, Vol. 17 (eds KM Smith, MA Lauffer, FB Bang). Academic Press, New York, pp. 129–186.

8. Krieg PA, Melton DA (1987) In vitro RNA synthesis with SP6 RNA polymerase. *Methods Enzymol.* **155**: 397–415.

Solutions to problems

Solutions to problems posed in Section 6.1

(i) The enzymes would make blunt ends, filling in recessed 3′ ends and digesting away overhanging 3′ ends. The replaced nucleotides are underlined:

```
(A) 5′– AATTGTACAGGCCCACCTGGGAATACATGA      –3′
    3′– TTAACATGTCCGGGTGGACCCTTATGTACT      –5′

(B) 5′– GATCTAAGAGCTACAGTTACTTACTACCCTCCGG   –3′
    3′– CTAGATTCTCGATGTCAATGAATGATGGGAGGCC   –5′

(C) 5′–      TATTACATACTTCGGAGAGTCAGGATAGCT   –3′
    3′–      ATAATGTATGAAGCCTCTCAGTCCTATCGA   –5′

(D) 5′–      GAGAGATCAGATCATGGGATCAGGAT       –3′
    3′–      CTCTCTAGTCTAGTACCCTAGTCCTA       –5′
```

It is a common mistake to think that a polymerase should 'fill in' recessed 5′ ends, such as those in the last example (D). However, a 5′ end cannot serve as a primer for a DNA polymerase.

(ii) If we are using a polymerase that has no 3′ to 5′ exonuclease, then recessed 3′ ends would be filled in but overhanging 3′ ends would not be digested back to give blunt ends:

(A) 5'- AATTGTACAGGCCCACCTGGGAATACATGATCAG -3'
    3'- TTAACATGTCCGGGTGGACCCTTATGTACT   -5'

(B) 5'- GATCTAAGAGCTACAGTTACTTACTACCCTCCGG -3'
    3'- CTAGATTCTCGATGTCAATGAATGATGGGAGGCC -5'

(C) 5'-   TATTACATACTTCGGAGAGTCAGGATAGCT -3'
    3'- AATTATAATGTATGAAGCCTCTCAGTCCTATCGA -5'

(D) 5'-   GAGAGATCAGATCATGGGATCAGGATAGCT -3'
    3'- GCGCCTCTCTAGTCTAGTACCCTAGTCCTA   -5'

(iii) A. The ends of the plasmid vector digested with *Eco*RI would have the following general structure:

5'-...NNNNNG   AATTCNNNNN...-3'
3'-...NNNNNCTTAA   GNNNNN...-5'

The ends of the DNA fragment digested with *Kpn*I would have the following general structure:

5'-...NNNNNGGTAC   CNNNNN...-3'
3'-...NNNNNC   CATGGNNNNN...-5'

Following treatment with Klenow enzyme and the dNTP substrates, the recessed 3' ends of the *Eco*RI fragment are filled in, and the overhanging 3' ends of the *Kpn*I fragment are digested, to make them blunt:

5'-...NNNNNGAATT   AATTCNNNNN...-3' (*Eco*RI)
3'-...NNNNNCTTAA   TTAAGNNNNN...-5'

5'-...NNNNNG   CNNNNN...-3' (*Kpn*I)
3'-...NNNNNC   GNNNNN...-5'

B. If dGTP had not been added to the reaction with Klenow enzyme, then the 3'-G nucleotide at the blunted *Kpn*I site would have been excised and not replaced (see *Figure 6.5D*), and the end would not be blunt.

5'-...NNNNN
3'-...NNNNNC

Depending on the composition of the 'N' nucleotides, the excision could be continued to make a longer 5' overhanging end.
   Furthermore, there would be a slight chance that the 3'-G of the *Eco*RI end would have been removed by a 3' to 5' exonuclease activity before the AATT could be synthesized to form the blunt end.

5'-...NNNNN
3'-...NNNNNCTTAA

Again, depending on the composition of the 'N' nucleotides the 5' overhanging end could be lengthened.
C. If the blunted *Eco*RI ends were joined to each other by ligation (after Klenow treatment), an *Eco*RI site (GAATTC) would not be regenerated:

(*Eco*RI) 5'-...NNNNNGAATT AATTCNNNNN...-3' (*Eco*RI)
          3'-...NNNNNCTTAA TTAAGNNNNN...-5'

D. If the blunted *Eco*RI and *Kpn*I ends were joined by ligation (after Klenow treatment), an *Eco*RI site (GAATTC) would be regenerated:

```
(EcoRI) 5'-...NNNNNGAATT CNNNNN...-3' (KpnI)
        3'-...NNNNNCTTAA GNNNNN...-5'
```

Solution to problem posed in Section 6.3

(iv) We need to design an oligonucleotide that will anneal to the following sequence, and generate a mutation (G to A) at the site of the lower-case 'g':

```
5'-...GCCGTTAGGAGTgAGTTACCCATCC...-3'
```

We can write the complementary sequence under this parental strand, taking care to substitute a 't' nucleotide for the 'C' that would base-pair with the lowercase 'g':

```
5'-...GCCGTTAGGAGTgAGTTACCCATCC...-3'
3'-   CGGCAATCCTCAtTCAATGGGTAGG   -5'
```

The oligonucleotide is the bottom strand, and written 5' to 3' it would read:

```
5'-GGATGGGTAACTtACTCCTAACGGC-3'
```

where the lowercase 't' indicates the point of mismatch between oligonucleotide and template. When the daughter strand is replicated, an 'A' nucleotide complementary to the 't' will be incorporated instead of the 'g' nucleotide, and the mutation will be established.

Solutions to problems posed in Section 6.4

(v) The first 10 chain-terminated extension products in the 'G' reaction would be as follows (the primer is written with uppercase letters, and the overall length is indicated at the end):

```
5'-GCTGACTCGCTCCAGAGCTGatg(23)
5'-GCTGACTCGCTCCAGAGCTGatgg(24)
5'-GCTGACTCGCTCCAGAGCTGatggg(25)
5'-GCTGACTCGCTCCAGAGCTGatgggaatcctg(32)
5'-GCTGACTCGCTCCAGAGCTGatgggaatcctgccg(35)
5'-GCTGACTCGCTCCAGAGCTGatgggaatcctgccgttcctactg(44)
5'-GCTGACTCGCTCCAGAGCTGatgggaatcctgccgttcctactgatccccatg(53)
5'-GCTGACTCGCTCCAGAGCTGatgggaatcctgccgttcctactgatccccatgg(54)
5'-GCTGACTCGCTCCAGAGCTGatgggaatcctgccgttcctactgatccccatggag(56)
5'-GCTGACTCGCTCCAGAGCTGatgggaatcctgccgttcctactgatccccatggagag(58)
```

(vi) A. Extending the primer at the 3' end would not affect the overall lengths of the products, because it is the 3' end that is being extended (primer is shown in uppercase):

```
5'-GCTGACTCGCTCCAGAGCTGAtg(23)
5'-GCTGACTCGCTCCAGAGCTGAtgg(24)
5'-GCTGACTCGCTCCAGAGCTGAtggg(25)
```

```
5'-GCTGACTCGCTCCAGAGCTGAtgggaatcctg(32)
5'-GCTGACTCGCTCCAGAGCTGAtgggaatcctgccg(35)
etc.
```

B. Extending the primer at the 5′ end would lengthen each product by one nucleotide; in the primer extension we would be 'pulling out our measuring tape' from a start point that is a little further away (primer is shown in uppercase):

```
5'-AGCTGACTCGCTCCAGAGCTGatg(24)
5'-AGCTGACTCGCTCCAGAGCTGatgg(25)
5'-AGCTGACTCGCTCCAGAGCTGatggg(26)
5'-AGCTGACTCGCTCCAGAGCTGatgggaatcctg(33)
5'-AGCTGACTCGCTCCAGAGCTGatgggaatcctgccg(36)
etc.
```

(vii) A. To obtain sequence information from the complementary strand, you need to design a primer that anneals to the end (primer is shown in uppercase):

```
151 catgatcttc caaaggctgt ggtcaaactc gagcccccgt ggatccaggt
                                   3'-CTCGGGGGCA CCTAGGTCCA-5'
```

The new sequencing primer (written 5′ to 3′) would be:

```
5'-ACCTGGATCCACGGGGGCTC-3'
```

B. The two templates are in agreement for the first six nucleotides, but beyond that point the products will be mixed (the primer is shown in uppercase):

```
5'-ACCTGGATCCACGGGGGCTCg
5'-ACCTGGATCCACGGGGGCTCga
5'-ACCTGGATCCACGGGGGCTCgag
5'-ACCTGGATCCACGGGGGCTCgagt
5'-ACCTGGATCCACGGGGGCTCgagtt
5'-ACCTGGATCCACGGGGGCTCgagttt
5'-ACCTGGATCCACGGGGGCTCgagtttg
5'-ACCTGGATCCACGGGGGCTCgagtttt
5'-ACCTGGATCCACGGGGGCTCgagtttga
5'-ACCTGGATCCACGGGGGCTCgagttttg
5'-ACCTGGATCCACGGGGGCTCgagtttgac
5'-ACCTGGATCCACGGGGGCTCgagttttga
5'-ACCTGGATCCACGGGGGCTCgagtttgacc
5'-ACCTGGATCCACGGGGGCTCgagttttgac
```

The sequencing ladder for this region is shown schematically in *Figure 6.17*.

(viii) We can determine by the clock that the first exposure amounted to 100 hours, or 4.17 days. If the amount of ^{33}P at noon on April 30 was taken to be 1 (arbitrary units), then after 4.17 days the fraction remaining will be $e^{-(0.02729)(4.17)} = 0.892$, using a constant shown in the inset formula in *Figure 6.10*. That is, $1-0.892$, or 0.108 of our arbitrary units of ^{33}P decayed during the first exposure. If we want

G A T C

Figure 6.17

Solution to a DNA sequencing ladder problem. The heterogeneous sequence is 5'-GAGTTT(G/T)(A/G)(C/A)C-3'.

five times that amount, or 0.540 units during the second exposure, then we want to stop the exposure process when the amount remaining is 0.892 − 0.540 = 0.352. That is, we need to solve the equation $e^{-(0.02729)(t)}$ = 0.352 for 't', where 't' is the elapsed time from noon on April 30. Taking the natural logarithm of both sides of the equation, and then solving algebraically, we find that:

$(-0.02729)(t) = \ln(0.352) = -1.044$
$t = 38.26$ days

Starting the clock at noon on April 30, the passage of 38.26 days would end at 6:15 PM on June 7. The first exposure was 100 hours, but the important point is that the second exposure was not 500 hours; it was a little more than 918 hours.

Solutions to problems posed in Section 6.8

(ix) A. On April 30 the concentration of γ-[^{32}P]-ATP in the vial was 10 μCi μl^{-1}. We can use the inset formula in *Figure 6.10* to determine that after 7 days the remaining concentration is $(10 \text{ μCi μl}^{-1})e^{-(0.04847)(7)}$ = 7.1 μCi μl^{-1}. Therefore, if we want to have exactly 50 μCi we need to use:

50 μCi ÷ 7.1 μCi μl^{-1} = 7.0 μl

B. The decayed γ-[^{32}P]-ATP is not biologically active, so let us assume that our 50 μCi of γ-[^{32}P]-ATP is close to being carrier-free, and at a specific activity of 6000 Ci mmol^{-1}. The 50 μCi is therefore:

5×10^{-5} Ci ÷ 6000 Ci mmol^{-1} = 8.3×10^{-9} mmol = 8.3 pmol ATP

If we want to add the same number of moles of DNA oligonucleotide to the reaction (so the oligonucleotide is labeled to a high specific activity), we need to figure out how many μl of the 5-μM solution would provide us with 8.3 pmol of oligonucleotide. If we remember that a 5-μM solution is always 5 pmol per μl, then we calculate that the answer is 8.3 pmol ÷ 5 pmol μl^{-1} = 1.7 μl. If we happen not to remember that little fact, then we can always write out the scientific notation and get the same answer:

8.3×10^{-12} mol ÷ 5×10^{-6} mol l^{-1} = 1.7×10^{-6} l = 1.7 μl

C. We added 8.3 pmol of oligonucleotide to the reaction, and now discover that 0.1% of it carries 4×10^4 dpm. The entire 8.3 pmol must carry 1000 times that amount, or 4×10^7 dpm. Our 8.3 pmol of oligonucleotide is:

8.3×10^{-12} mol $\times 9900$ g mol^{-1} = 8.2×10^{-8} g = 8.2×10^{-2} μg

and the specific activity is therefore:

4×10^7 dpm ÷ 8.2×10^{-2} μg = 4.9×10^8 dpm μg^{-1}

We could also calculate that the 4×10^7 dpm is:

4×10^7 dpm ÷ 2.2×10^{12} dpm Ci^{-1} = 1.8×10^{-5} Ci

and the specific activity of the oligonucleotide is therefore:

1.8×10^{-5} Ci ÷ 8.3×10^{-12} mol = 2.2×10^6 Ci mol^{-1}

and

2.2×10^6 Ci mol^{-1} = 2200 Ci mmol^{-1}

We had started the reaction with γ-[^{32}P]-ATP that was approximately 6000 Ci mmol^{-1} but the theoretical maximum specific activity for ^{32}P is approximately 8800 Ci mmol^{-1}, as discussed in Section 6.8. Our product is about 2200 Ci mmol^{-1}. This tells us that the percentage of oligonucleotide labeled with ^{32}P, compared to the theoretical maximum of 100%, is approximately 2200/8800 or 25%.

(x)   A. The calculation of specific activity depends only on the amount of DNA being labeled (in μg) and the amount of ^{32}P that is incorporated. We conducted the labeling reaction with 50 ng (0.050 μg) and the total ^{32}P incorporated was 1000×5000 dpm, or 5×10^6 dpm. Therefore, the specific activity of the probe is:

5×10^6 dpm ÷ 0.050 μg = 1×10^8 dpm μg^{-1}.

That is a low specific activity, and the probe would probably be too weak to use in a Southern blot of human DNA (genome size 3×10^9 bp). The amount of DNA that would need to be loaded on the gel to establish a detectable signal on the Southern blot would probably exceed the capacity of the gel. However, the organism in this case (see part B) has a smaller genome of 5×10^8 bp, so each μg of DNA has more copies of the genome.

B. We can start the calculation by imagining that we have 100 dpm of probe hybridized to a 5-kbp band on the Southern blot. From there, we can work our way back to the amount of target DNA that would need to be loaded to obtain the required signal. We calculate that the 100 dpm of probe is:

100 dpm ÷ 1×10^8 dpm μg^{-1} = 1×10^{-6} μg = 1×10^{-12} g probe.

We assumed a hybridization efficiency of 5%, meaning that only 5% of the target molecules on the blot are bound by probe. If the efficiency were 100%, the amount of probe hybridized to the target would be 20 times greater, or 2×10^{-11} g. However, only 1 kbp of the probe covers the target, so the target is five times larger than the probe; the amount of target in the 5-kbp band on the Southern blot would have to be five times 2×10^{-11} g, or 1×10^{-10} g.

How much genomic DNA would we need to load on the gel to have 1×10^{-10} g in that one 5-kbp fragment? The 5-kbp DNA fragment is 1/100 000 of the whole genome, because

$$5 \times 10^3 \text{ bp} \div 5 \times 10^8 \text{ bp} = 1 \times 10^{-5}.$$

Therefore, we would need to load 1×10^5 times 1×10^{-10} g, or 1×10^{-5} g of genomic DNA on the gel. This equals 10 µg, which will not exceed the capacity of an agarose gel if the electrophoresis is conducted at low voltage.

The polymerase chain reaction

7

7.1 Overview

In the polymerase chain reaction (PCR), a segment of a template DNA is copied repetitively using a DNA polymerase, and its concentration increases geometrically as the reaction proceeds (1,2). The outcome may remind us of chain letters or chain e-mails that we sometimes receive: 'Copy this letter every day and send it to a different friend,' the message might say, with an added promise that we will be lucky and win a lot of money in about 3 weeks if we do so. Suppose we all actually followed these instructions and copied the letter. If 1000 of us gave the letter to a friend on April 1, then 2000 people would have the letter at the end of the day, 4000 would have the letter by the end of the day on April 2, and 8000 would have the letter by the end of the day on April 3. Assuming that no person is counted as a friend by more than one person, and nobody is friendless, then we can watch the amplification and spread of the letter, day-by-day. By the end of the day on April 10, 1 024 000 would have the letter; by the end of the day on April 20, 1.05 billion people would have the letter. If we could continue the chain letter to the end of the month, then by the end of the day on April 30, approximately 1.07 trillion people would have the letter; or they would, except that we actually run out of people on Earth around 7:45 PM on April 23, after nearly 23 cycles of letter writing are completed. Now, if these letter recipients would each just send the 1000 original authors one dollar, euro, peso, rupee, or yuan, along with their best wishes, the promise of immense wealth would be fulfilled!

This story of a chain letter helps to explain the power of PCR, and also some of its difficulties in planning. If we're going to write 5.36×10^{11} new letters on April 30, for example we're going to need to have 1.07 billion reams of paper in stock, and an amount of ink that is proportional to the planned length of each letter. Similarly, with PCR if we're going to make 5.36×10^{11} copies of a template in the 30th cycle, a number of copies that incidentally would be approximately 100 ng of a 170-nt DNA fragment, we're going to consume 0.89 pmol of each primer and an amount of dNTP substrate that depends on the length of each PCR product. In the world of chain letters there are always some lazy writers who don't make the copies they are supposed to, and this reduces the efficiency of the geometric amplification. In PCR there are copies that are not fully extended and cannot be used as templates in the next cycle, and this similarly reduces the efficiency so that the number of templates is not actually doubled in each cycle.

During the process of PCR, each cycle starts with a heat denaturation step, usually at 94°C to 98°C, which converts the double-stranded DNA fragments into single strands. Subsequently, the oligonucleotide primers are allowed to anneal to the DNA strands at an appropriate lower temperature, usually 50°C to 65°C, and finally the temperature is adjusted to approximately 72°C so that the DNA polymerase can conduct a primer extension reaction (discussed in Chapter 6). During this final part of each cycle, the DNA templates are copied and the newly synthesized DNA is added to the template that is available in the next cycle. The components of the PCR are added to the sample tube at the start of the reaction, and are gradually consumed over the course of many cycles; there is usually no need to open the tube and refresh the supply of any component. The polymerase chain reaction is the topic of another book in this series (3).

7.2 Primer design

Designing PCR oligonucleotide primers is not difficult, provided that a few matters are systematically addressed. When sitting down to design primers for PCR it is important to know what PCR product is desired, because the placement of primers on the template establishes the composition of the product. Next, the primer sequences must be designed so that they anneal to the correct strands; that is, the primers must be on the right strands, and pointed in the right direction for synthesis. Finally, the researcher must give some consideration to the annealing temperatures of the primers, and the possibility of unwanted side reactions; this may necessitate revising the original plan. There are software programs that can help a researcher with planning oligonucleotides, but this chapter will be written with the assumption that the researcher is not using one of them.

The 3′ end of the oligonucleotide is the primer

The right direction, the left direction, and the wrong direction

The 3′ ends of the oligonucleotide primers are the ends that are extended by a DNA polymerase during PCR, so they must each point in the intended direction of synthesis. If we write out a double-stranded DNA sequence on paper, with the top strand running 5′ to 3′ from left to right, and the bottom, complementary strand running 5′ to 3′ from right to left, then it is straightforward to mark off the primer sequences that would generate such a DNA fragment (see *Figure 7.1*). We can circle some of the DNA sequence at the left end of the top strand, and mark it overhead with an arrow pointed to the right. Then, we can circle some of the sequence at the right end of the bottom strand, and underline it with an arrow pointed to the left. We are going to delay, for the moment, any discussion of how many nucleotides need to be included in each primer; the important thing at this stage is simply to get the oligonucleotide sequences running in the right directions. The two arrows should point towards each other, because we want the polymerase to synthesize the DNA between the arrowheads.

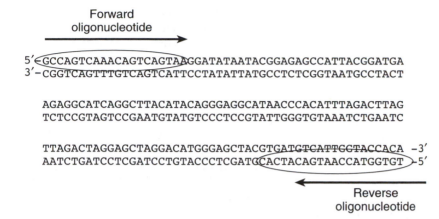

Figure 7.1

Designing oligonucleotide primers for PCR. The sequence in the upper left will become the content of the forward oligonucleotide primer, and the sequence in the lower right will become the content of the reverse oligonucleotide primer. The 5′ ends of the oligonucleotide primers determine the boundaries of a PCR product.

Now we should give some thought to the sequences of the two oligonucleotides that might be ordered. The circled sequence on the top strand reads 5′ to 3′, from left to right, and that makes it easy. For example, if we circled a sequence reading 'GCCAGTCAAACAGTCAGTAA' on the top strand (see *Figure 7.1*), we might order an oligonucleotide primer with the sequence

```
1. Forward: 5′-GCCAGTCAAACAGTCAGTAA-3′
```

The 'A' nucleotide at the right end carries the 3′ hydroxyl group that will be extended by the polymerase during PCR, in the direction indicated by the arrowhead. We call this oligonucleotide '1' the 'forward' primer, because the arrow points to the right (see *Figure 7.1*).

On the other hand, the circled sequence on the bottom strand reads 5′ to 3′ from right to left, because double-stranded DNA consists of two antiparallel strands. If we circled a sequence reading 'CACTACAGTAAC-CATGGTGT' on the bottom strand, that sequence could be annotated as 3′-CACTACAGTAACCATGGTGT-5′, with the arrowhead pointing to the left (see *Figure 7.1*). However, if we wanted to order it as an oligonucleotide we would have to rewrite it so that we follow the convention of placing the 5′ end on the left:

```
2. Reverse: 5′-TGTGGTACCAATGACATCAC-3′
```

Rewriting it in the reverse direction does not change the sequence, provided that we change the 5′ and 3′ markings accordingly; we only need to rewrite the sequence so that the oligonucleotide manufacturer will understand the ordering instruction. We call this oligonucleotide '2' the 'reverse' primer, because it is pointed in the opposite direction to the forward primer.

A word should be said about the common mistakes that can be made in planning primers, because studying mistakes can help us to learn how to do it right. If the researcher does not make sure that the 3′ ends of the primers face each other, then the polymerization will not converge to generate a product that can be amplified. For example, if a researcher needs to order the forward and reverse primers 5′-GCCAGTCAAACAGTCAGTAA-3′ and 5′-TGTGGTACCAATGACATCAC-3′, but instead orders their complementary sequences 5′-TTACTGACTGTTTGACTGGC-3′ and 5′-GTGATGTCATTG-GTACCACA-3′, then the DNA polymerases will be sent in the wrong direction. Another type of mistake can occur if a researcher becomes confused about how to rewrite the reverse oligonucleotide for ordering. If we want to rewrite the sequence 3′-CACTACAGTAACCATGGTGT-5′ with the 5′ end on the left, we have to rewrite both the sequence and the labels; switching only the labels gives us an incorrect sequence: 5′-CACTACAGTAACCATGGTGT-3′ that will not anneal to the template. Finally, if a researcher stares at the problem for too long, he may become confused about the template strand for each primer and make some rash and unwarranted change. Here is a simple explanation of how the primers will be used in the reaction: the forward primer, the sequence that we circled in the upper-left corner of our diagram of the double-stranded DNA sequence, will use the bottom strand as a template; the reverse primer, the sequence that we circled in the lower-right corner of our diagram, will anneal to the top strand.

A unique primer binding site

Now that we have a clear idea about the direction each primer must point, which is a purely mechanical step in the planning, we need to think more deeply about whether the primers are going to provide the reaction product that we want. For example, if we are using PCR to isolate a fragment of DNA from a genome, we need to ask whether the primer sequences are present as a single copy in the genome or are part of a repetitive DNA element. For example, a gene of interest may be duplicated in the genome, or part of a gene family that contains a similar sequence motif. If a primer can anneal to many DNA sequences, then there are opportunities for the amplification of spurious PCR products.

If genome sequence information is available for an organism, a researcher may take advantage of this information to search for sites at which the planned oligonucleotide may have a partial 3′ end match. Depending on the extent of the unwanted match, the oligonucleotide design could be modified to step around the problem, or a slightly higher annealing temperature could be used to avoid generation of a nonspecific product. For example, if we test the oligonucleotide

```
1.  Forward:  5′-GCCAGTCAAACAGTCAGTAA-3′
```

that we have been using as model, using a BLAST search of the GenBank® (U.S. Department of Health and Human Services) database, we obtain 130 hits and find that at the 3′ end there is a match with 13 to 18 nt of specific sequence from the human and zebrafish genomes. This oligonucleotide sequence '1' was generated at random, without any plan to match a pre-existing sequence; however, it would very likely prime synthesis of DNA from those human and zebrafish sequences and many others. We may expect background annealing between almost any oligonucleotide and a complex genome, but this is only going to generate a problem in PCR if there is a second downstream site nearby, at which one of the oligonucleotides can prime the reverse direction of synthesis.

A 3′ region of exact match

The 3′ ends of the oligonucleotide primers must match a template perfectly if they are to be extended by a DNA polymerase, as discussed in Chapter 6. For example, when we use oligonucleotide primers for mutagenesis there is a mismatch between the primer and a parental template. However, we place the mismatched regions towards the 5′ end of the oligonucleotide so that the bases at the 3′ end are annealed. How many nucleotides of perfect match are required at the 3′ end of a primer for the PCR to work? There is no single answer to the question, because annealing is temperature-dependent. Furthermore, the problem of mismatch is gradually relieved after the first few cycles of PCR, if there is some successful synthesis, because the primer sequences are copied into the template and become a perfect match thereafter. The most cautious approach involves designing a primer that has a generous base-pairing region at the 3′ end, perhaps 18 nt, to ensure successful annealing during the first few PCR cycles. Once the PCR is underway, the full length of the primer will anneal to the newly made templates.

The 5′ end boundaries define the product

The 5′ ends of the oligonucleotide primers become the 5′ ends of the PCR product. Therefore, we must give some thought to the precise boundaries of the desired product when designing oligonucleotides. This is particularly important if we are preparing a fragment for cloning into an expression vector, because we may want to maintain a particular spacing of regulatory elements or construct an in-frame fusion protein gene.

Suppose that we have a 1-kbp sequence, and that we are interested in generating a PCR product that stretches from nt 101 of a specific sequence to nt 400. Where would our oligonucleotide primers start? The forward primer would have nt 101 as its 5′ end, and might extend to nt 118 at its 3′ end if the overall oligonucleotide is 18 nt in length. The reverse oligonucleotide would be based on the complementary strand, and would have a 5′ end complementary to nt 400. If the reverse oligonucleotide was also 18 nt in length, then the 3′ end would be complementary to nt 383. We would get exactly the same product if we extended each oligonucleotide at its 3′ end while leaving the 5′ end the same. This is because the 5′ ends of the primers define the boundaries of the product, and adding to the oligonucleotide at its 3′ end simply means that there is less work for the polymerase to do during synthesis.

Predicted annealing temperature

When a primer anneals to a template, it is forming two hydrogen bonds for every AT base pair and three for every GC base pair. There is a relationship between the total number of hydrogen bonds formed and the optimal annealing temperature of an oligonucleotide primer, in that having more hydrogen bonds stabilizes the interaction and raises the optimal annealing temperature. When we perform a PCR with a pair of oligonucleotides, we try to predict the optimal annealing temperatures for each primer so we can set the thermocycler accordingly in the second step of each cycle. The annealing temperature may be lower than the melting temperature T_M by a few degrees.

Given a choice, we would rather use an oligonucleotide with a higher annealing temperature, in the range of 55 to 65°C, because it can be used with greater specificity than an oligonucleotide with an annealing temperature of 45°C. The oligonucleotide

```
1.  Forward:  5'-GCCAGTCAAACAGTCAGTAA-3'
```

has a predicted annealing temperature of about 50°C, but may anneal to the wrong target at 40°C, for example a hypothetical pairing in which 13 out of 20 bases are annealed:

```
            5'- GCCAGTCAAACAGTCAGTAA-3'
                 | |||||  |  |  |||||
3'-GGATCATACATTAAGAGAATTAGTTTCTTACTCATTGGACAGTAC-5'
```

With lower specificity, more unwanted side reactions occur. We may increase the annealing temperature of an oligonucleotide by increasing its length at the 3' end, which will not have any effect on the 5' boundaries of the PCR product.

Primer length and GC content

Increasing the length of a primer, or increasing its GC content, will increase the number of hydrogen bonds formed with a template and therefore increase its optimal annealing temperature. A simple algorithm for predicting the T_M of an oligonucleotide longer than 14 nt is this: count the number of G and C bases in the primer and the total number of nucleotides in the primer, then use the formula

$$T_M = 64.9°C + 41°C \times (\#GC - 16.4)/n$$

where #GC is the number of G or C nucleotides and n is the length of the primer. For example, if we are planning oligonucleotides with sequences

```
1.  Forward:  5'-GCCAGTCAAACAGTCAGTAA-3'
2.  Reverse:  5'-TGTGGTACCAATGACATCAC-3'
```

then each will have 9 G or C bases and 20 nucleotides total. From the formula we have a predicted T_M of 50°C, and the optimal annealing temperature might be a few degrees below that. We could set a temperature of 50°C for the annealing step of each PCR cycle, and if necessary make changes to find the best annealing temperature empirically.

However, suppose that our second oligonucleotide had 11 GC bases and 9 AT bases; in that case, the predicted T_M would be approximately 54°C. Would we set the machine differently to account for having two different T_M values? Probably not, because if we raised the annealing temperature above 50°C then the first primer might have a more difficult time forming a stable interaction with the template. We want to ensure that both primers have an opportunity to anneal, so we attend to the requirements of the primer that has the lower T_M. If we are using two oligonucleotides with widely different T_M values, for example an 18-mer with a predicted T_M of 48°C and a 30-mer with a predicted T_M of 68°C, then the annealing of the shorter oligonucleotide would be the focus of the annealing step of the PCR and might be conducted at 45°C to 50°C. The annealing of the longer oligonucleotide could happen concurrently with the extension of the primers, during the next step in the cycle.

Software programs are available that use more sophisticated algorithms to predict T_M, some using empirical data such as the nearest neighbor model to account for the order of nucleotides. However, these calculations are not definitive and only provide a general guide to the researcher. The work of optimizing a PCR may still require careful experimentation.

Primer complexity and spurious annealing

The two oligonucleotide sequences

```
1.  Forward:  5'-GCCAGTCAAACAGTCAGTAA-3'
3.  Forward:  5'-GGGGAAAAAAAATTTCCCCC-3'
```

have the same sequence composition and length, and taken one nucleotide at a time they are equally likely as random sequences. However, sequence '3' has less complexity and is more commonly found in GenBank®, generating 613 hits in a BLAST search, substantially more than the 130 hits generated with '1'. Large genomes contain tracts of low-complexity sequence, often in repeated units that may generate spurious PCR products. Redundant sequences such as 'AGAGAGAGA', or 'CCTCCTCCT', or 'GATCGATCGATC' are therefore best avoided in oligonucleotide design. A sequence that has a reasonable balance of the four nucleotide bases is more likely to be complex than a sequence that makes scant use of one or more of the bases.

For example, the oligonucleotide

```
1.  Forward:  5'-GCCAGTCAAACAGTCAGTAA-3'
```

has the same number of G+C and A+T bases as

```
4.  Forward:  5'-GCGAGTGAAAGAGTGAGTAA-3'
```

but oligonucleotide '4' is slightly less complex than '1' because all but one of the cytidylate nucleotides have been changed to guanylate nucleotides, making the 'C' nucleotides under-represented. Indeed, a BLAST search using this C-deficient sequence '4' generates 179 hits, instead of 130 hits obtained with '1', the more complex sequence. Taking this a step further and changing all but one of the adenylate nucleotides to thymidylate nucleotides, the oligonucleotide sequence

```
5.  Forward:  5'-GCGTGAGTTTGTGTGTGTTT-3'
```

has under-representation of both A and C, and generates 361 hits in a BLAST search. This sequence '5' is reasonably balanced between purines and pyrimidines. However, had we instead changed all but one of the thymidylate nucleotides into adenylate nucleotides then the resulting sequence

6. Forward: 5'-GCGAGTGAAAGAGAGAGAAA-3'

would have only two pyrimidine bases and 18 purine bases. In a BLAST search, this purine-rich sequence '6' generated 673 hits, highlighting the problem with oligonucleotides lacking complexity.

A primer can be a template

Oligonucleotide primers can involve themselves in side reactions, even without any DNA being added to a PCR. For example, if we modified the last nucleotide of the oligonucleotide

1. Forward: 5'-GCCAGTCAAACAGTCAGTAA-3'

changing the A-3' ending to a C-3' end, then the slightly revised oligonucleotide would read

7. Forward: 5'-GCCAGTCAAACAGTCAGTAC-3'

We can imagine two molecules of oligonucleotide '7', with GTAC-3' ends, base-pairing with each other and being extended by a DNA polymerase.

```
5'-GCCAGTCAAACAGTCAGTAC-3'
              3'-CATGACTGACAAACTGACCG-5'
```

This would not be a very common annealing reaction at a high temperature, but it would only have to occur once to create a template molecule with the sequence

```
5'-GCCAGTCAAACAGTCAGTACtgactgtttgactggc-3'
3'-cggtcagtttgtcagtCATGACTGACAAACTGACCG-5'
```

where the lowercase letters indicate the sequence added by the DNA polymerase and the uppercase letters indicate the sequence coming from oligonucleotide '7'. This 36-nt product is a dimer of the primers, and represents a class of unwanted side reactions called primer dimers. Once made in a reaction, it base-pairs perfectly to the oligonucleotide and can readily anneal at the normal annealing temperature of the PCR. Primer dimer formation consumes oligonucleotide that was intended for use on the legitimate template, and may compete with that reaction, preventing it from working well.

Primer dimers can be homodimers, forming between two copies of a single oligonucleotide, or heterodimers, forming between two different oligonucleotides. Suppose we start a PCR with the oligonucleotide primers

1. Forward: 5'-GCCAGTCAAACAGTCAGTAA-3'
8. Reverse: 5'-GAATTCACCAATTACTTCAC-3'

The AGTAA-3' end of primer '1' can base-pair with the internal sequence 5'-TTACT-3', nucleotides 12 to 16 of primer '8':

```
(primer '1') 5'-GCCAGTCAAACAGTCAGTAA-3'
                                 |||||
                    3'-CACTTCATTAACCACTTAAG-5' (primer '8')
```

and the first primer could use a portion of the second as a template, generating a 31-nt extended sequence

```
5'-GCCAGTCAAACAGTCAGTAAttggtgaattc-3'
                  ||||||||||||||||
              3'-CACTTCATTAACCACTTAAG-5'
```

where the new synthesis is shown in lowercase. This dimer would not be amplified geometrically, because the second primer would not be able to anneal its 3' end to the dimer and use it as a template. However, the five nucleotides 5'-gtgaa-3', from nt 24 to 28 of the dimer, could base-pair to the TTCAC-3' end of the second oligonucleotide:

```
5'-GCCAGTCAAACAGTCAGTAAttggtgaattc-3'
                        |||||
                 3'-CACTTCATTAACCACTTAAG-5'
```

generating a different primer dimer (bottom strand)

```
5'-GCCAGTCAAACAGTCAGTAAttggtgaattc-3'
   |||||||||||||||||||||||||||||||
3'-cggtcagtttgtcagtcattaacCACTTCATTAACCACTTAAG-5'
```

This 43-nt dimer would contain both oligonucleotide sequences and if synthesis of the top strand happened to be incomplete, or the three 3' nucleotides were removed by exonuclease activity, the dimer could be amplified geometrically:

```
5'-GCCAGTCAAACAGTCAGTAAttggtgaagtaattggtgaattc-3'
3'-cggtcagtttgtcagtcattaacCACTTCATTAACCACTTAAG-5'
```

We should also consider the possibility that the previous 31-nt primer dimer, while not being amplified geometrically, could still be generated in impressive amounts due to the high concentration of the two oligonucleotide reactants. The 3' end of that primer dimer is a symmetric sequence gaattc-3', and two of these primer dimers could pair

```
5'-GCCAGTCAAACAGTCAGTAAttggtgaattc-3'
                        ||||||
              3'-cttaagtggttAATGACTGACAAACTGACCG-5'
```

and be extended to form a tetramer of 56 nucleotides:

```
5'-GCCAGTCAAACAGTCAGTAAttggtgaattcaccaattactgactgtttgactggc-3'
3'-cggtcagtttgtcagtcattaaccacttaagtggttAATGACTGACAAACTGACCG-5'
```

This tetramer product could be amplified geometrically using only primer '1' (shown in uppercase in the product).

In short, a bit of cross-priming between oligonucleotides can lead to side reaction products that are unwanted, and that consume primers that should have been used in synthesis of the desired product. Some of the reaction products may be amplified linearly, and others geometrically, but a characteristic of primer dimers is that they can form in the absence of any template being added to the reaction. If a PCR is plagued by the production of primer dimers, a researcher may wish to initiate the PCR using a hot start method, as discussed in Section 7.8, to minimize side reactions. If this does not solve the problem, the primers might need to be redesigned in a way that will minimize primer dimer formation.

A template can be a primer

There are some interesting cases in which DNA synthesized during a PCR cycle may act as a primer. For example, two sequences can be joined in a PCR if there is overlapping sequence between them, and this is a common approach to mutagenesis that is discussed in Section 7.5. Suppose two oligonucleotides, A and B, are used to generate one DNA fragment by PCR, and two different oligonucleotides, C and D, are used to generate another DNA fragment by PCR (see *Figure 7.2A*), and suppose the 5′ end of oligonucleotide B and the 5′ end of oligonucleotide C overlap 15 nt in sequence. If the two DNA fragment products are put in solution together, denatured, and cooled, then they too will overlap by 15 nt and may anneal (see *Figure 7.2B*). In pairings in which the 3′ ends of the single-stranded DNA fragments overlap, these may be extended by a DNA polymerase to make a combined fragment containing the sequences of both DNA fragments (see *Figure 7.2C*). In essence, the product of one reaction serves as a primer for another.

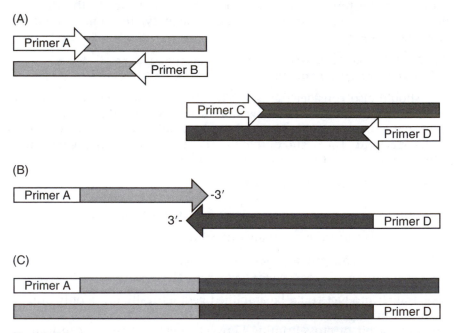

Figure 7.2

A template can act as a primer during PCR. (A) Two PCR reactions are shown; primer A and primer B generate the product on the left, and primer C and primer D generate the product on the right. Primer B and primer C have complementary sequences at their ends. (B) If the two reaction products are mixed, the 3′ ends may base-pair and serve as primers for each other (arrowheads). (C) Extension of the 3′ ends results in a fusion product that can be amplified with primer A and primer D.

Experimental results can be misleading if products act as primers. Suppose a researcher is using RT-PCR to uncover a pattern of alternative mRNA splicing for a particular gene (see *Figure 7.3A*). The researcher uses an exon 5 primer with the mRNA in a reverse transcriptase reaction, then

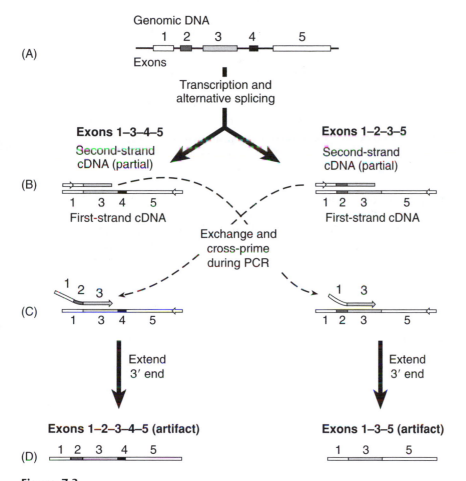

Figure 7.3

Possible generation of artifacts during RT-PCR mapping of alternatively spliced transcripts. (A) The genomic structure of a gene with five exons. (B) cDNA copies of two alternatively spliced forms of the transcript. One form contains exons 1, 3, 4, and 5 (on the left) and the other contains exons 1, 2, 3, and 5 (on the right). The first-strand cDNA synthesis was initiated from a specific oligonucleotide primer in exon 5, and is complete. The second-strand synthesis from a specific oligonucleotide in exon 1 is shown as incomplete, with extension paused in exon 3. (C) During PCR the incompletely synthesized strands are denatured and may anneal to a different template (dashed lines). The exon 1–2–3 product may anneal to the 1–3–4–5 template and the exon 1–3 product may anneal to the 1–2–3–5 template. (D) Extension of these partially completed strands on a different template leads to cross-priming artifacts: a cDNA containing all of the exons 1–2–3–4–5 (on the left) and a cDNA containing only exons 1–3–5 (on the right). These cDNAs do not represent naturally occurring transcripts.

follows that with a PCR using the exon 5 primer and a forward primer directed to exon 1. In the example provided in *Figure 7.3*, the gene transcript is alternatively spliced to yield one mRNA with exons 1, 2, 3, and 5, and another with exons 1, 3, 4, and 5. The PCR synthesis of these alternative cDNAs will lead to some full-length product but also to the accumulation of partially completed cDNAs containing only exons 1–2–3 or 1–3 (see *Figure 7.3B*). During subsequent cycles, these partially completed products may anneal to the exon 3 sequence of a different cDNA template and be extended (see *Figure 7.3C, D*). An exon 1–2–3 fragment that is extended on an exon 1–3–4–5 cDNA template will become a product containing exons 1–2–3–4–5. An exon 1–3 fragment that is extended on an exon 1–2–3–5 cDNA template will become a product containing exons 1–3–5 only. These latter two products are chimeras that are not reflective of the alternatively spliced mRNA pool, which only contains exon 1–2–3–5 and exon 1–3–4–5 transcripts. In other words, structural characterizations of nucleic acids can be erroneous if PCR is involved, and that is a reason to be very cautious.

The general principle is that if two templates overlap, they may cross-prime during PCR and make a fused template. A fragment with structure 'ABC' can be put together from 'AB' and 'BC', if a researcher has added a forward primer that matches 'A' and a reverse primer that base-pairs to 'C'. If segment 'B' is a repetitive sequence that exists in many copies in the genome, then the joining of segments 'A' and 'C' in 'ABC' may be a complete artifact. This fusion can even occur if one of the original templates is RNA, because the widely used DNA polymerase *Taq* has a reverse transcriptase activity.

The products made by a polymerase are at substantially lower molar concentrations than the oligonucleotide primers for most of the PCR cycles, and this probably prevents most cross-priming effects. Furthermore, the 3' ends of completed PCR products are usually covered by the 5' ends of the complementary oligonucleotide primer and not available for use as a primer. However, if a template is subjected to a large number of PCR cycles, then rare cross-primed templates may rise to the forefront of detectable products.

Degenerate primers

Degenerate primers are mixtures of primers that show variation in sequence; despite the name 'degenerate', they are not primers found lacking in some moral quality! A degenerate primer mixture is often used when a researcher is uncertain of the exact template sequence to which the primer must bind, or wishes to incorporate combinatorial variation into the product mixture.

Reverse translation

Here is an example of how degenerate primers might be used. If we isolate a protein from a cell and perform Edman degradation and peptide sequencing, we may find that one part of the protein contains the peptide Gly-Asp-Lys-Trp-Thr-Gln-Thr-Cys, and another part of the protein contains the peptide Ala-Glu-Thr-Gly-Gly-Asn-Tyr-Val. We don't know which of the two peptides comes first in the protein sequence, nor do we know the exact mRNA sequence that encodes these peptides because of the degeneracy of the genetic code. However, we may represent these two peptides as sequences of degenerate triplet codons:

```
Peptide 1.  N-Gly-Asp-Lys-Trp-Thr-Gln-Thr-Cys-C
            5'-GGN-GAY-AAR-UGG-ACN-CAR-ACN-UGY-3'

Peptide 2.  N-Ala-Glu-Thr-Gly-Gly-Asn-Tyr-Val-C
            5'-GCN-GAR-ACN-GGN-GGN-AAY-UAY-GUN-3'
```

where 'N' represents any of the four nucleotides, 'Y' represents any pyrimidine, and 'R' represents any purine. We might attempt to make and isolate a cDNA containing these two sequences by RT-PCR, and so would order a set of forward and reverse oligonucleotide primers for each:

```
 9.  1-forward:  5'-GGNGAYAARTGGACNCARACNTG-3'
10.  1-reverse:  5'-CANGTYTGNGTCCAYTTRTCNCC-3'
11.  2-forward:  5'-GCNGARACNGGNGGNAAYTAYGT-3'
12.  2-reverse:  5'-ACRTARTTNCCNCCNGTYTCNGC-3'
```

The last nucleotide of the degenerate codon sequence was not included in the oligonucleotides, because it adds no specificity, so the overall length of each oligonucleotide is 23 nt. The 'N', 'R', and 'Y' codes for degeneracy are widely used, and a researcher can use these letters when ordering degenerate primers from a commercial source. The RT-PCR would be conducted on mRNA in two separate trial reactions, because we don't know the order of the peptides in the protein sequence; these would be primers '9' and '12' together in one reaction, and primers '11' and '10' together in a different reaction. If the order of peptide sequences in the protein is 'peptide 1' followed by 'peptide 2', then the first reaction will generate an RT-PCR product, and if the order of peptide sequences is 'peptide 2' followed by 'peptide 1', then the second reaction will generate an RT-PCR product.

Some of the oligonucleotides in a degenerate mixture will be effective as primers, because they are complementary to the template, and others will be inert because they do not base-pair at their 3' ends. In the examples provided, the primers '9' and '10' had three positions containing 'N' nucleotides and three positions containing either 'R' or 'Y' nucleotides. There are four possible nucleotides for each 'N' position, and two possible nucleotides for each 'R' or 'Y' position, so the combinatorial possibilities for the 1-forward and 1-reverse primer are $(4)(4)(4)(2)(2)(2) = 512$. The primers '11' and '12' were slightly more degenerate, with four positions containing 'N' nucleotides and three positions containing either 'R' or 'Y' nucleotides; they would each have $(4)(4)(4)(4)(2)(2)(2) = 2048$ possible sequences.

In any PCR, only one of the primers in each degenerate set will match the template sequence exactly. Some of the oligonucleotides that are an imperfect match may still participate in occasional reactions, particularly if the mismatch between primer and template is near the 5' end of the primer and the annealing conditions are not stringent. However, the majority of the oligonucleotides will have a mismatch at the 3' end that prevents their use as primers, and the amount of oligonucleotide added to the RT-PCR may need to be increased to account for the limited activity of the overall mixture of degenerate primers.

Conserved protein sequence information

Degenerate PCR primers are also useful when a researcher is attempting to clone a gene from an organism of interest, using sequence information gathered from a group of closely related organisms. The researcher may create an alignment of homologous protein sequences from these related organisms, and search for peptide segments that appear to be highly conserved. The amino acids that are the most informative about mRNA sequence are Met and Trp, with only one codon specifying each, and Asp, Asn, Cys, Glu, Gln, His, Lys, Phe, and Tyr with only two codons specifying each. Ile is specified by three codons, and Ala, Gly, Pro, Thr, and Val are specified by four, and the amino acids Arg, Leu, and Ser are least informative, with six codons specifying each.

To take Arg, Leu, and Ser as the 'worst possible' scenario, the peptide sequence Arg-Leu-Ser would be encoded by the 9-nt sequence MGNYT-NWSN, where M represents either A or C, W represents either A or T, and S represents either G or C. The combinatorial possibilities would include $(2)(4)(2)(4)(2)(4) = 512$ different oligonucleotides; however, many of these would not even code for the desired sequence. For example, AGATTAAGC encodes Arg-Leu-Ser, but AGCTTCACC encodes Ser-Phe-Thr and would also be part of the degenerate oligonucleotide mixture.

With conserved sequences in mind, degenerate primers can be designed for use in a PCR to isolate the gene from the organism of interest. In this application the order of peptide segments in the gene is known and only one degenerate forward and reverse primer need be made for any conserved segment. The researcher may decide to design oligonucleotides in a way that Arg, Leu, and Ser codons are avoided entirely, because they are of little help in making the oligonucleotide specific. Furthermore, the 3′ ends of oligonucleotides may be planned around codons for amino acids such as Asp and Tyr, which have limited degeneracy, or Met and Trp, which are not degenerate.

Codons that are rarely used in the genes of a particular organism are sometimes excluded from degenerate oligonucleotides, and tables of codon usage may be consulted for information on how frequently each codon is used. There is some risk that a rare codon will be missed and a PCR won't work if the researcher takes too many chances in designing a degenerate oligonucleotide. For example, the percentage use of each of the alanine codons in human genes is 26% GCU, 40% GCC, 23% GCA, and 11% GCG; a researcher might design an oligonucleotide with only the first three possible endings, GCT, GCC, and GCA, and accept the 11% risk that the oligonucleotide will not base-pair to a GCG template. If the alanine codon were part of a reverse oligonucleotide, the anticodon sequences AGC, GGC, and TGC might be included in the oligonucleotide, omitting the more rare possibility of CGC. However, the researcher may take some solace from the fact that guanine and thymine bases can form a single hydrogen bond when they are mispaired, and so the TGC anticodon may have some limited ability to anneal to both GCA and GCG codons.

Combinatorial libraries

Degenerate oligonucleotides can be used in the construction of combinatorial libraries, which may be useful in mapping the function of individual amino acid residues or nucleotides. For example, a researcher might take an interest in the amino acid sequence of an enzyme, focusing in particular on a peptide Gly-Ala-Thr-His-Ser-Tyr-Glu-Gly-Ser-Leu that is near the catalytic site. Let us suppose that the codon sequence of this part of the gene is known to be

5′-GGG-GCC-ACA-CAT-AGT-TAC-GAG-GGA-AGC-CTT-3′

and that the researcher wishes to generate different versions of the gene in which the His codon is changed so that it specifies any of the 20 amino acids. The researcher might design the forward oligonucleotide

13. Forward: 5′-GGG-GCC-ACA-NNK-AGT-TAC-GAG-GGA-AGC-CTT-3′

in which the His codon has been replaced with the triplet NNK, and where 'K' represents either the G or T nucleotide. This would allow the three-nucleotide degenerate sequence to become any of 32 codons, representing each of the 20 amino acids at least once. The oligonucleotide '13' could be used in PCR, as part of a site-directed mutagenesis procedure (discussed in Section 7.5) and the 20 different versions of the enzyme tested for activity.

Degenerate oligonucleotide concentration and T_M

It is difficult to estimate the T_M of a degenerate oligonucleotide mixture that is being used in PCR, because by their very nature the sequences will have different percentages of G, A, T, and C nucleotides. However, the most GC-rich and most AT-rich versions of the degenerate sequence might be compared, to determine the probable upper and lower boundaries of the T_M. Software programs are available to assist a researcher with modeling the problem, but the solution may lie in empirical testing. This is a time at which a researcher might consider ramping up the temperature of the annealing step during PCR, or using a thermocycler that allows for a temperature gradient in the sample block, to account for all of the possible T_M values. If stringency of annealing is desirable, the optimal annealing temperature can be approached 'from the top', using a touchdown PCR in which sequential cycles of PCR use slightly lower annealing temperatures (discussed in Section 7.8). However, if a broad representation of degenerate products is desired for combinatorial studies, then a lower, less stringent annealing temperature may be warranted.

Problems

The solutions to these problems are provided at the end of the chapter.

(i) Suppose that you want to design oligonucleotide primers to generate by RT-PCR the 5′ untranslated region of the mouse caspase I cDNA (see *Figure 7.5*). That 200-nt segment of the cDNA sequence is shown below:

```
  1 CTTCCAAGTG TTGAAGAAGA ATCATTTCCG CGGTTGAATC CTTTTCAGAC TTGAGCATTT
 61 AAACCTAACT TTAATTAGGG AAAGAAACAT GCGCACACAG CAATTGTGGT TATTTCTCAA
121 TCTGTATTCA CGCCCTGTTG GAAAGGAACT AACAATATGC TTTCAGTTTC AGTAGCTCTG
181 CGGTGTAGAA AAGAAACGCC
```

A. If the forward and reverse oligonucleotide primers are each 25 nt in length, what will be their sequences?

B. What are the (estimated) T_M values of the primers, using the formula provided below?

$$T_M = 64.9°C + 41°C \times (\#GC - 16.4)/n$$

C. Which of the two oligonucleotides would be used with reverse transcriptase to generate a first-strand cDNA, before the PCR step?

(ii) The 200-nt sequence shown below is the reverse-complement of the sequence shown in the previous problem.

```
  1 GGCGTTTCTT TTCTACACCG CAGAGCTACT GAAACTGAAA GCATATTGTT AGTTCCTTTC
 61 CAACAGGGCG TGAATACAGA TTGAGAAATA ACCACAATTG CTGTGTGCGC ATGTTTCTTT
121 CCCTAATTAA AGTTAGGTTT AAATGCTCAA GTCTGAAAAG GATTCAACCG CGGAAATGAT
181 TCTTCTTCAA CACTTGGAAG
```

Suppose that you wanted to design 25-nt oligonucleotide primers to generate this fragment from a cloned cDNA. Would you need to design a different set of primers from the ones you used in the previous problem? If so, explain how they would be different.

(iii) A. The five oligonucleotide sequences below all have the same 5′ ends, but are of different lengths.

5′-GGCGTTTCTTTTCTACACCGCAGAGC
5′-GGCGTTTCTTTTCTACACCGCAGAGCT
5′-GGCGTTTCTTTTCTACACCGCAGAGCTA
5′-GGCGTTTCTTTTCTACACCGCAGAGCTAC
5′-GGCGTTTCTTTTCTACACCGCAGAGCTACT

Which of these oligonucleotides has the greatest potential of forming a primer dimer with itself during PCR synthesis?

B. Which of the five oligonucleotide sequences (shown above) has the greatest potential of forming a primer dimer with this new oligonucleotide (below), during PCR synthesis?

5′-CAGGACTACGATTTAGTAGCGGTAGT-3′

C. Would the primer dimer generated in the previous part of the problem be amplified geometrically during PCR?

7.3 Cycle efficiency

PCR is usually explained as a reaction in which the amount of template is doubled in each cycle, leading to geometric growth of 1 template to 2, then 4, 8, 16, 32, 64, 128, 256, 512, and finally 1024 after 10 cycles. That story is a bit of fiction, because the efficiency of PCR is not 100% in each cycle, and there are changing conditions that affect the initiation and completion of synthesis. For example, at the start of a PCR the DNA polymerase is added fresh, and it gradually loses activity as the cycles progress; some thermostable enzymes have a half-life of about 30 to 60 min at 95°C, and depending on the timing of the denaturation step in each cycle the enzyme may have lost half its activity by the 30th cycle. That loss of enzymatic activity, and the depletion of substrates towards the end of the PCR may lead to a steady drop in polymerase efficiency. On the other hand, the annealing reaction may be less efficient in earlier cycles, because the template concentration is low; it may increase in efficiency as the number of templates increases, and finally decrease as the primer concentration becomes limiting.

The myth of a perfect cycle

A 100% efficiency for PCR cycles, were it to exist, would lead to doubling of template in each cycle, and the amount of amplification of a single template molecule after n cycles could be expressed as 2^n. If we started the PCR with an initial amount of template that is A_0 copies, then after n cycles we would have $A_0 2^n$ copies. We could call that final amount A, and write the equation for overall yield:

$$A = A_0 2^n$$

We could do a little test to see if that formula is predictive in the laboratory, and we would quickly find that it was not. For example, if we conducted a PCR to isolate a unique 500-bp DNA sequence from the human genome, which is 3×10^9 bp in size and has a formula weight of 1.98×10^{12} Da, we might initiate the reaction with 50 ng of genomic DNA template, which is 2.5×10^{-20} mol of the genome. Let's suppose that we conduct the PCR using 35 cycles; how much product should we expect? If the efficiency of PCR is 100%, then we expect

$$(2.5 \times 10^{-20} \text{ mol}) \ 2^{35} = 8.6 \times 10^{-10} \text{ mol}$$

The 500-bp product would have a formula weight of

$$(500 \text{ bp})(660 \text{ Da bp}^{-1}) = 3.3 \times 10^5 \text{ Da}$$

because the average base pair is about 660 Da, so our 8.6×10^{-10} moles would be about 280 µg! That is more than 1000 times what we actually obtain, and well beyond the theoretical yield from the reactants in the tube. We can back-calculate the net efficiency over 35 cycles by knowing what amount of template went into the reaction and measuring our actual yield. If we assume that we make 100 ng of the 500-bp product in 35 cycles, a respectable band on a gel and a yield that is fairly typical, then our molar yield is about 3×10^{-13} mol. Our efficiency formula might then be written as:

$$(2.5 \times 10^{-20} \text{ mol}) \ P^{35} = 3 \times 10^{-13} \text{ mol}$$

where the factor P is no longer 2, but a number between 1 and 2 that represents the geometric mean productivity of each cycle in the PCR. We can solve the equation and determine that $P^{35} = 1.2 \times 10^6$, the actual amount of net amplification, and P = 1.59. That means 59 out of every 100 templates are copied in any given cycle, yielding 1.59 times as much template in the next cycle. Rather than model the geometric growth of

PCR products over n cycles as 1, 2, 4, 8, 16, ..., 2^n, we may do better by modeling them as 1, 1.59, 2.53, 4.02, 6.39, ..., 1.59^n.

A simple predictive calculation

It is important to guess, or better yet know the productivity factor P, because it can help a researcher in designing and troubleshooting a method. What would be a good guess for P? If the reaction is very straight-forward and well optimized, then P may be in the neighborhood of 1.60. If the reaction is more difficult, perhaps because it leads to a long synthetic product or is being performed under conditions that are not yet optimized, then P = 1.40 to P = 1.55 may be a better guess.

We also need to know how to calculate molar amounts of DNA fragments. This is not difficult if we keep in mind that the average nucleotide base is 330 Da and the average base pair is 660 Da. The formula weight of a DNA molecule is simply 660 times the number of base pairs, or 330 times the number of nucleotide bases if the DNA is single-stranded. An 18-mer oligonucleotide of average composition would have a formula weight of about $(18 \text{ nt})(330 \text{ Da nt}^{-1}) = 5940$ Da; a double-stranded DNA product of 250 bp would have a formula weight of

$$(250 \text{ bp})(660 \text{ Da bp}^{-1}) = 1.65 \times 10^5 \text{ Da}$$

a plasmid of 4.5 kbp would have a formula weight of

$$(4500 \text{ bp})(660 \text{ Da bp}^{-1}) = 2.97 \times 10^6 \text{ Da}$$

and a genome of 3000 Mbp would have a formula weight of

$$(3 \times 10^9 \text{ bp})(660 \text{ Da bp}^{-1}) = 1.98 \times 10^{12} \text{ Da}$$

We need to know how to calculate formula weights, because the formula weight is the conversion factor that relates μg and μmol, ng and nmol, and pg and pmol. For example, 10 ng of a 4.5-kbp plasmid is

$$(10 \text{ ng}) \div (2.97 \times 10^6 \text{ ng nmol}^{-1}) = 3.37 \times 10^{-6} \text{ nmol}$$

which we may also call 3.37×10^{-15} mol, or 3.37 fmol. We may also calculate the number of copies of a DNA molecule using Avogadro's number: In this case, 10 ng of a 4.5-kbp plasmid is

$$(3.37 \times 10^{-15} \text{ mol})(6.02 \times 10^{23} \text{ copies mol}^{-1}) = 2.03 \times 10^9 \text{ copies}$$

Suppose we start with 50 ng of human DNA as template in a PCR, leading to the production of a 500-bp product. It is interesting to compare the yield of product in a PCR of 35 cycles with productivity factors (P) of 1.45, 1.50, or 1.55. In each case the equation for PCR yield may be written as $A = A_0 P^n$, where A is the molar yield and A_0 is the initial molar amount of template, P is the productivity factor, and n is the number of cycles. As we calculated earlier, 50 ng of human genomic DNA is 2.5×10^{-20} mol of genome. The three molar yields with different P factors would be

$$(2.5 \times 10^{-20} \text{ mol})(1.45)^{35} = 1.1 \times 10^{-14} \text{ mol} = 11 \text{ fmol}$$
$$(2.5 \times 10^{-20} \text{ mol})(1.50)^{35} = 3.6 \times 10^{-14} \text{ mol} = 36 \text{ fmol}$$
$$(2.5 \times 10^{-20} \text{ mol})(1.55)^{35} = 1.1 \times 10^{-13} \text{ mol} = 110 \text{ fmol}$$

In a previous calculation, we found that a 500-bp product has a formula weight of 3.3×10^5 Da. Therefore the mass yields of the three results are, respectively:

$(1.1 \times 10^{-14}\ \text{mol})(3.3 \times 10^5\ \text{g mol}^{-1}) = 3.7 \times 10^{-9}\ \text{g} = 3.7\ \text{ng}$ [where P = 1.45]
$(3.6 \times 10^{-14}\ \text{mol})(3.3 \times 10^5\ \text{g mol}^{-1}) = 1.2 \times 10^{-8}\ \text{g} = 12\ \text{ng}$ [where P = 1.50]
$(1.1 \times 10^{-13}\ \text{mol})(3.3 \times 10^5\ \text{g mol}^{-1}) = 3.8 \times 10^{-8}\ \text{g} = 38\ \text{ng}$ [where P = 1.55]

We also calculated previously that a 100-ng yield would be consistent with a P factor of 1.59; we now see that a P factor of 1.55 would reduce the yield to 38 ng, still easily detectable on a gel stained with ethidium bromide. However, if the P factor drops to 1.50 or 1.45, the yield will be 12 ng or 3.7 ng, respectively, and the latter may not be easily detectable on an ethidium-bromide-stained gel. The lesson from these calculations is that a very slight change in PCR efficiency in each cycle can mean the difference between seeing a band on a gel, and not seeing one. A change from 1.55 to 1.45 is a drop of only 6.5%, but if that 6.5% drop is compounded geometrically through 35 cycles then it leads to more than a 90% drop in overall yield.

Refining a guess

We can take a guess at our P factor for a PCR, and use a bit of initial success in the laboratory to refine our guess. Suppose that we are performing site-directed mutagenesis on a gene and are using a plasmid copy as the template. Let us suppose further that the plasmid's overall size is 5.5 kbp and that the PCR product will be 1 kbp. If we want to start our PCR with 1 pg of plasmid template, how many cycles would we need to achieve a yield of 100 ng? First, we calculate that our starting material, 1 pg of the 4.5-kbp plasmid, is 2.75×10^{-19} mol, so A_0 is 2.75×10^{-19} mol. Next, we calculate that our desired yield, 100 ng of a 1-kbp product, is 1.51×10^{-13} mol, so $A = 1.51 \times 10^{-13}$ mol. We decide to assume a P factor of 1.45 (because these calculations are making us feel a bit depressed) and we do a calculation to predict the number of cycles n that will be needed:

$(1.51 \times 10^{-13}\ \text{mol}) = (2.75 \times 10^{-19}\ \text{mol})(1.45)^n$

so $(1.45)^n = 5.5 \times 10^5$. We can solve for n using logarithms:

$\log_{10}(1.45)^n = n \log_{10}(1.45) = n(0.1614)$
$\log_{10}(5.5 \times 10^5) = 5.74$
$n(0.1614) = 5.74$
$n = 5.74/0.1614 = 35.6$

and $n = 36$ cycles, rounded up to the next integer.

Let's suppose that we conduct the PCR with the 1 pg of plasmid template, and 36 cycles, but rather than getting a 1-kbp band of 100 ng we get a broad smear of stained DNA that could amount to as much as 1 µg, if taken together. Time to recalculate, because our guess that P = 1.45 was obviously too pessimistic. If we assume that our yield of product was 1 µg, including all of the poorly resolved DNA on the gel, then our molar yield was about 1.5×10^{-12} mol (if we assume the same formula weight as before). A very rough estimate for P can be calculated:

$$P^{36} = (A/A_0) = (1.5 \times 10^{-12} \text{ mol}/2.75 \times 10^{-19} \text{ mol}) = 5.45 \times 10^6$$
$$P = (5.45 \times 10^6)^{1/36} = 1.54$$

However, a word of caution is in order: if the efficiency dropped significantly in the last few cycles of the method this may be an underestimate. The change in efficiency during the end of a PCR may be dramatic, and this is a topic discussed in Section 7.7.

With this empirical estimate for P, we can set up additional PCR samples in a more informed manner. For example, if we decided that we would really rather have 200 ng of the 1-kbp product, and would rather start with 10 pg of plasmid, then the number of cycles necessary to do that amplification is easily calculated. The 10 pg of plasmid template will be 2.75×10^{-18} mol, and the yield will be 3.0×10^{-13} mol, so

$$(1.54)^n = (3.0 \times 10^{-13} \text{ mol})/(2.75 \times 10^{-18} \text{ mol}) = 1.09 \times 10^5$$

and $n = 27$ cycles (rounded up to the next integer). Alternatively, if we want to start the reaction with 100 pg of plasmid template, or 10 times as much, then the denominator is adjusted to 2.75×10^{-17} mol so that

$$(1.54)^n = (3.0 \times 10^{-13} \text{ mol})/(2.75 \times 10^{-17} \text{ mol})$$

and n = 22 cycles (rounded up to the next integer).

Optimization

Why is PCR so fickle?

The polymerase chain reaction is very susceptible to slight drops in efficiency because the net efficiency is compounded over many cycles. For example, if we compare the yield between two reaction tubes, one of which has a value of P = 1.50, and the other of which has a value of P = 1.48, after 30 cycles the first tube will have 50% more product than the second. That is, $(1.50)^{30}$ is approximately 50% greater than $(1.48)^{30}$. As the efficiency of each cycle drops, the PCR yield deteriorates quickly; a reaction tube with a value of P = 1.40 would have only one-eighth the yield of the first tube (P = 1.50) after 30 cycles.

These huge swings in PCR yield may baffle a researcher, because there is such a fine line between success and failure. The road to success is partly a matter of thoughtfully designing the oligonucleotides and reaction conditions in advance, and doing the necessary calculations. It is also a matter of setting up reactions using consistent, high quality reagents and good technique. For example, if a pipetting device lacks precision because it needs repair, or the operator is not using the device properly, then a key component of the reaction may not be delivered to the reaction tubes consistently and this may lead to changes in reaction efficiency. To take another example, if a researcher thaws a frozen stock tube of $MgCl_2$ each day but does not vortex it after thawing, then the stock concentration of that critical component may change over time because the freezing process makes the solute inhomogeneous. The experiment-to-experiment yield may drop over a period of weeks or months, even though the researcher is ostensibly using the very same stock reagent tubes.

Consumables

With PCR being a bit of a fickle reaction, we should start the discussion of optimization from the beginning with a review of the necessary components. In particular, we should calculate the appropriate amounts of each component to make sure that they are present in the reaction at an appropriate concentration, realizing that several of the component requirements may vary depending on the size of the desired product.

The oligonucleotides used in PCR define the 5′ ends of the reaction product, so one copy of each oligonucleotide needs to be incorporated into every double-stranded product. How much oligonucleotide do we need, then? We anticipate a certain yield of double-stranded product, in grams, and divide that by the formula weight of the product to determine the molar yield. That molar yield tells us exactly how many moles of oligonucleotide will be incorporated into the product. For example, if we expect to generate 200 ng of a 2-kbp product in PCR tube 'A', and 200 ng of a 150-bp product in PCR tube 'B', then the formula weights of the products are

$$(2000 \text{ bp})(660 \text{ Da bp}^{-1}) = 1.32 \times 10^6 \text{ Da}$$

in tube 'A', and

$$(150 \text{ bp})(660 \text{ Da bp}^{-1}) = 9.9 \times 10^4 \text{ Da}$$

in tube 'B'. The respective molar yields would be $2.00 \times 10^{-7}/1.32 \times 10^6 = 1.51 \times 10^{-13}$ mol (0.151 pmol) in tube 'A', and $2.00 \times 10^{-7}/9.9 \times 10^4 = 2.02 \times 10^{-12}$ (2.02 pmol) in tube 'B'. If we added 1 pmol of each oligonucleotide to PCR tube 'A' while setting up the reaction, we would have a 5.6-fold excess of oligonucleotide at the end of the PCR because we only consumed 0.15 pmol. On the other hand, if we added 1 pmol of each oligonucleotide to PCR tube 'B' then we would fall short, and have less than half of what we needed.

Most recipes and protocols call for the addition of 0.5 to 10 pmol of each oligonucleotide primer, but we have to think about our needs before we follow these instructions. Smaller PCR products consume more oligonucleotide per nanogram of yield than larger PCR products. Could we just dodge the calculation by always resolving to add 5 to 10 pmol of oligonucleotide, regardless of product size? No, because if too much oligonucleotide is added, it may cause spurious priming of template or primer dimer formation, and unwanted side reactions. Typical working stock concentrations of oligonucleotide primers are 1 µM, 5 µM, and 10 µM; and 1 µl of each of these stocks would contain 1 pmol, 5 pmol, and 10 pmol of primer, respectively. If these were incorporated into a 50-µl PCR, the final concentrations of oligonucleotide would be 20 nM, 120 nM, and 200 nM, respectively.

Slightly higher concentrations of oligonucleotide (100 nM to 500 nM) may be warranted if the thermostable DNA polymerase has a 3′ to 5′ exonuclease (proofreading) activity, because some of the primer may be degraded during the natural course of the PCR. The presence of a proofreading activity must also be considered in the order of assembly of a reaction, because primers may be degraded if the polymerase is added before the dNTP substrates. Overall, the 3' to 5' exonuclease improves the fidelity of the enzyme by allowing it to excise its mistakes, but it does require some accommodation on the part of the experimental design.

The dNTP substrates are the components of PCR that are largely responsible for the mass of the product. The dNTP substrate concentrations in a reaction may be in the range of 50 μM to 250 μM, and in a 50 μl reaction the respective total amounts would be

$$(5 \times 10^{-5} \text{ l})(5 \times 10^{-5} \text{ mol l}^{-1}) = 2.5 \times 10^{-9} \text{ mol} = 2.5 \text{ nmol}$$

in the case of 50 μM dNTP, to 12.5 nmol in the case of 250 μM dNTP. If these amounts could be fully incorporated into PCR product, and we assume the average incorporated nucleotide has a formula weight of 330 Da, then the yields would be in the range of

$$(2.5 \times 10^{-8} \text{ mol})(330 \text{ g mol}^{-1}) = 8.2 \times 10^{-6} \text{ g} = 8.2 \text{ μg}$$

to five times that amount, or 4.1×10^{-5} g (41 μg). The lower end of the range represents a substantial theoretical maximum yield; we need not worry that we will fall short of dNTP substrate to incorporate for most applications.

This is more than we can incorporate, but is it a wasteful amount? Probably not, and for some enzymes the 50 μM concentration may not be sufficient for optimal reaction conditions because the Michaelis–Menten constant K_M for thermostable DNA polymerases varies considerably. *Taq* DNA polymerase has a K_M for dNTP of approximately 13 μM, and Deep VentR® enzyme has a K_M for dNTP of approximately 50 μM. If these enzymes followed Michaelis–Menten kinetics, then a dNTP substrate concentration of 50 μM would lead to polymerase reaction velocities that were (50/(50 + 13)) = 79% and (50/(50 + 50)) = 50% of V_{MAX}, respectively. In the case of *Pfu* polymerase, for example, a concentration of 100 to 250 μM dNTP is recommended for the reaction mixture.

Could we make the enzyme work harder by adding a huge amount of dNTP substrate, say 5 to 10 mM? Probably not, because pushing a reaction out to the far end of a Michaelis–Menten substrate diagram to achieve the perfect state of V_{MAX} is more of a fantasy of extrapolation than a reality. Enzymes may be slightly inhibited by excessive amounts of substrate, perhaps for reasons that are entirely nonspecific, and a slight inhibition in each PCR cycle may be compounded into a significantly lower net yield. Furthermore, higher concentrations of substrate may decrease the fidelity of the enzyme; the rate at which it misincorporates nucleotides.

Nonconsumables

The salt and buffer conditions for PCR, and in particular the concentration of magnesium ions, are critically important parameters. Commercially available polymerases are usually distributed along with a 10 × buffer cocktail that may or may not contain Mg^{2+} salts, and the buffer cocktail is intended for use at a 1 × final concentration. For example, if the researcher is setting up a 50-μl PCR, then 5 μl of that 50 μl, or one part in 10 should be the $10 \times$ buffer. The final concentration of Mg^{2+} in the reaction is adjusted by the researcher so that it falls within the range of 1 mM to 5 mM, using a stock of $MgCl_2$ or $MgSO_4$ that is approximately 10 to 50 mM. The typical buffer conditions for use of *Taq* DNA polymerase in PCR would be 10 mM Tris-Cl, pH 8.3, 50 mM KCl, and 1.5 to 5.0 mM $MgCl_2$. However, different thermostable enzymes have different optimal conditions; *Pfu* DNA polymerase is used under conditions of 20 mM Tris-Cl, pH 8.8, 10 mM $(NH_4)_2SO_4$, 10 mM KCl, 2.0 to 3.0 mM $MgSO_4$, 0.1% (v/v) Triton® X-100, and 0.1 mg ml^{-1} BSA.

Testing a reaction using different Mg^{2+} concentrations is one of the more important first steps in development of a new PCR method. The Mg^{2+} ion has a strong effect on DNA annealing because of its ability to provide electrostatic shielding of the phosphodiester backbones. If the concentration of Mg^{2+} is insufficient, then the reaction may not generate an acceptable yield, and if it is excessive, then it may encourage nonspecific annealing and side reactions or decrease the activity or fidelity of the polymerase.

A researcher using *Taq* DNA polymerase might test the reaction with a series of different $MgCl_2$ concentrations, from 1 mM to 6 mM, in 1 mM, or even 0.5 mM steps, to see what range is most favorable. In the case of *Pfu* DNA polymerase, $MgSO_4$ is recommended in place of $MgCl_2$. For example, if a 50-µl reaction is being assembled using a $10 \times$ buffer for *Pfu* polymerase, one that already contains 20 mM $MgSO_4$ and would yield a final concentration of 2 mM $MgSO_4$ upon dilution, a final concentration of 3 mM $MgSO_4$ could be arranged by combining 5 µl of the $10 \times$ *Pfu* buffer and 2 µl of a 25 mM $MgSO_4$ stock into the final volume. On the other hand, if a 50-µl reaction is being assembled using a $10 \times$ *Pfu* buffer that lacks $MgSO_4$, then a final reaction condition of 2 to 3 mM $MgSO_4$ could be arranged by combining 5 µl of the Mg-free $10 \times$ *Pfu* buffer with 4 to 6 µl of a 25 mM $MgSO_4$ stock.

As with most enzymes used in the laboratory, thermostable DNA polymerases are typically distributed by suppliers in a solution of 50% glycerol, to prevent freezing under the $-20°C$ storage conditions. The stock tubes of enzyme should be kept cold at all times to preserve the activity of the enzyme, and should be handled carefully to prevent contamination (a matter that is discussed in Section 7.8). The enzyme may be distributed at a stock concentration of 1 to 5 units per µl, with the recommendation that 1 to 2.5 units of enzyme be used in a single reaction. The 'unit' of activity is defined as an amount of enzyme that can incorporate 10 nmol of dNTP in 30 min, under a specific solution and temperature condition; however, the template used in that unit definition may be a nicked genomic DNA, with the result that the unit definition is actually based on a rate of nick translation rather than a typical PCR test. An incorporation of 10 nmol of dNTP in 30 min would be equivalent to a rate of approximately 0.1 µg DNA synthesis per minute, or 25 ng per 15-second time interval.

Many PCR methods require longer extension times than that for full efficiency, and this is particularly the case for generation of long PCR products in the range of 1 to 10 kbp. For example, a supplier may recommend an extension time of 1 to 2 min per kbp of synthesis for long PCR products, which means that in the last cycles of a PCR it might take 5 to 10 min to generate 100 ng of a 5-kbp product. That seems about 10 times slower than what was promised in the unit definition: why does the unit definition not match the actual behavior of the enzyme? The unit definition is based on enzyme activity on a nicked genomic template in which the nicks may be closely spaced and the enzyme is not limited by the number of available 3' ends to extend. On the other hand, PCR synthesis is initiated from a single forward or reverse primer, and with long PCR products the overall density of priming sites may be low. The efficiency of PCR is limited by the progress of the enzyme on long templates, and 1 to 2 min per kbp is equivalent to an incorporation rate of about 10 to 20 nt per s.

The thermostable DNA polymerase is not consumed during the PCR; however, it does have a finite lifespan under the extreme conditions of thermal cycling. Some companies report that the half-lives of their enzymes are approximately 30 to 60 min at 95°C, meaning that if a PCR starts with an initial 5-min period of denaturation and then adds 30 s of denaturation at the beginning of each cycle, a 40-cycle reaction would amount to a high temperature exposure of at least 25 min. Half as much enzyme activity might be present in the last cycle, as compared to the first cycle. If the denaturation temperature used is slightly higher, perhaps in the range of 96°C to 98°C, then the half-life of the enzyme is likely to be somewhat less, and if the thermocycler is slow to change temperatures, then exposure of the enzyme to temperatures above 90°C may be a consideration as well. A gradual loss of polymerase activity may be a limiting factor in some methods; one solution is to extend the synthesis time in later cycles to give the remaining polymerase a bit of extra time to complete the job, and another is to add extra polymerase at the beginning to allow for a sufficient activity at the end.

The DNA template can be added to a PCR in varying amounts, but a typical method would be based on approximately 10^4 to 10^5 copies of the template, or 1.7×10^{-20} to 1.7×10^{-19} mol. Multiplying this number of moles by the formula weight of the starting template gives us the starting number of grams; for example, the human genome has a formula weight of approximately 2×10^{12} Da, and so 10^4 copies is

$$(1.7 \times 10^{-20} \text{ mol})(2 \times 10^{12} \text{ g mol}^{-1}) = 3.2 \times 10^{-8} \text{ g} = 32 \text{ ng}$$

On the other hand, 10^4 copies of the *Drosophila* genome would only be about 1.3 ng, because the genome size is 1.2×10^8 bp, and the genome formula weight is therefore approximately 8×10^{10} Da. In both cases, 100 ng of a 500-bp product could be generated from the 10^4 templates, over the course of 40 cycles, assuming a productivity factor P = 1.52. This calculation is based on the formula $A = A_0 P^n$, where A_0 is 1.7×10^{-20} mol and P^n is $(1.52)^{40}$ or approximately 1.8×10^7, and the formula weight of product fragment is

$$(500 \text{ bp})(660 \text{ Da bp}^{-1}) = 3.3 \times 10^5 \text{ Da}$$

As more DNA template copies are added to a reaction, fewer cycles are needed to generate a desired yield, and there are fewer opportunities for accidental incorporation of mutations into the product. However, adding more than 200 ng of any template to a PCR is inadvisable, because an excessive concentration of DNA may encourage the formation of spurious annealing reactions with the oligonucleotide primers. These lead to side reactions that may cause a mixture of wanted and unwanted products.

Let's consider how a much smaller-sized template would be used. A 5-kbp plasmid would have a formula weight of approximately 3.3×10^6 Da, and so 10^4 copies would amount to only 5.5×10^{-14} g (55 fg). However, 55 fg of DNA would be challenging to handle in solution, without a carrier to keep it from adsorbing to surfaces, and so we would probably use considerably more copies of a small plasmid template and compensate by calculating the number of cycles necessary for the desired yield. For example, if we started with 10^7 copies of the plasmid as template, an amount equal to 55 pg, then only 24 cycles would be needed to generate 100 ng of a 500-bp product, assuming the same productivity factor $P = 1.52$. Again, this is based on the formula $A = A_0 P^n$, where the starting template A_0 is 1.7×10^{-17} mol, and the amount of amplification P^n is 1.52^{24} or approximately 1.8×10^4-fold.

Finally, we sometimes wish to take a small PCR product and amplify it in a second round of PCR, using the same primers or another set that lies nested inside the boundaries of the first set. We could start with 1 ng of a 500-bp product and amplify it so that we have a yield of 100 ng. If we assume $P = 1.52$, then only 11 cycles would be needed for this 100-fold amplification, because 100 is approximately 1.52^{11}.

Adjuvants, adjuncts, and additives

Besides the basic components of a PCR, the template, substrates, primers, buffer, and salts, there are a few compounds that are sometimes used to make a specific reaction more favorable. These compounds may help to stabilize the enzyme, or moderate the annealing or melting of template and primer, or increase the activities of solutes in the reaction. They may be particularly helpful with templates that are unusually GC-rich, or other templates with potentially interfering secondary structures. Examples of additives include BSA and nonionic detergents such as Triton® X-100, dimethylsulfoxide (DMSO), formamide, and proprietary cocktails that are commercially available.

In some cases these additives may give a slight boost to the efficiency in each cycle, and that can mean the difference between success and failure. However, these additives are probably not necessary for most applications, provided that the researcher has optimized the Mg^{2+}, oligonucleotide primer, dNTP, and template concentrations, and has properly estimated the number of cycles that would be needed for success.

Problems

The answers to these problems are provided at the end of the chapter.

(iv) The *Fugu* genome is approximately 4×10^8 bp. Suppose that you want to use *Fugu* DNA as a template in a PCR, isolating a unique gene sequence of 2 kbp. If you add 20 ng of genomic DNA to each reaction, and 1 µl of a 5 µM stock of each primer, what is the molar ratio of primer to template in the very first cycle? You may assume that the average base pair is 660 Da.

(v) What is the theoretical maximum µg yield of product, in the previous problem, assuming that you have excess dNTP substrates and are limited by the amount of oligonucleotide primer?

(vi) If you obtained a yield of 200 ng of product, during a PCR of 30 cycles, what is the overall amount of template amplification and what is the productivity factor P?

(vii) Suppose that your 200-ng DNA yield (from the previous problem) is now gel purified, and in a solution of 1 ml. If you take 1 µl of this sample and set up a new PCR using the same oligonucleotides, how many cycles would be needed to generate 100 ng of product, assuming that the productivity factor P is still the same?

7.4 Setting up a reaction

Reagents and plasticware

Slight variations in PCR cycle efficiency may be compounded to give wild fluctuations in product yield, so it is important that the reagents used to assemble the reaction be of high quality, and consistently so. The dNTP stocks and primer solutions should be stored frozen in small aliquots at $-20°C$ or $-70°C$, and mixed thoroughly after thawing. The principal stocks of reagents should be diluted to make working stocks, rather than being used on a day-to-day basis, to protect them from contamination. The concentrations of DNA and oligonucleotide stock solutions can be determined or confirmed by UV spectrophotometry at 260 nm.

For example, a concentrated primer stock of 100 µM could be prepared by taking 23 nmol of a desiccated primer supplied by a company, and resuspending it in 230 µl of a diluent such as sterile distilled water or 10 mM Tris-Cl, 1 mM EDTA, pH 8.0. A 5-µM working stock could be prepared from the 100-µM primer stock by a 1:20 dilution, and 1 µl of the 5-µM working stock would contain 5 pmol of primer. As a point of reference, a 5-µM stock of a 20-mer primer would be approximately 33 µg ml^{-1}, depending on its nucleotide composition, and would have an absorbance at 260 nm (A_{260}) of a little more than 1.0 (assuming the spectrophotometer cuvette has a path length of 1 cm). However, the exact absorbance will vary, depending on base composition and sequence, and will be somewhat higher if the sequence is purine-rich and lower if the sequence is pyrimidine-rich. If an oligonucleotide has X_g guanylate nucleotides, X_a adenylate nucleotides, X_t thymidylate nucleotides, and X_c cytidylate nucleotides, then the expression

$$(A_{260})/((11.5)X_g + (15.4)X_a + (8.7)X_t + (7.4)X_c)$$

will yield the approximate µM concentration of oligonucleotide. For example, if a sample of the oligonucleotide

```
1. Forward 5'-GCCAGTCAAACAGTCAGTAA-3'
```

(having 4 Gs, 8 As, 3 Ts and 5 Cs) has an absorbance at 260 nm of 0.250, then

$$(0.250)/((11.5)(4) + (15.4)(8) + (8.7)(3) + (7.4)(5)) = 0.0011 \text{ mM}$$

so the solution is approximately 1.1 µM. Dinucleotide information in the sequence can be used to improve the estimate, and web-based software tools are available to help a researcher calculate the exact extinction co-efficient for an oligonucleotide.

A 25-mM concentrated stock of dNTP can be purchased as a pre-made mix, or prepared from high quality reagents in a 10-mM Tris-Cl, 1-mM EDTA, pH 8.0 solution. In any case, 25-mM dNTP means that each of the component nucleotides dGTP, dATP, dTTP, and dCTP are individually at concentrations of 25 mM. From this concentrated stock, a 2.5-mM working stock of dNTP could be prepared by a 1:10 dilution into the Tris and EDTA buffer. The 2.5-mM working stock could be treated as a $10 \times$ or $25 \times$ stock for reaction assembly, depending on whether the researcher wishes to have a final concentration of 250-µM or 100-µM dNTP in the reaction. The molar extinction coefficients of each nucleotide are different, as are their absorbance maxima at pH 7.0. For example, a 25-µM stock of dGTP would have an absorbance of 0.34 at 253 nm, a 25-µM stock of dATP would have an absorbance of 0.38 at 259 nm, a 25-µM stock of dTTP would have an absorbance of 0.24 at 267 nm, and a 25-µM stock of dCTP would have an absorbance of 0.23 at 271 nm.

A DNA template may be stored in a 10-mM Tris-Cl, 1-mM EDTA, pH 8.0 solution, at 4°C or frozen at −20°C or −70°C but, if frozen, the solution should be thoroughly mixed after thawing, and before a sample is taken. A concentrated DNA stock might be in the range of 100 µg ml^{-1} to 1 mg ml^{-1}, and these concentrations are also equivalent to 100 ng µl^{-1} and 1 µg µl^{-1}, respectively. A working stock of DNA template that is used for reaction assembly might be prepared as a 1:10 dilution of the concentrated stock in Tris-EDTA buffer, in the range of 10 ng µl^{-1} to 100 ng µl^{-1}, so that 1 µl of the working stock is 10 to 100 ng. For solutions of double-stranded DNA, single-stranded DNA (not oligonucleotide length), or single-stranded RNA, an A_{260} of 1.0 corresponds to approximate concentrations of 50 µg ml^{-1}, 37 µg ml^{-1}, and 40 µg ml^{-1}, respectively. As a point of reference, a 10-ng µl^{-1} solution of double-stranded DNA will have an absorbance of 0.20 at 260 nm, and a 10-ng µl^{-1} solution of single-stranded DNA will have an absorbance of 0.25 at 260 nm.

The salts used in a PCR buffer should be prepared from high-grade reagents, and stored as $10 \times$ stock solutions at −20°C, in small aliquots. Once thawed, these solutions should be thoroughly mixed so that they are homogeneous. Salts have a tendency to be excluded from ice during freezing of water, and so they can become concentrated in one part of a tube. They drop to the bottom of the tube upon thawing due to their density, and if the researcher does not mix the tube before taking a sample then an incorrect amount of solute will be delivered.

The reagent source of Mg^{2+} ion is a particular concern, because we use $MgCl_2$ with some thermostable enzymes and $MgSO_4$ with others. Furthermore, magnesium salts can be purchased as anhydrous (dry) reagents or hydrated crystals, for example $MgCl_2(6H_2O)$ or $MgSO_4(7H_2O)$, but the important point is that these salts are hygroscopic and can vary in water content, depending on whether the reagent bottle has been

tightly capped during storage. This means that a researcher may anticipate variation in the true formula weights of crystalline reagents. For
consistency, a researcher may wish to prepare a substantial stock of a
1.0-M $MgSO_4$ or $MgCl_2$, storing it tightly capped at room temperature,
and use that stock to prepare working stocks of 25 mM by a 1:40
dilution.

Order of addition and master mixes

A suggested recipe for a single polymerase chain reaction of 50 µl is as
follows, with the order of components indicating the order of addition to
the reaction tube:

- 29 µl sterile, distilled H_2O
- 5 µl 10 × PCR buffer without Mg (final concentration 1 ×)
- 5 µl 25-mM $MgSO_4$ (final concentration 2.5 mM)
- 5 µl 2.5-mM dNTP mix (final concentration 250 µM)
- 1 µl 5-µM Forward oligonucleotide primer (final concentration 0.1 µM)
- 1 µl 5-µM Reverse oligonucleotide primer (final concentration 0.1 µM)
- 3 µl 10-µg ml^{-1} human genomic DNA (final concentration 0.6 ng µl^{-1})
- 1 µl 1-u µl^{-1} *Pfu* DNA polymerase (final concentration 0.02 u µl^{-1}).

An important point about the order of addition is that the buffer and
salts, particularly Mg^{2+}, are mixed together before the more sensitive
components such as nucleic acids and enzymes are added. Nucleic acids
may be precipitated by Mg^{2+} at high concentration, and so this order of
addition prevents precipitation from occurring. The enzyme is added last
of all, and may even be added after the reaction tube is pre-heated if a
hot start protocol is being followed (discussed in Section 7.8). The *Pfu*
enzyme has a 3' to 5' exonuclease activity, and in the absence of dNTP
substrates it may degrade the oligonucleotides and template; so it is important that the dNTP substrates be added to the reaction prior to the
Pfu DNA polymerase. In the recipe given, the supplier's 2.5-u µl^{-1} stock of *Pfu*
DNA polymerase was diluted 1:2.5 in 1 × reaction buffer to obtain a
1-u µl^{-1} working stock, and this working stock is made fresh for each
experiment.

If we were setting up 10 reactions, all under identical conditions but
with different DNA templates, we might consider setting up a master mix
of the common components so that the reactions were consistent. Setting
up the 10 reactions with eight components each would require 80 pipetting steps, with all of the accompanying problems of pipette tip waste,
pipetting precision, and researcher fatigue. For example, we could scale
the single reaction up 10.5 times, the extra 5% being added so that we do
not run out of the master mix on the 10th reaction due to loss of some
of the solution in the wetted and discarded pipette tips.

Master mix recipe (total volume 493.5 µl):

- 311.0 µl sterile, distilled H_2O
- 52.5 µl 10 × PCR buffer without Mg
- 52.5 µl 25-mM $MgSO_4$
- 52.5 µl 2.5-mM dNTP mix

- 10.5 µl 5-µM forward oligonucleotide primer
- 10.5 µl 5-µM reverse oligonucleotide primer
- 4.2 µl 2.5-u µl^{-1} *Pfu* DNA polymerase.

We could then combine 47 µl of this master mix with 3 µl of DNA template in each reaction tube, or substituting 3 µl of H_2O for the DNA if we are conducting a negative control test, and the reaction volume would be 50 µl.

Alternatively, we could establish a master mix containing only 2 mM $MgSO_4$, and then test a range of different Mg^{2+} levels by supplementing the reactions with 0, 1, 2, 3, 4, 5, or 6 µl of the 25-mM $MgSO_4$ stock, adding H_2O as needed to make the final reaction volume 50 µl in each case. These seven reactions would span the range of 2 mM to 5 mM $MgSO_4$, in 0.5 mM steps, and would be consistent in all other components because they were included in the master mix.

Many researchers seal the top of a reaction solution with a thin layer of mineral oil or a surface wax, to prevent evaporation of the reaction volume during thermocycling. The oil or wax does not mix with the aqueous volume and so does not change the concentrations of any of the solutes; however, it does add mass to the tube that will slightly slow down the rate of heating and cooling. Waxes are sometimes distributed as solid pellets that can be added to each reaction tube, and the wax melts as the temperature first rises. Some wax pellets are sold with a pre-measured amount of magnesium salt embedded, so that a reaction can be delayed until the wax has melted and the aqueous components combined; a type of hot start protocol (discussed in Section 7.8).

It is common for researchers to use thin-walled plastic tubes, tube strips, or reaction plates in a thermocycler to allow fast heating and cooling, and to use a thermocycler with a heated lid to prevent condensation of liquid at the top of the tube. This ostensibly replaces the need for mineral oil or wax, but may create a different problem because the heated lids may soften the top of the thin-walled tube and cap, and allow leakage of the pressurized water vapor during the high-temperature denaturation steps. The result is that the reactions may vary in volume at the end of the PCR, having lost different amounts of water vapor, and the reactions may have inconsistent solute concentrations. If this occurs, the continued use of wax or mineral oil might be advisable, or a better sealing system for the tubes may be implemented. Researchers often don't like to use mineral oil or wax because it is a bit messy, but traces of wax or mineral oil can be effectively removed from aqueous reaction products by rolling the sample over a waxy surface such as Parafilm®; the residue of oil or wax adsorbs to the Parafilm® leaving a purely aqueous sample.

Contamination

PCR is such a sensitive reaction that a little bit of contamination can go a long way towards spoiling an experiment. The researcher's own DNA may be inadvertently introduced into an experiment through contamination with skin flakes, and many researchers who work with human DNA samples will wear gloves, hairnets, masks, and other protective clothing to prevent this type of contamination. The disposable pipette tips that are used in reaction assembly may contain a filter disk to prevent contamination of a sample through the pipette shaft, and researchers may assign one area of a lab for clean PCR set up, and stock it with plasticware and equipment that is only to be touched with gloved hands. If available, a laminar flow hood with a high-efficiency particulate air (HEPA) filter may also be used during PCR set up, to help keep a sample free of airborne contaminants.

The most profound source of contamination may be a successful prior experiment, as a simple calculation reveals. Let us suppose that on Monday morning a researcher sets up a series of five polymerase chain reactions using a set of oligonucleotide primers, and 10^4 copies of a template in each tube. On Tuesday morning the researcher loads the five reactions on an agarose gel, discarding the loading tips in a sharps' container on the bench, and runs the gel. At the conclusion of the electrophoresis the gel buffer is poured down the drain, and the gel is stained in SYBR® Green dye before being photographed on a UV transilluminator screen. Let us suppose that the experiment is a resounding success, and that each of the five reactions generates 100 ng of a 500-bp PCR product. The gel is discarded in the trash, the staining solution is discarded down the drain, and a bottle of Champagne is opened in celebration of success.

On Wednesday morning, when the researcher comes back to repeat the experiment, the success of the previous day will weigh heavily on the day's events. Let us remember that the five successful reactions on Tuesday generated a total of 500 ng of a 500-bp PCR product, which is 1.5 pmol, or, to say it in a more menacing way, nearly 10^{12} copies! That yield is 100 million times the amount that was used to initiate each reaction on Monday, and since it was handled casually during analysis on Tuesday it will be difficult to keep cross-contamination levels below 0.000001% on Wednesday. The pipette tip sharps' container, the pipetting device used to load the gel, the gel apparatus, the sink, the UV transilluminator, the camera, the waste basket, the telephone, and the bench and floor around these areas are contaminated with PCR product, and probably enough to serve as a template for 100 000 new reactions. The researcher may have even fallen asleep in his clothes after the party, and worn some of the PCR product back to the lab the next day! On Wednesday morning, when the new reaction tubes are assembled, the stock solutions may become contaminated with the PCR product, and from then on every experiment with these solutions may result in amplification of a contaminant from the first successful experiment, rather than *de novo* amplification from an added template.

If the researcher is running a sufficient number of negative controls, reaction tubes into which no template is ostensibly being added, the contamination might be detected immediately because the 'template-free' reaction products would look as if template had been added. The problem is a highly personal one to the researcher; the contaminant will be product-specific, and will have no effect on experimental results of other researchers who are using primers that do not anneal to the contaminant. By Thursday morning the contamination may be recognized, and the researcher may take several steps to mitigate the problem.

First, the researcher may discard the working stocks of template, oligonucleotides, dNTPs, 10 × buffers, and salts, and prepare fresh working stocks from the (hopefully) uncontaminated original stocks. This fresh dilution would have to be performed in an uncontaminated laboratory, perhaps a friendly neighboring lab. Second, it would be best if all future reactions using these stocks were also assembled in a different laboratory room, using an uncontaminated set of pipetting devices, and with the researcher wearing a clean lab coat and gloves. The contaminated lab can still be used for gel electrophoresis of products, and, indeed, it would be a good place to do that because the location in which PCR products are handled and analyzed should always be physically separated from the location in which PCR samples are assembled. Traffic of reagents and equipment between the reaction analysis lab and the reaction assembly lab should be minimized. Third, the PCR products from future reactions, including the contaminated pipette tips, gel, and electrophoresis running buffer, can be destroyed by autoclaving or bleaching to control most of the 10^{12} copies of potential contaminant. By Friday, the experiment will be working again, without contamination of the negative controls, and a second bottle of Champagne ought to be delivered as a gift of thanks to the neighboring lab – the one that came to the rescue by providing clean space and pipetting devices!

One step that laboratories may take to try to control cross-contamination of experiments is to use dUTP in place of dTTP in the dNTP mixes. The deoxyuridylate nucleotide would not be a part of the template in most situations, and could be specifically destroyed as a contaminant by treatment of a reaction with uracil N-glycosylase enzyme. However, some thermostable DNA polymerases cannot use dUTP substrates, and template DNA from a *dut− ung−* strain of E. *coli* would contain uridine bases, so a researcher must be aware of the limitations of this method.

Thermocycling

Thermocyclers, the instruments that are used to change the incubation temperature of PCR tubes, have different programming interfaces. However, thermocyclers typically allow a researcher to specify a series of temperature steps and times at each temperature, and to specify that a programming segment be repeated for a variable number of cycles. For example, a researcher might set up a thermocycler to have four programming segments (see *Figure 7.4*).

- Segment A (run once): hold the temperature at 96°C for 5 min (see *Figure 7.4A*).
- Segment B (repeat cycle 35 times): hold the temperature at 96°C for 30 s, then at 55°C for 30 s, then at 72°C for 1 min (see *Figure 7.4B*; 12 repeats are shown).
- Segment C (run once): hold the temperature at 72°C for 5 min (see *Figure 7.4C*).
- Segment D (run once): hold the temperature at 4°C indefinitely (see *Figure 7.4D*).

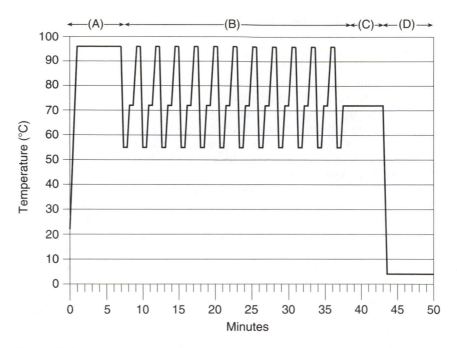

Figure 7.4

The temperature profile of a PCR. The temperature of the reaction is shown as a function of time. (A) An initial temperature rise to 96°C for complete denaturation of the template. (B) Repetitive cycling between 96°C, 55°C, and 72°C, pausing for 30 s at each temperature. Only 12 cycles are shown. (C) A segment during which the temperature is held at 72°C to allow for completion of products being synthesized in the last cycle. (D) Refrigeration of the samples at 4°C until they are removed from the thermocycler.

The first segment is used to denature the template the very first time, which because of its length may take a few minutes. Segment B consists of 35 repeated cycles and represents the part of the PCR during which amplification takes place. Segment C is included at the end of the cycling to allow completion of unfinished products, and segment D is included so that the tubes will be chilled after the reaction is done.

The thermocycler does not change the temperatures instantly, although some are very fast, and so the stated times for each temperature step do not include the seconds that it takes to move from one temperature hold to the next. Segment B would not take 2 min per cycle, the sum of 30 s plus 30 s plus 60 s, for example, and might require 3 min or more per cycle if the thermocycler is slow to change temperatures. The rate of temperature change is a characteristic of a machine, and a method optimized for one thermocycler may not perform as well on another, using the same program.

Slow temperature changes effectively extend the times for annealing and synthesis, and can permit annealing over a T_M range as the temperature gradually ramps up and down. This can be advantageous when degenerate primers are being used, and there is a lack of certainty about the

T_M. Some thermocyclers allow for programming of ramp times between temperatures; for example, a researcher could program segment B in the previous example so that it instructed the thermocycler to take at least 2 min to raise the temperature from 55°C to 72°C.

Temperatures and times

How does a researcher decide upon a thermocycling program? The denaturation temperature is usually 94°C to 98°C, so there is not much variation during that step. The annealing temperature will depend on the composition and length of the oligonucleotides (as discussed in Section 7.2), and 30 s is usually adequate unless there is reason to believe that there is a low concentration of functional oligonucleotides; for example, if the oligonucleotides are degenerate. The extension temperature is usually in the range of 72°C to 75°C, depending on the optimal conditions for the thermostable polymerase, and the synthesis time depends on the length of the product. For that determination, the polymerase supplier's instructions should be consulted, but 1 to 2 min per kbp of product would be typical. If an oligonucleotide has a relatively high annealing temperature that is close to the optimal temperature of the enzyme, then the annealing step and the synthesis step might be combined to make a simple two-step cycle of 96°C and 72°C.

Cycle number

The number of cycles required in the PCR will depend on the amount of starting template, the desired amount of final product, and the cycle efficiency (as discussed in Section 7.3). In brief, if A is the molar amount of final product that is desired and A_0 is the molar amount of starting template, then $A = A_0 P^n$, where P is the productivity factor, a number that is usually between 1.40 and 1.60, and n is the number of cycles. We may rearrange the formula to calculate that $n = \log(A/A_0)/\log(P)$, using either base 10 logarithms or natural logarithms. For example, if we start with 10^4 template copies and want to generate 2×10^{11} product copies, and we assume that P = 1.55, then

$$n = \log_{10}(2 \times 10^{11}/1 \times 10^4)/\log_{10}(1.55) = 7.30/0.190 = 38.4$$

so 39 cycles would be the appropriate number. Note that we did not divide the copies by Avogadro's number, because moles and copies are proportional so we saved ourselves the trouble of converting each 'copy' number to a 'mole' number. However, to put this yield in perspective, 2×10^{11} copies of a 1-kbp product is about 220 ng, and 2×10^{11} copies of a 250-bp product is about 55 ng.

Is there a problem with having too many cycles? For example, could we round up the 39 cycles in the previous example to 40 cycles, just to make it an even number? We could, but there are indeed problems associated with over-cycling and so it is better to limit the number of cycles to just what is needed. In the final cycles of a PCR, for example the 39th cycle in the previous example, there may be 2×10^{11} copies of double-stranded product, about 0.33 pmol. If we extend the PCR to a 40th cycle, then after the denaturation step many of the 2×10^{11} single-stranded product copies will anneal with each other because they are finally at a high concentration and can compete with the oligonucleotide for binding. The annealing could sometimes involve three or four products that form a ragged 'T' or a cruciform structure, or a more elaborate tangle. These oddly shaped products could contribute to a smear during gel electrophoresis and that may conceal the expected product. Furthermore, if there have been side reactions such as primer dimer formation that have consumed a significant part of the oligonucleotide pool, the extra cycles may not be productive and there may be single-stranded products that are not converted to double-stranded products, or are only partially copied.

One strategy for optimizing the cycle number is to set up a PCR with a series of identical tubes, and to stand by the thermocycler so that tubes can be pulled out at defined cycle numbers. For example, a researcher might pull identical reactions out of a thermocycler at 30, 32, 34, 36, 38, and 40 cycles, just to see when the product first appears and when its production is optimal. In each case the tube should be pulled out of the thermocycler at the end of the synthetic step and before the DNA is denatured for the subsequent cycle.

Analysis by gel electrophoresis

Once the reaction is run, the next step is often to analyze the sample by gel electrophoresis. Fragments larger than about 1 kbp are readily separated on an agarose gel, and fragments smaller than 1 kbp are easily resolved on a polyacrylamide gel. A lane of known size markers is included so that the sizes of unknown fragments can be estimated by comparison. In studying the results of electrophoresis, it is important to pay attention to all of the data. In addition to a desired band there may be extra bands above or below the right one. There may be stained material at the top of the gel or still stuck in the well, there may be a faint or intense smear of stained nucleic acid, and there may be a glow of unincorporated oligonucleotide primer or primer dimer near the bottom of the gel. Not all of these are pathological signs, and the matter of extra bands is discussed in more detail in Section 7.8.

For some applications it may be desirable to excise a band from the gel and elute the DNA for further analysis or handling. In the case of an agarose gel, a low melting agarose can be liquified by raising the temperature to 65°C, and after melting it can remain in a liquid state down to at least 37°C. A normal 'high melting point' agarose gel fragment can be dissolved in a sodium iodide solution, or treated with the enzyme agarase to liquify it and release the DNA. A polyacrylamide gel fragment can be crushed and soaked in a 50-mM Tris-Cl, 1-mM EDTA, 100-mM NaCl, pH 8.0 solution to elute the DNA, and the solution then combined with 2.5 volumes of ethanol to precipitate the DNA.

Polymerase fidelity

Thermostable polymerases vary in their ability to copy sequences accurately, and the conditions of a reaction can have an effect on the error rates of polymerases. The polymerases with 3′ to 5′ exonuclease (proof-reading) activities tend to have a lower rate of incorporation of errors, probably because a mismatched 3′ nucleotide can be excised and replaced with the correct complementary nucleotide.

Error rates are usually described in terms of errors per nucleotide synthesized; for example, a *Taq* polymerase that has an error rate of 2×10^{-5} errors per nt would, on average, produce an error every 5×10^{4} nt during synthesis. That does not mean that the product of the last cycle of synthesis will share that average level of accuracy, because errors that occur early in the PCR will be fixed in the template population and perpetuated through later cycles. For example, if we are generating a 1-kbp product using a polymerase that has an error rate of 2×10^{-5}, then approximately 2% of the 1000-bp products have an error at the end of the first cycle, and approximately 40% have at least one error at the end of cycle 35. Error rates can be minimized by using error correcting polymerases such as *Pfu*, and following manufacturers' recommendations for salt conditions. Error rates tend to increase when Mg^{2+} or dNTP concentrations are higher than recommended in a reaction.

If a researcher is relying on PCR to generate templates for sequencing, then some caution needs to be exercised when interpreting the results of a single experiment. Different clone candidates derived from a single pool of PCR products may contain different errors, and so the true sequence of the original template cannot be determined from only two examples; a third independent example would be needed to settle any disagreements between the first two sequences, and more confirmatory examples would usually be desirable.

7.5 Mutagenesis and engineering

The polymerase chain reaction is particularly useful for making changes in a DNA sequence for experimental purposes, or for simply making a fragment of DNA easier to insert into a plasmid vector. The topic of site-specific mutagenesis was introduced in Chapter 6, and the principles of primer design were also discussed in Section 7.2 of this chapter. The two important lessons from those discussions are that the 3′ end of the primer must base-pair effectively with the template, and that the 5′ end defines the boundaries of the PCR fragment. In this section, we will broaden the discussion to demonstrate the full power of PCR as an engineering tool.

Adding a restriction site, revisited

It is a simple thing to add a restriction endonuclease recognition sequence to the 5′ ends of oligonucleotide primers, and when these are used to generate a PCR fragment then the restriction endonuclease recognition sequence will be incorporated into the product. Suppose that we want to clone the coding sequence for a mouse caspase I gene into a cloning vector. We might start by looking at the published sequence for the mouse caspase I gene, which includes a coding sequence of 402 codons, starting with an ATG start codon at nt 201 and ending with a TAA stop codon at nt 1409 (see boxed or underlined sequence in *Figure 7.5*).

```
   1 CTTCCAAGTG  TTGAAGAAGA  ATCATTTCCG  CGGTTGAATC  CTTTTCAGAC  TTGAGCATTT
  61 AAACCTAACT  TTAATTAGGG  AAAGAAACAT  GCGCACACAG  CAATTGTGGT  TATTTCTCAA
 121 TCTGTATTCA  CGCCCTGTTG  GAAAGGAACT  AACAATATGC  TTTCAGTTTC  AGTAGCTCTG
 181 CGGTGTAGAA  AAGAAACGCC  ATGGCTGACA  AGATCCTGAG  GGCAAAGAGG  AAGCAATTTA
 241 TCAACTCAGT  GAGTATAGGG  ACAATAAATG  GATTGTTGGA  TGAACTTTTA  GAGAAGAGAG
 301 TGCTGAATCA  GGAAGAAATG  GATAAAATAA  AACTTGCAAA  CATTACTGCT  ATGGACAAGG
 361 CACGGGACCT  ATGTGATCAT  GTCTCTAAAA  AAGGGCCCCA  GGCAAGCCAA  ATCTTTATCA
 421 CTTACATTTG  TAATGAAGAC  TGCTACCTGG  CAGGAATTCT  GGAGCTTCAA  TCAGCTCCAT
 481 CAGCTGAAAC  ATTTGTTGCT  ACAGAAGATT  CTAAAGGAGG  ACATCCTTCA  TCCTCAGAAA
 541 CAAAGGAAGA  ACAGAACAAA  GAAGATGGCA  CATTTCCAGG  ACTGACTGGG  ACCCTCAAGT
 601 TTTGCCCTTT  AGAAAAAGCC  CAGAAGTTAT  GGAAAGAAAA  TCCTTCAGAG  ATTTATCCAA
 661 TAATGAATAC  AACCACTCGT  ACACGTCTTG  CCCTCATTAT  CTGCAACACA  GAGTTTCAAC
 721 ATCTTTCTCC  GAGGGTTGGA  GCTCAAGTTG  ACCTCAGAGA  AATGAAGTTG  CTGCTGGAGG
 781 ATCTGGGGTA  TACCGTGAAA  GTGAAAGAAA  ATCTCACAGC  TCTGGAGATG  GTGAAAGAGG
 841 TGAAAGAATT  TGCTGCCTGC  CCAGAGCACA  AGACTTCTGA  CAGTACTTTC  CTTGTATTCA
 901 TGTCTCATGG  TATCCAGGAG  GGAATATGTG  GGACCACATA  CTCTAATGAA  GTTTCAGATA
 961 TTTTAAAGGT  TGACACAATC  TTTCAGATGA  TGAACACTTT  GAAGTGCCCA  AGCTTGAAAG
1021 ACAAGCCCAA  GGTGATCATT  ATTCAGGCAT  GCCGTGGAGA  GAAACAAGGA  GTGGTGTTGT
1081 TAAAAGATTC  AGTAAGAGAC  TCTGAAGAGG  ATTTCTTAAC  GGATGCAATT  TTTGAAGATG
1141 ATGGCATTAA  GAAGGCCCAT  ATAGAGAAAG  ATTTTATTGC  TTTCTGCTCT  TCAACACCAG
1201 ATAATGTGTC  TTGGAGACAT  CCTGTCAGGG  GCTCACTTTT  CATTGAGTCA  CTCATCAAAC
1261 ACATGAAAGA  ATATGCCTGG  TCTTGTGACT  TGGAGGACAT  TTTCAGAAAG  GTTCGATTTT
1321 CATTTGAACA  ACCAGAATTT  AGGCTACAGA  TGCCCACTGC  TGATAGGGTG  ACCCTGACAA
1381 AACGTTTCTA  CCTCTTCCCG  GGACATTAAA  CGAAGAATCC  AGTTCATTCT  TATGTACCTA
1441 TGCTGAGAAT  CGTGCCAATA  AGAAGCCAAT  ACTTCCTTAG  ATGATGCAAT  AAATATTAAA
1501 ATAAAACAAA  ACAAAAAAAA  AAAAAAAAAA  AAA
```

Figure 7.5

A cDNA sequence of the mouse caspase I gene. The ATG start codon is located at nt 201 and the TAA stop codon is located at nt 1407. The coding sequence is underlined, and the locations of oligonucleotide primer sequences are additionally boxed (the primer at nt 1407 would be the reverse-complement of the strand shown). (GenBank® locus NM_009807)

If we wanted to generate the coding sequence and stop codon for the mouse caspase I gene by RT-PCR, a technique that was introduced in Chapter 6, we might prime the reverse transcription reaction with an oligonucleotide that is antisense to the mRNA, with a 5' end complementary to nt 1409. That is, we might consider a 'reverse' oligonucleotide complementary to the 18-nt segment 1392-1409 which would be

```
14.  Reverse (caspase I): 5'-TTAATGTCCCGGGAAGAG-3'
```

(see boxed sequence, *Figure 7.5*). The 'forward' oligonucleotide for second-strand cDNA synthesis would start with the ATG start codon and extend into the coding sequence, for example the 18-nt sequence

```
15.  Forward (caspase I): 5'-ATGGCTGACAAGATCCTG-3'
```

(see boxed sequence, *Figure 7.5*). Using these reverse and forward oligonucleotide primers '14' and '15', and a suitable mouse tissue that expresses caspase I mRNA, we could generate a fragment of caspase I cDNA stretching from nt 201 to nt 1409 by RT-PCR.

What are our options for cloning this cDNA? If we performed the PCR with an enzyme such as *Taq* DNA polymerase, many of the fragments would have an untemplated adenylate nucleotide added to the 3' ends because of the enzyme's limited terminal transferase activity (discussed in Chapter 6). In that case we could use a TOPO TA Cloning® vector, or a similar vector that takes advantage of the 3' adenylate nucleotide. Alternatively if we performed the PCR with an enzyme that leaves blunt ends, such as *Pfu* polymerase, we could use a cloning vector such as pZErO®-1 (discussed in Chapter 4), and prepare it by digestion with an enzyme that cleaves the vector in the multiple cloning site and leaves a blunt end; for example, the enzyme *Eco*RV (GAT/ATC).

However, if neither of these options were satisfactory, we might consider adding restriction endonuclease sequences to the 5' ends of the oligonucleotides and then cleaving the PCR product with that specific restriction enzyme to make ends that are cohesive and compatible with a cloning site in a vector. We would need to select a restriction endonuclease that does not cleave the cDNA fragment internally, and in this specific example of mouse caspase I we could make use of the fact that there are no internal *Bam*HI recognition sequences (G/GATCC). We could extend our original oligonucleotides, adding restriction endonuclease sites for *Bam*HI (G/GATCC), shown in lowercase letters:

```
16. Forward (caspase I): 5'-gggatccATGGCTGACAAGATCCTG-3'
17. Reverse (caspase I): 5'-gggatccTTAATGTCCCGGGAAGAG-3'
```

When we complete the PCR, the *Bam*HI restriction endonuclease sequences would be incorporated into the ends of the fragment, and would have the following general structure:

```
5'-GGGATCCATGGCTGACAAGATCCTGAGGGC...TCTACTCTTCCCGGGACATTAAGGATCCC-3'
3'-CCCTAGGTACCGACTGTTCTAGGACTCCCG...AGATGAGAAGGGCCCTGTAATTCCTAGGG-5'
```

When the DNA product is digested with the enzyme *Bam*HI, the enzyme will generate 5'-GATC overhanging ends, suitable for cloning into the *Bam*HI site of pZErO®-1 (or another vector with a *Bam*HI cloning site):

```
5'-  GATCCATGGCTGACAAGATCCTGAGGGC...TCTACTCTTCCCGGGACATTAAG     -3'
3'-      GTACCGACTGTTCTAGGACTCCCG...AGATGAGAAGGGCCCTGTAATTCCTAG -5'
```

The observant reader may have noticed that we added an extra nucleotide to the 5' end of each of oligonucleotides '16' and '17', beyond the recognition sequence of the *Bam*HI enzyme. As was discussed in Chapter 5, restriction endonucleases vary in their ability to cleave DNA sequences close to the end of a fragment. *Bam*HI is an example of an enzyme that will cleave DNA fragments successfully if there is at least one extra nucleotide at the 5' end of the GGATCC recognition sequence. While the nucleotide was added to improve the efficiency of the enzyme, it is excised during cleavage with the enzyme and does not become a part of the cloned product.

Matching a triplet reading frame

Let's make the cloning problem a little more sophisticated by supposing that we not only wish to clone the mouse caspase I cDNA, but that we wish to have it be expressed from an *E. coli* protein expression vector such as pET-50b(+) (4). The general principle of protein expression vectors, and the use of tags and fusion proteins was discussed in Chapter 4. The pET-50b(+) vector allows for foreign protein expression as a fusion protein with N-terminal Nus·Tag™ and His·Tag® coding sequences (see *Figure 7.6*). There is a unique *Bam*HI restriction endonuclease sequence that appears in the vector, after the nucleotide sequence encoding the human rhinovirus protease (HRV 3C) cleavage site (see site marked '1' in *Figure 7.6*).

We can make use of this *Bam*HI site for cloning, but the restriction endonuclease recognition sequence G/GATCC must also be considered in the context of the triplet reading frame of the ribosome if we want the mouse caspase I coding sequence to be expressed as a fusion protein following the Nus·Tag™ and His·Tag® sequences.

```
             BamHI
...GGG TAC CAG GAT CCG AAT TCT...
...Gly Tyr Gln Asp Pro Asn Ser...
```

Were we to join the mouse caspase I sequence to the vector using the embedded *Bam*HI site, the result would be a mismatch between the original Nus·Tag™ and His·Tag® triplet reading frame and the caspase I reading frame.

```
             BamHI
...GGG TAC CAG GAT CCA TGG CTG ACA AGA TCC TG ...
...Gly Tyr Gln Asp Pro Trp Leu Thr Arg Ser    ...
        caspase I    Met Ala Asp Lys Thr Leu ...
```

Rather than express the underlined caspase I nucleic acid sequence as the polypeptide Met-Ala-Asp-Lys-Thr-Leu ..., the inserted sequence is expressed as Trp-Leu-Thr-Arg-Ser ... because the ribosome is frame-shifted. In our preliminary design for adding a *Bam*HI site to the PCR fragment, we happened to set it up so that the 5' end of the caspase I coding sequence

Figure 7.6

A portion of the sequence of pET-50b(+) (Novagen®). Nucleotides 1–180 (above the dashed line) include a T7 RNA polymerase promoter used to control expression, and the site of translation initiation, followed by a His·Tag® sequence and the start of a Nus·Tag™ sequence (white on black lettering). The central part of the Nus·Tag™ is not shown (dashed line). Nucleotides 1561–1920 (below the dashed line) include the end of the Nus·Tag™, a second His·Tag®, protease cleavage sites for HRV 3C and thrombin (the triangles show the points of cleavage), and S·Tag™. The amino acid sequence encoded by the DNA is shown below the strand. Selected restriction endonuclease recognition sequences are shown above the strand. Specific sites discussed in the text are marked with circled numbers (1–7). Oligonucleotide primers are indicated by line segments, with black lines indicating 'forward' and gray lines indicating 'reverse' oligonucleotides, and arrowheads indicating 3' ends. The nucleotide index numbers do not match the numbering system of Novagen® for this vector.

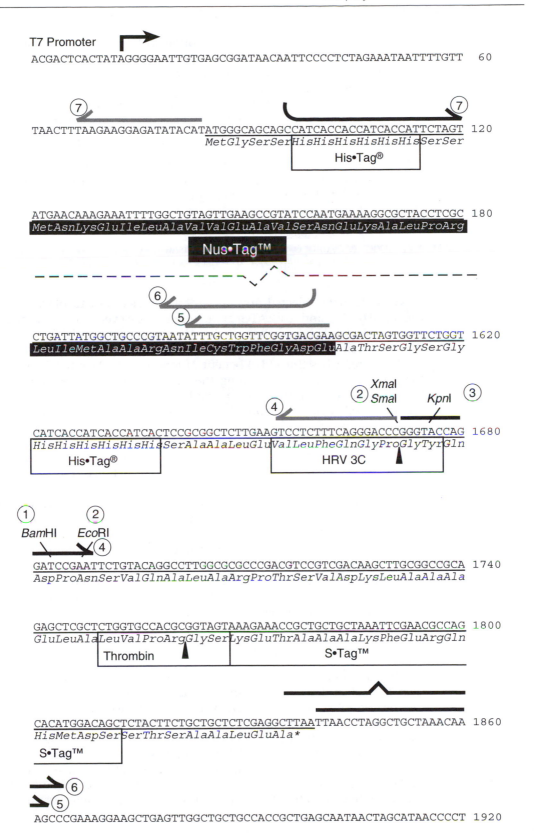

T7 Promoter

ACGACTCACTATAGGGGAATTGTGAGCGGATAACAATTCCCCTCTAGAAATAATTTTGTT 60

TAACTTTAAGAAGGAGATATACATATGGGCAGCAGCCATCACCACCATCACCATTCTAGT 120
                MetGlySerSerHisHisHisHisHisHisSerSer

His•Tag®

ATGAACAAAGAAATTTTGGCTGTAGTTGAAGCCGTATCCAATGAAAAGGCGCTACCTCGC 180
MetAsnLysGluIleLeuAlaValValGluAlaValSerAsnGluLysAlaLeuProArg

Nus•Tag™

CTGATTATGGCTGCCCGTAATATTTGCTGGTTCGGTGACGAAGCGACTAGTGGTTCTGGT 1620
LeuIleMetAlaAlaArgAsnIleCysTrpPheGlyAspGluAlaThrSerGlySerGly

*Xma*I
*Sma*I
*Kpn*I

CATCACCATCACCATCACTCCGCGGCTCTTGAAGTCCTCTTTCAGGGACCCGGGTACCAG 1680
HisHisHisHisHisHisSerAlaAlaLeuGluValLeuPheGlnGlyProGlyTyrGln

His•Tag®   HRV 3C

*Bam*HI   *Eco*RI

GATCCGAATTCTGTACAGGCCTTGGCGCGCCCGACGTCCGTCGACAAGCTTGCGGCCGCA 1740
AspProAsnSerValGlnAlaLeuAlaArgProThrSerValAspLysLeuAlaAlaAla

GAGCTCGCTCTGGTGCCACGCGGTAGTAAAGAAACCGCTGCTGCTAAATTCGAACGCCAG 1800
GluLeuAlaLeuValProArgGlySerLysGluThrAlaAlaAlaLysPheGluArgGln

Thrombin   S•Tag™

CACATGGACAGCTCTACTTCTGCTGCTCTCGAGGCTTAATTAACCTAGGCTGCTAAACAA 1860
*HisMetAspSerSerThrSerAlaAlaLeuGluAla**

S•Tag™

AGCCCGAAAGGAAGCTGAGTTGGCTGCTGCCACCGCTGAGCAATAACTAGCATAACCCCT 1920 needs to be lengthened by one nucleotide so that it will fuse seamlessly with the triplet reading frame of the fusion protein in pET50-b(+). Without the extra nucleotide, the ribosome will translate gibberish.

The matter is easily solved by adding one nucleotide to the forward oligonucleotide, immediately after the *Bam*HI sequence and before the ATG start codon of the caspase I. For example, the oligonucleotide

18. Forward (caspase I): 5'-GGGATCCcATGGCTGACAAGATCCTG-3'

is lengthened by one nucleotide, in lowercase, and when the PCR product is digested with *Bam*HI and inserted into the *Bam*HI site of pET-50b(+), the fusion of reading frames would be perfect (the inserted nucleotide is again shown in lowercase):

<u>BamHI</u>
```
...GGG TAC CAG GAT CCc ATG GCT GAC AAG ATC CTG...
...Gly Tyr Gln Asp Pro Met Ala Asp Lys Ile Leu...
                       caspase I
```

We could have selected any of the four nucleotides to place between the *Bam*HI site and the ATG start codon; however, we chose the 'C' nucleotide to fill the gap because it is the same nucleotide that precedes the ATG start codon in the cDNA, at nt 200 of the caspase I sequence. Furthermore, it has the added benefit of retaining an *Nco*I (C/CATGG) site that might be useful in screening the candidate clones after transformation. Having added a nucleotide matching the cDNA, we may also shorten our proposed forward oligonucleotide '18' at its 3' end so that it once again only has 18 nt of sequence matching the cDNA (see '19' below).

It happens that we are not concerned about matching a reading frame at the 3' end of the caspase I sequence, because it was our plan to retain the TAA stop codon from the gene. Had we not included that stop codon, we might face a similar issue of reading frame as the ribosome passes out of the caspase I sequence and back into the pET-50b(+)-derived sequence. In all likelihood, we would want to add a sufficient number of nucleotides so that the frame of the ribosome rejoined that of the vector tag sequences. The alternative to rejoining the tag sequence reading frame would be to add a bit of arbitrary amino acid sequence to the carboxyl end of the caspase I; a design that would probably not be wise for most applications.

At this point we should take a look at the two oligonucleotides and consider their technical merits and flaws.

19. Forward (caspase I): 5'-GGGATCCCATGGCTGACAAGATCCT-3'
17. Reverse (caspase I): 5'-GGGATCCTTAATGTCCCGGGAAGAG-3'

The forward and reverse oligonucleotides are 25 nt in length, and have 14 G or C nucleotides and 11 A or T nucleotides. The 3' 18 nucleotides of the reverse oligonucleotide '17' will base-pair to the mRNA and be extended using reverse transcriptase. The 3' 18 nucleotides of the forward oligonucleotide '19' will base-pair to the first-strand cDNA, and be extended using a thermostable DNA polymerase. For these initial annealing steps of 18 nt at the 3' end, both oligonucleotides have 9 G or C nucleotides and 9 A or T nucleotides, and are predicted to have a T_M of approximately 50°C to 54°C and an optimal annealing temperature a few degrees below that. However, that annealing temperature would only be applicable to the first rounds of PCR when the original cDNA template is being copied, because once the PCR is underway the entire 25 nt of each oligonucleotide will anneal to the replicated template. The T_M of the 25-nt oligonucleotides would be approximately 61°C, so a higher annealing temperature might be appropriate after the first few cycles of PCR.

In evaluating these oligonucleotides, we should notice a lack of complexity of the 3' end of the reverse oligonucleotide '17'; the last 8 nt of the sequence (GGGAAGAG) are made up of only 'G' and 'A' nucleotides. We cannot move the 5' end of the oligonucleotide because it provides the stop codon for the coding sequence, and it is just bad luck that the non-complex sequence lies right where it does. Specificity may not be a problem using an mRNA template, but if there were a problem then the complexity problem could be addressed by extending the 3' end to include additional sequence. For example, the oligonucleotide

```
20. Reverse (caspase I): 5'-
GGGATCCTTAATGTCCCGGGAAGAGgtagaaac-3'
```

is extended by eight nucleotides shown in lowercase.

Restriction sites, unrestricted

In the previous example, we modeled the cloning of a mouse caspase I coding sequence into the *Bam*HI site of the vector pET-50b(+), by adding *Bam*HI recognition sequences to the 5' ends of the oligonucleotides used in the generation of caspase I. We chose *Bam*HI as a cloning enzyme in part because the caspase I coding sequences lack *Bam*HI sites and would not be fragmented internally during cloning. Suppose that we had wanted to introduce the caspase I cDNA into the pET-50b(+) using the enzymes *Sma*I (CCC/GGG) and *Eco*RI (G/AATTC) (see sites marked '2' in *Figure 7.6*). It happens that there are internal sites in the caspase I gene for *Sma*I and *Eco*RI, so the method used with *Bam*HI would not work. The PCR product digested with *Sma*I and *Eco*RI would be fragmented.

Creating an overhanging end

One elegant method for cloning the PCR product into the *Sma*I and *Eco*RI sites involves using a DNA polymerase with a 3' to 5' exonuclease activity, to make a compatible end. Let's suppose that we have a vector pET-50b(+) that is digested with *Sma*I (CCC/GGG) and *Eco*RI (G/AATTC), and therefore has one blunt end and one 5' overhanging end (see sites marked '2' in *Figure 7.6*). The digestion of the vector sequence

```
              SmaI                        EcoRI
...GTCCTCTTTCAGGGACCCGGGTACCAGGATCCGAATTCTGTACAGGCCTTGGC...
...CAGGAGAAAGTCCCTGGGCCCATGGTCCTAGGCTTAAGACATGTCCGGAACCG...
...ValLeuPheGlnGlyProGlyTyrGlnAspProAsnSerValGlnAlaLeuAla...
```

yields these ends:

```
              SmaI                 EcoRI
...GTCCTCTTTCAGGGACCC          AATTCTGTACAGGCCTTGGC...
...CAGGAGAAAGTCCCTGGG              GACATGTCCGGAACCG...
```

Suppose that we modify the 5' ends of the previous oligonucleotides so that they no longer contain a *Bam*HI site, but rather have adaptable ends. In this example, the nucleotides in uppercase are caspase I sequence and the nucleotides in lowercase are extensions to the 5' ends of the oligonucleotide.

```
21. Forward (caspase I): 5'-gggATGGCTGACAAGATCCT-3'
22. Reverse (caspase I): 5'-aattcTTAATGTCCCGGGAAGAG-3'
```

A PCR product that was generated using these oligonucleotides would have the following general structure:

```
5'-GGGATGGCTGACAAGATCCTGAGGGC...TCTACCTCTTCCCGGGACATTAAGAATT-3'
3'-CCCTACCGACTGTTCTAGGACTCCCG...AGATGGAGAAGGGCCCTGTAATTCTTAA-5'.
```

Upon treatment with Klenow enzyme or T4 DNA polymerase, and only the substrates dCTP and dGTP, the proofreading activity of the enzyme would subtract nucleotides from the 3' ends until a 'C' or 'G' base was uncovered. The end derived from the forward oligonucleotide would experience no net change, but the end derived from the reverse oligonucleotide would be converted to a 5' overhanging end compatible with an *Eco*RI-generated end.

```
5'-GGGATGGCTGACAAGATCCTGAGGGC...TCTACCTCTTCCCGGGACATTAAG    -3'
3'-CCCTACCGACTGTTCTAGGACTCCCG...AGATGGAGAAGGGCCCTGTAATTCTTAA-5'
```

This enzymatic treatment left one blunt end and one cohesive end in the PCR product, and no restriction enzymes were used to make the compatible ends. When the end left by *Sma*I in the pET-50b(+) is fused with the end left by the forward oligonucleotide in the PCR product, the triplet reading frame is preserved (the caspase I sequence is underlined):

```
...ValLeuPheGlnGlyProGlyMetAlaAspLysIleLeuArgAla...
...GTCCTCTTTCAGGGACCCGGGATGGCTGACAAGATCCTGAGGGC...
```

Furthermore, this is an example of forced cloning. The PCR fragment may only be inserted into the pET-50b(+) vector in one orientation since the ends are not compatible in the reverse orientation.

Digging a bit deeper into the problem

Ligating a *Sma*I-generated end with another blunt end may seem a bit artless, and we might prefer to have two cohesive ends involved in the ligation. We could change the design a bit by digesting the pET-50b(+) with *Eco*RI (G/AATTC) and *Xma*I (C/CCGGG), which is an isoschizomer of *Sma*I (CCC/GGG) that leaves a 5' overhanging end (see sites marked '2' in *Figure 7.6*). Then we could treat the digested vector with Klenow enzyme or T4 DNA polymerase, and only the substrates dCTP and dATP, which would fill in two nucleotides at each end (shown in lowercase) and leave 5' overhanging ends of two nucleotides.

```
         XmaI                          EcoRI
...GTCCTCTTTCAGGGACcc            AATTCTGTACAGGCCTTGGC...
...CAGGAGAAAGTCCCTGGGCC          aaGACATGTCCGGAACCG...
```

The oligonucleotides could be designed so that the PCR product can also be treated with Klenow enzyme or T4 DNA polymerase to generate 5′ overhanging ends of 2 nt that are compatible with the vector. Again, the 5′ extensions of the caspase I sequence are shown in lowercase:

```
23. Forward (caspase I): 5′-ggcATGGCTGACAAGATCCT-3′
24. Reverse (caspase I): 5′-aaTTAATGTCCCGGGAAGAG-3′
```

The structure of the PCR product would be

```
5′-GGCATGGCTGACAAGATCCTGAGGGC...TCTACCTCTTCCCGGGACATTAATT-3′
3′-CCGTACCGACTGTTCTAGGACTCCCG...AGATGGAGAAGGGCCCTGTAATTAA-5′
```

and after treatment with Klenow enzyme or T4 DNA polymerase and only the substrates dGTP and dATP it would be

```
5′-GGCATGGCTGACAAGATCCTGAGGGC...TCTACCTCTTCCCGGGACATTAA  -3′
3′-  GTACCGACTGTTCTAGGACTCCCG...AGATGGAGAAGGGCCCTGTAATTAA-5′
```

When joined to the vector by ligation the reading frame of the vector-derived tag sequences would match that of the caspase I gene (underlined below):

```
...ValLeuPheGlnGlyProGlyMetAlaAspLysIleLeuArgAla...
...GTCCTCTTTCAGGGACCCGGCATGGCTGACAAGATCCTGAGGGC...
```

The GGC codon preceding the ATG start codon is synonymous with the original GGG glycine codon, so there is no coding change.

Alternatively, we might digest the pET-50b(+) vector with only *Sma*I:

```
...GTCCTCTTTCAGGGACCC       GGGTACCAGGATCCGAATTCTGTACAGGCCTTGGC...
...CAGGAGAAAGTCCCTGGG       CCCATGGTCCTAGGCTTAAGACATGTCCGGAACCG...
```

and then treat those ends with Klenow enzyme or T4 DNA polymerase and dATP substrate, leaving 3-nt overhanging ends.

```
...GTCCTCTTTCAGGGA         GGGTACCAGGATCCGAATTCTGTACAGGCCTTGGC...
...CAGGAGAAAGTCCCTGGG           ATGGTCCTAGGCTTAAGACATGTCCGGAACCG...
```

Compatible ends to the vector could be generated in the PCR product if the oligonucleotides had 3-nt extensions of CCC:

```
25. Forward (caspase I): 5′-cccATGGCTGACAAGATCCT-3′
26. Reverse (caspase I): 5′-cccTTAATGTCCCGGGAAGAG-3′
```

The PCR product would have the following structure:

```
5′-CCCATGGCTGACAAGATCCTGAGGGC...TCTACCTCTTCCCGGGACATTAAGGG-3′
3′-GGGTACCGACTGTTCTAGGACTCCCG...AGATGGAGAAGGGCCCTGTAATTCCC-5′
```

and after treatment with Klenow enzyme or T4 DNA polymerase and only the substrates dATP and dTTP it would be:

```
5′-CCCATGGCTGACAAGATCCTGAGGGC...TCTACCTCTTCCCGGGACATTAA   -3′
3′-   TACCGACTGTTCTAGGACTCCCG...AGATGGAGAAGGGCCCTGTAATTCCC-5′
```

The fused reading frames of the vector sequence and caspase I sequence would match, the only difference being the loss of a vector-derived glycine codon:

```
...ValLeuPheGlnGlyProMetAlaAspLysIleLeuArgAla...
...GTCCTCTTTCAGGGACCCATGGCTGACAAGATCCTGAGGGC...
```

This method would require only a single restriction endonuclease, but would not permit forced cloning because the PCR product could be inserted in either orientation.

These examples are meant to illustrate two principles: first, that a compatible end can be generated without the use of a restriction endonuclease on a DNA fragment; and second, that a restriction enzyme used on a vector creates a point of entry for cloning but does not lock the design into the use of a specific overhanging end. Examples of enzymes leaving blunt ends (*Sma*I) or 5′ overhanging ends (*Xma*I, *Eco*RI) have been provided; however, an additional example that makes use of a 3′ overhanging end may be interesting as well. The enzyme *Kpn*I (GGTAC/C) has a recognition sequence that overlaps with the previously used *Sma*I (CCCGGG) recognition sequence, but it leaves a 3′ overhanging end after cleavage (see site marked '3' in *Figure 7.6*). When a *Kpn*I-digested vector is treated with Klenow enzyme or T4 DNA polymerase, the 3′ overhanging end is always removed, regardless of its sequence content (for reasons that are discussed in Chapter 6).

If we were to treat pET-50b(+) with *Kpn*I, and follow that by treating the vector with Klenow enzyme or T4 DNA polymerase and only the substrate dATP, we would be left with 5′ overhanging ends of 4 nt and 5 nt. The vector has this structure at the *Kpn*I site:

*Kpn*I
```
...GTCCTCTTTCAGGGACCCGGGTACCAGGATCCGAATTCTGTACAGGCCTTGGC...
...CAGGAGAAAGTCCCTGGGCCCATGGTCCTAGGCTTAAGACATGTCCGGAACCG...
...ValLeuPheGlnGlyProGlyTyrGlnAspProAsnSerValGlnAlaLeuAla...
```

3′ overhanging ends are left following digestion with *Kpn*I:

*Kpn*I
```
...GTCCTCTTTCAGGGACCCGGGTAC          CAGGATCCGAATTCTGTACAGGCCTTGGC...
...CAGGAGAAAGTCCCTGGGC               CCATGGTCCTAGGCTTAAGACATGTCCGGAACCG...
...ValLeuPheGlnGlyProGlyTyr          GlnAspProAsnSerValGlnAlaLeuAla...
```

Upon treatment with Klenow enzyme or T4 DNA polymerase and only the substrate dATP, the 3′ overhanging ends are removed and digestion continues until a 3′ 'A' is uncovered:

```
...GTCCTCTTTCAGGGA                   CAGGATCCGAATTCTGTACAGGCCTTGGC...
...CAGGAGAAAGTCCCTGGGC               AGGCTTAAGACATGTCCGGAACCG...
...ValLeuPheGlnGlyPro                GlnAspProAsnSerValGlnAlaLeuAla...
```

The caspase I PCR fragment could be generated using a revised set of oligonucleotides with 6-nt extensions:

```
27. Forward (caspase I): 5′-cccgaaATGGCTGACAAGATCCT-3′
28. Reverse (caspase I): 5′-tcctgaTTAATGTCCCGGGAAGAG-3′
```

yielding a structure

```
5′-CCCGAAATGGCTGACAAGATCCTGAGGGC...TCTACCTCTTCCCGGGACATTAATCAGGA-3′
3′-GGGCTTTACCGACTGTTCTAGGACTCCCG...AGATGGAGAAGGGCCCTGTAATTAGTCCT-5′
```

After treatment with Klenow enzyme or T4 DNA polymerase and only the substrate dTTP, a set of compatible ends could be generated:

```
5'-CCCGAAATGGCTGACAAGATCCTGAGGGC...TCTACCTCTTCCCGGGACATTAAT      -3'
3'-    TTTACCGACTGTTCTAGGACTCCCG...AGATGGAGAAGGGCCCTGTAATTAGTCCT-5'
```

When joined to the vector by ligation the reading frame of the vector-derived tag sequences would match that of the caspase I gene (underlined below):

```
...ValLeuPheGlnGlyProGluMetAlaAspLysIleLeuArgAla...
```

```
...GTCCTCTTTCAGGGACCCGAAATGGCTGACAAGATCCTGAGGGC...
```

The reader may note that this design led to an additional codon for glutamic acid (GAA) before the ATG start codon of the caspase I gene, perhaps an unnecessary sacrifice; however, the point is made that a 3' overhanging end, such as that left by *Kpn*I, can be converted to a 5' overhanging end by Klenow enzyme or T4 DNA polymerase.

There are many other solutions to the caspase I cloning problem using restriction endonucleases and DNA polymerases, and readers may wish to amuse themselves by finding the most artful ones. This particular example was constrained by the need to preserve the triplet reading frame of the ribosome through the fusion protein junction, and more general cloning applications would be a bit easier to design. However, the use of restriction endonucleases against the cloning vector does limit the points of entry into the sequence, and the 5' ends generated by the restriction endonucleases are fixed points into which the PCR fragment must be fitted.

Kicking the restriction enzyme habit

We can take the art of joining DNA segments a step further by using PCR to generate both the DNA insert and the vector. The vector is a circular sequence, which in the case of pET-50b(+) is 6733 nt, but it can be linearized using a method called inverse PCR in which two 'outward facing' oligonucleotide primers regenerate the whole sequence. We will illustrate this by extending the example previously given, of the pET-50b(+) vector (*Figure 7.6*) and the mouse caspase I cDNA (*Figure 7.5*).

Inverse PCR

We are used to seeing oligonucleotide designs in which a 'forward' oligonucleotide copies the 'top strand' sequence, and lies to the left of a 'reverse' oligonucleotide that copies the 'bottom strand' sequence. However, suppose we break that rule and design oligonucleotides for the pET-50b(+) sequence in which the 'reverse' oligonucleotide lies to the left (see sites marked '4', *Figure 7.6*):

```
29. Reverse (pET): 5'-GGGTCCCTGAAAGAGGAC-3'
30. Forward (pET): 5'-GGGTACCAGGATCCGAAT-3'
```

These anneal to the template as follows:

```
                    29. Reverse (pET)
              3'-CAGGAGAAAGTCCCTGGG-5'
...GCGGCTCTTGAAGTCCTCTTTCAGGGACCCGGGTACCAGGATCCGAATTCTGTACAGGCCT...
...CGCCGAGAACTTCAGGAGAAAGTCCCTGGGCCCATGGTCCTAGGCTTAAGACATGTCCGGA...
                          5'-GGGTACCAGGATCCGAAT-3'
                            30. Forward(pET)
```

It might seem that we've made a mistake because the 3' ends don't point toward each other, but in fact we haven't, and they do! The 3' ends point toward each other going the long way around the vector, just as an airplane flying on an Earth meridian will eventually reach the point where it started, if it doesn't run out of fuel. The structure of the pET-50b(+) vector, generated as a linear PCR product, will be:

```
5'–GGGTACCAGGATCCGAATTCTGTACAGGCCT...GCGGCTCTTGAAGTCCTCTTTCAGGGACCC–3'
3'–CCCATGGTCCTAGGCTTAAGACATGTCCGGA...CGCCGAGAACTTCAGGAGAAAGTCCCTGGG–5'
```

The reader may note that the ends of the PCR product are identical in sequence to those that would have been left by the enzyme *Sma*I (CCC/GGG), except that the free 5' ends are oligonucleotide-derived and may not have a 5' phosphate.

That analogy of running out of fuel while circumnavigating the Earth raises the related question of whether it is feasible to generate a 6.7-kbp PCR product from a circular template. It is feasible, but it requires a polymerase that has high fidelity, making few mistakes. There are commercially available kits and polymerase buffer formulations for performing long PCR, as this is sometimes called; however, an error correcting polymerase such as *Pfu* can go the distance if it is given at least 1 to 2 min of synthesis time for every 1 kbp of fragment size. A 6.7-kbp product might be achieved using a 14-min synthesis step at 72°C. When conducting inverse PCR for the purposes of mutagenesis or genetic engineering it is important to start the reaction with the smallest amount of template DNA possible; the original plasmid template is a 'parental' type that may persist through the PCR and be recovered during DNA transformation among the candidate clones.

Making your own cloning site

We have absolute control over the point of entry into the vector, because we specify the 5' ends of the linearized vector during inverse PCR. Let's return to the example of the pET-50b(+) vector, and use inverse PCR to organize our own cloning site (see the two sites marked '5', *Figure 7.6*). Suppose that we want to clone the mouse caspase I coding sequence immediately downstream of the Nus·Tag™ sequence (the white-on-black lettering in *Figure 7.6*), and in the process eliminate the second His·Tag® coding sequence, the multiple cloning site, and the S·Tag™ sequence at the carboxyl terminus. The sequence at the 3' end of the Nus·Tag™ sequence is

```
5'–ATTTGCTGGTTCGGTGACGAA–3'
```

and for inverse PCR an oligonucleotide that is complementary to this sequence could be used:

```
31. Reverse (pET): 5'–TTCGTCACCGAACCAGCAAAT–3'
```

The 'forward' oligonucleotide could be based on sequence immediately following the TAA stop codon that lies 9 codons beyond the end of the S·Tag™ sequence.

```
32. Forward (pET): 5'–TTAACCTAGGCTGCTAAACAAAG–3'
```

These two oligonucleotides '31' and '32' are 21 and 23 nt, respectively, and would have a predicted T_M of approximately 52°C.

If a PCR were conducted using pET-50b(+) as a template and these oligonucleotides '31' and '32' as primers, a linear product would be generated that deletes 237 nt of the vector, and has the general structure:

```
5'-TTAACCTAGGCTGCTAAACAAAGCCCGAAA...TGCCCGTAATATTTGCTGGTTCGGTGACGAA-3'
3'-AATTGGATCCGACGATTTGTTTCGGGCTTT...ACGGGCATTATAAACGACCAAGCCACTGCTT-5'
```

We might then treat this fragment with Klenow enzyme or T4 DNA polymerase, and only the substrate dCTP, to generate unique 5' overhanging ends of 8 nt and 3 nt at the forward and reverse oligonucleotide ends:

```
5'-TTAACCTAGGCTGCTAAACAAAGCCCGAAA...TGCCCGTAATATTTGCTGGTTCGGTGAC    -3'
3'-          CCGACGATTTGTTTCGGGCTTT...ACGGGCATTATAAACGACCAAGCCACTGCTT-5'
```

This is just a starting plan, and we may have to make changes to '31' and '32' in order to make a perfect connection with the coding sequence of mouse caspase I. It should be noted at the outset that the limited terminal transferase activities of some thermostable polymerases, such as *Taq* DNA polymerase, will not have any effect on the outcome of this method; if an untemplated 3' adenylate nucleotide is added by the thermostable polymerase, it will be immediately removed by the 3' to 5' exonuclease of Klenow enzyme or T4 DNA polymerase.

At the 5' end, the caspase I PCR fragment could be generated using a revised oligonucleotide with a 3-nt extension:

```
33. Forward (caspase I): 5'-gaaATGGCTGACAAGATCCT-3'
```

yielding a structure:

```
        caspase I sequence
        MetAlaAspLysIleLeuArgAla...
5'-GAAATGGCTGACAAGATCCTGAGGGC...
3'-CTTTACCGACTGTTCTAGGACTCCCG...
```

We would now like to connect the Nus·Tag™ sequence to the caspase I coding sequence with the following compatible ends:

```
    Nus•Tag™ sequence                      caspase I sequence
...ArgAsnIleCysTrpPheGlyAspGlu             MetAlaAspLysIleLeuArgAla...
...CGTAATATTTGCTGGTTCGGTGAC    -3'   5'-GAAATGGCTGACAAGATCCTGAGGGC...
...GCATTATAAACGACCAAGCCACTGCTT-5'    3'-    TACCGACTGTTCTAGGACTCCCG...
```

Unfortunately, we cannot make the indicated compatible end for the caspase I cDNA that has a 3' overhanging end and connects the Glu codon of the Nus·Tag™ directly to the Met codon of the caspase I. However, we could generate a 4-nt overhanging end of 5'-GAAA, by treating with Klenow enzyme or T4 DNA polymerase and only dATP as a substrate:

```
        caspase I sequence
        MetAlaAspLysIleLeuArgAla...
5'-GAAATGGCTGACAAGATCCTGAGGGC...
3'-    ACCGACTGTTCTAGGACTCCCG...
```

We can modify our plan for the inverse PCR of the pET-50b(+), altering '31' so that the ends will fit together. The replacement oligonucleotide for the inverse PCR would be

    34. Reverse (pET): 5'-tTTCGTCACCGAACCAGCAAAT-3'

with the only difference between '31' and '34' being that the revised oligonucleotide would be extended at its 5' end by an extra 'T' base (shown in lowercase). The PCR product generated using '32' and '34' could be treated with Klenow enzyme or T4 DNA polymerase, and only the substrate dCTP, to generate unique 5' overhanging ends of 8 nt and 4 nt:

```
5'-TTAACCTAGGCTGCTAAACAAAGCCCGAAA...TGCCCGTAATATTTGCTGGTTCGGTGAC    -3'
3'-          CCGACGATTTGTTTCGGGCTTT...ACGGGCATTATAAACGACCAAGCCACTGCTTT-5'
```

Now the pET-50b(+) and caspase I ends can be joined using compatible 4-nt ends:

```
        Nus·Tag™ sequence                       caspase I sequence
     ArgAsnIleCysTrpPheGlyAspGlu            MetAlaAspLysIleLeuArgAla
 ...CGTAATATTTGCTGGTTCGGTGAC    -3'    5'-GAAATGGCTGACAAGATCCTGAGGGC...
 ...GCATTATAAACGACCAAGCCACTGCTTT-5'    3'-    ACCGACTGTTCTAGGACTCCCG...
```

When joined by ligation the reading frame of the vector-derived Nus·Tag™ sequences would match that of the caspase I gene (underlined below):

    ...ArgAsnIleCysTrpPheGlyAspGluMetAlaAspLysIleLeuArgAla...
    ...CGTAATATTTGCTGGTTCGGTGACGAAATGGCTGACAAGATCCTGAGGGC...

Now let us design the joint at the 3' end of the caspase I gene. We want to have an overhanging end complementary to the planned 5'-TTAACCTA overhanging end of the vector.

```
5'-TTAACCTAGGCTGCTAAACAAAGCCCGAAA...
3'-          CCGACGATTTGTTTCGGGCTTT...
```

Ideally, this would entail a 3' end for caspase I that could generate an 8-nt overhanging end matching the pET-50b(+) overhanging end.

```
...PheTyrLeuPheProGlyHisStop
...TTCTACCTCTTCCCGGGACATTAA      -3'
...AAGATGGAGAAGGGCCCTGTAATTAATTGGAT-5'
```

Once again, we have to make some adjustment in the design because it is not feasible to generate an 8-nt end stopping on the last 'A' of the TAA stop codon of caspase I. One way of resolving this would be to change oligonucleotide '32' in the inverse PCR so that it leaves a different overhanging end in the pET-50b(+) vector. We don't mind changing the vector sequence after the stop codon, because it is a noncoding sequence. We could redesign the forward oligonucleotide '32' for inverse PCR so that it reads:

    35. Forward (pET): 5'-TTccCCTtGGCTGCTAAACAAAGccc-3'

where the changed oligonucleotides are in lowercase. These changed nucleotides at the 5' end of '35' do not match the template; we're just making up an extended sequence that is free of 'G' nucleotides so that Klenow enzyme or T4 DNA polymerase can remove nucleotides on the complementary strand. The introduction of mismatched nucleotides at the 5′ end of '35' causes us to extend the oligonucleotide at the 3′ end with 'ccc' so that there are 18 nt at the 3′ end of '35' that match the vector template perfectly. The inverse PCR product generated using '34' and '35' would have a revised end with the following structure:

```
5′-TTCCCCTTGGCTGCTAAACAAAGCCCGAAA...
3′-AAGGGGAACCGACGATTTGTTTCGGGCTTT...
```

After treatment with Klenow enzyme or T4 DNA polymerase and only the substrate dCTP, an 8-nt overhanging end would be generated:

```
5′-TTCCCCTTGGCTGCTAAACAAAGCCCGAAA...
3′-      CCGACGATTTGTTTCGGGCTTT...
```

The reverse oligonucleotide '28' used previously to generate the caspase I gene would then be redesigned at its 5′ end:

```
36. Reverse (caspase I): 5′-aaggggaaTTAATGTCCCGGGAAGAG-3′
```

where the lowercase letters indicate the segment that will become a compatible overhanging end. The caspase I PCR product using '33' and '36' would have a 'reverse' end with the structure:

```
    PheTyrLeuPheProGlyHisStop
...TTCTACCTCTTCCCGGGACATTAATTCCCCTT-3′
...AAGATGGAGAAGGGCCCTGTAATTAAGGGGAA-5′
```

Upon treatment with Klenow enzyme or T4 DNA polymerase, and only the substrate dATP, the caspase I PCR product would have the following structure at the 'reverse' end:

```
    PheTyrLeuPheProGlyHisStop
...TTCTACCTCTTCCCGGGACATTAA        -3′
...AAGATGGAGAAGGGCCCTGTAATTAAGGGGAA-5′
```

This caspase I 'reverse' end is compatible with the previously described pET-50b(+) 'forward' end, and upon ligation the fused ends would have the following structure:

```
...PheTyrLeuPheProGlyHisStop
...TTCTACCTCTTCCCGGGACATTAATTCCCCTTGGCTGCTAAACAAAGCCCGAAA...
```

(caspase I sequence is shown underlined).

The modified ends of caspase I and pET-50b(+) can be joined in only one way; the vector cannot close on itself without the insert, and the insert can only enter the vector in one orientation.

Now that we have completed a design for linking the Nus·Tag™ directly to the caspase I sequence, let's improve the design by inserting a factor Xa protease cleavage site between the Nus·Tag™ and the caspase I, so that the two polypeptide segments can be separated after synthesis. The factor Xa protease cleaves a protein sequence immediately after the arginine of the tetrapeptide sequence Ile Glu Gly Arg. There are several ways that we could add the sequence, but one way would be to modify the reverse oligonucleotide used in inverse PCR of the pET-50b(+) to read

```
37. Reverse (pET): 5′-ccttcgatTTCGTCACCGAACCAGCAAAT-3′
```

where the lowercase letters represent new factor Xa cleavage site coding sequence, and the uppercase letters represent Nus·Tag™ coding sequence. After the inverse PCR of the pET-50b(+) vector, and treatment with Klenow enzyme or T4 DNA polymerase and the substrate dCTP, the vector would have the following 5-nt overhanging structure at the 'reverse' end:

```
 AlaArgAsnIleCysTrpPheGlyAspGluIle
...TGCCCGTAATATTTGCTGGTTCGGTGACGAAATC        -3'
...ACGGGCATTATAAACGACCAAGCCACTGCTTTAGCTTCC-5'
```

(where the underlined sequence is the end of the Nus·Tag™). The forward oligonucleotide used in second-strand cDNA synthesis of the caspase I gene could be modified to

```
38. Forward (caspase I): 5'-gaaggtcgcATGGCTGACAAGATCCTG-3'
```

where the lowercase letters again represent new factor Xa cleavage site coding sequence, and the uppercase letters represent caspase I coding sequence. Following PCR using oligonucleotides '38' and '36', and treatment of the fragment with Klenow enzyme or T4 DNA polymerase and dATP, the 'forward' end of the caspase I fragment would have the following structure:

```
    Xa         caspase I sequence
    GluGlyArgMetAlaAspLysIleLeuArgAla
5'-GAAGGTCGCATGGCTGACAAGATCCTGAGGGC...
3'-        AGCGTACCGACTGTTCTAGGACTCCCG...
```

(where the underlined sequence is the beginning of caspase I). The 5-nt overhanging ends are compatible, and upon ligation the Nus·Tag™ sequence and caspase I sequence (underlined) would be separated by a factor Xa site (Ile Glu Gly Arg):

```
    Nus·Tag™               (factor  Xa)  caspase I sequence
... AlaArgAsnIleCysTrpPheGlyAspGluIleGluGlyArgMetAlaAspLysIleLeuArgAla
... TGCCCGTAATATTTGCTGGTTCGGTGACGAAATCGAAGGTCGCATGGCTGACAAGATCCTGAGGGC..
```

The work of adding the four codons was split between the two oligonucleotides, and we did not alter the design for the ligation at the 3' end of the caspase I gene.

Kicking the DNA ligase habit

A plasmid carrying nicks on different strands can be transformed into *E. coli* successfully, and provided that the nicks are widely spaced, the plasmid DNA duplex will be held together by hydrogen-bonding of the intervening nucleotides. We may therefore dispense with ligation altogether, if we design a sufficiently long cohesive end and use a method called ligation-independent cloning (5).

A site for ligation-independent cloning may be devised either by searching for, or constructing, patches of sequence that are devoid of one of the four nucleotides. The lack of one nucleotide base in a sequence allows for an overlapping end to be developed by treating with Klenow enzyme or T4 DNA polymerase. For example, the sequence encoding the Gly Asp

Glu Ile Glu Gly Arg that joins the Nus·Tag™ coding sequence to the factor Xa cleavage site coding sequence could be modified to a series of synonymous codons lacking cytidylate nucleotides until the very last codon: GGT GAT GAA ATA GAA GGT CGA. We could design a reverse oligonucleotide for inverse PCR of the pET-50b(+) vector that starts with the complementary sequence to the first six codons, and finishes with sequence matching the pET-50b(+) template at the 3′ end (see sites marked '6' in *Figure 7.6*):

```
39. Reverse (pET): 5'-accttctatTTCaTCACCGAACCAGCAAATATTA-3'
```

The uppercase letters represent sequence that matches the pET-50b(+) template and the lowercase letters represent sequence that is complementary to sequence encoding Gly Asp Glu Ile Glu Gly. When this oligonucleotide is used in inverse PCR, and the end of the pET-50b(+) vector is treated with Klenow enzyme or T4 DNA polymerase and the substrate dCTP, an overhanging end of 18 nt will be established.

```
...ATGGCTGCCCGTAATATTTGCTGGTTC              -3'
...TACCGACGGGCATTATAAACGACCAAGCCACTACTTTATCTTCCA -5'
```

We could also design a forward oligonucleotide for the caspase I cDNA:

```
40. Forward (caspase I):
5'-ggtgatgaaatagaaggtcgaATGGCTGACAAGATCCTG-3'
```

to incorporate a cohesive overhanging end, and in which the uppercase letters represent caspase I sequence and the lowercase letters represent the sequence encoding the peptide Gly Asp Glu Ile Glu Gly Arg. Following RT-PCR of the caspase I cDNA using oligonucleotide '40' (and a yet to be designed reverse oligonucleotide '41'), and treatment with Klenow enzyme or T4 DNA polymerase and the substrate dGTP, a matching overhanging end of 18 nt would be established at the 5′ end of caspase I:

```
5'-GGTGATGAAATAGAAGGTCGAATGGCTGACAAGATCCTGAGGGCAAAG...
3'-                     GCTTACCGACTGTTCTAGGACTCCCGTTTC...
```

After annealing of the cohesive ends, the triplet reading frame of the Nus·Tag™, factor Xa, and caspase I sequence are contiguous, and the nicks in the duplex (underlined nucleotides) are separated by 18 base pairs.

```
  Nus·Tag™              (factor Xa)   caspase I sequence
...IleCysTrpPheGlyAspGluIleGluGlyArgMetAlaAspLysIleLeuArgAla...
...ATTTGCTGGTTCGGTGATGAAATAGAAGGTCGAATGGCTGACAAGATCCTGAGGGCA...
...TAAACGACCAAGCCACTACTTTATCTTCCAGCTTACCGACTGTTCTAGGACTCCCGT...
```

A ligation-independent cloning site at the 3′ end of the caspase I sequence may be designed with fewer constraints, since there is no triplet reading frame or amino acid coding sequence to maintain. However, the dNTP substrates used to generate the overhanging ends of the vector and caspase I cDNA must follow the precedents we have established at the first end: dCTP for the vector (see '39'), and dGTP for the caspase I cDNA (see '40'), along with Klenow enzyme or T4 DNA polymerase.

We could design the caspase I reverse oligonucleotide so that it starts with 18 nt of sequence complementary to the pET-50b(+) vector (with two minor nucleotide changes):

41. Reverse (caspase I): 5' agg<u>g</u>taggttaattaagcTTAATGTCCCGGGAAGAG-3'

The lowercase letters represent pET-50b(+) sequence, the reverse complement of sequence flanking the *Pac*I (TTAAT/TAA) restriction endonuclease site, with the two mutated nucleotides underlined, and the uppercase letters represent the reverse complement of the 3' end of the caspase I coding sequence. Following RT-PCR using '41' and '40' the caspase I 'reverse' end would have the structure:

```
...PheTyrLeuPheProGlyHisStop
...TTCTACCTCTTCCCGGGACATTAAGCTTAATTAACCTACCCT-3'
...AAGATGGAGAAGGGCCCTGTAATTCGAATTAATTGGATGGGA-5'
```

Upon treatment of the product with Klenow enzyme or T4 DNA polymerase and dGTP as the substrate, a 17-nt overhanging end will be established.

```
...PheTyrLeuPheProGlyHisStop
...TTCTACCTCTTCCCGGGACATTAAG                 -3'
...AAGATGGAGAAGGGCCCTGTAATTCGAATTAATTGGATGGGA-5'
```

A forward oligonucleotide for inverse PCR of the pET-50b(+) vector could be designed to generate a complementary overhanging end, with the sequence

42. Forward (pET): 5'-CTTAATTAACCTAccCTGCTAAACAAAGCCC-3'

where the two lowercase letters indicate mismatch with the pET-50b(+) template (see sites marked '6' in *Figure 7.6*). Following inverse PCR using '42' and '39' the 'forward' end of the pET-50b(+) product would have the structure:

```
5'-CTTAATTAACCTACCCTGCTAAACAAAGCCCGAAA...
3'-GAATTAATTGGATGGGACGATTTGTTTCGGGCTTT...
```

Upon treatment with Klenow enzyme or T4 DNA polymerase and dCTP as a substrate, a 17-nt overhanging end would be established:

```
5'-CTTAATTAACCTACCCTGCTAAACAAAGCCCGAAA...
3'-           CGATTTGTTTCGGGCTTT...
```

The cohesive 17-nt ends of the pET-50b(+) and caspase I can be annealed, and the nicks in the duplex (underlined nucleotides) are separated by 17 base pairs.

```
    caspase I                    pET-50b(+)
... TyrLeuPheProGlyHisStop
... TACCTCTTCCCGGGACATTAA<u>G</u>CTTAATTAACCTACCCTGCTAAACAAAGCCC...
... ATGGAGAAGGGCCCTGTAATTCGAATTAATTGGATGGG<u>AC</u>GATTTGTTTCGGG...
```

The predicted T_M of the 18-nt cohesive end at the 5' end of the caspase I sequence is approximately 43°C, and the predicted T_M of the 17-nt cohesive end at the 3' end of the caspase I sequence is similar, approximately 40°C. The caspase I and pET-50b(+) fragments could be annealed to each other at that temperature, prior to transformation into competent *E. coli* cells. The length of time needed for annealing would depend on molar concentration.

Let us consider an example: 50 ng of the pET-50b(+) PCR product and 9.3 ng of the caspase I RT-PCR product would each be approximately

12 fmol of DNA, and if combined in an annealing reaction of 5 µl (prepared in 10 mM Tris-Cl, 100 mM NaCl, pH 8.0) they would have molar concentrations of approximately 2.4 nM. By comparison, a PCR oligonucleotide might be used in an annealing reaction at a concentration of 1000 fmol per 50 µl, or about 20 nM, and towards the end of the thermocycling reaction there might be a similar concentration of template target. We usually allow about 30 s for the annealing step in PCR. Assuming that the annealing rate is proportional to the products of the concentrations of reactants, then we ought to allow about 30 min for the annealing of the caspase I cDNA and vector PCR products.

Kicking the annealing habit

Fragments of DNA can be joined by cross-priming, a matter that was introduced and explained as a problem in Section 7.2. Here the cross-priming effect becomes useful, and we use it to join two templates to simplify a cloning method. Let us return to the example of the caspase I cloning in pET-50b(+), as introduced above, and change the design problem as well. Suppose that we wish to clone the mouse caspase I coding sequence as an N-terminal fusion protein, meaning that the ribosome will start with the translation of the caspase I and then add additional amino acid sequences to the carboxyl terminus. For this design we will omit the stop codon that we have been using up to this point, with the caspase I cDNA. Let us also suppose that we wish to add a thrombin protease cleavage site (Leu Val Pro Arg Gly Ser) after the caspase I amino acid sequence, then immediately enter the first His·Tag of the pET-50b(+) sequence. We design forward and reverse oligonucleotides for RT-PCR generation of the caspase I cDNA (see *Figure 7.7A*, left side). The forward oligonucleotide is simply the previously described '15', the first 18 nt of the caspase I coding sequence:

```
15. Forward (caspase I): 5'-ATGGCTGACAAGATCCTG-3'
```

The reverse oligonucleotide is complementary to the last 18 nt of the caspase I coding sequence, plus an extension of 18 nt that forms the junction with the thrombin cleavage site (shown in lowercase):

```
43. Reverse (caspase I) 5'-actaccgcgtggcaccagATGTCCCGGGAAGAGGTA-3'
```

Upon completion of RT-PCR using '43' and '15', the overall structure of the cDNA ends would be:

```
5'-ATGGCTGACAAGATCCTGAGGGC...TCTACCTCTTCCCGGGACATCTGGTGCCACGCGGTAGT-3'
3'-TACCGACTGTTCTAGGACTCCCG...AGATGGAGAAGGGCCCTGTAGACCACGGTGCGCCATCA-5'
```

We design the forward oligonucleotide used for inverse PCR of the pET-50b(+) vector so that it overlaps 16 nt with the 3' end of the sense strand of the cDNA product (underlined in the sequences, above and below), and ends with 24 nt of sequence matching the first His·Tag of the pET-50b(+) sequence (see sites marked '7' in *Figure 7.6*, and the right side of *Figure 7.7A*):

```
44. Forward (pET): 5'-GGTGCCACGCGGTAGTCATCACCACCATCACCATTCTAGT-3'
```

The His·Tag amounts to only 18 nt of coding sequence, but by extending the match to 24 nt we ensure that it is the first His·Tag in the vector and not the second that serves as a template in the inverse PCR. The reverse oligonucleotide used for inverse PCR of the pET-50b(+) vector is simply sequence that is complementary to the 18 nt just preceding the ATG start

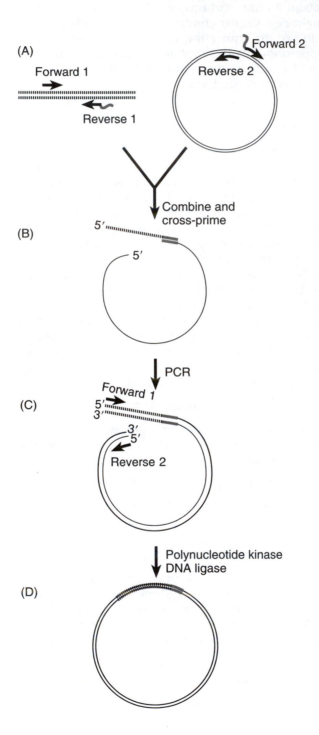

codon that would normally be used by the ribosome for expression from the vector:

```
45.  Reverse (pET): 5′-ATGTATATCTCCTTCTTA-3′
```

Upon completion of inverse PCR using '44' and '45', the overall structure of the pET-50b(+) vector would be:

```
5′-GGTGCCACGCGGTAGTCATCACCACCATCAC...TAACTTTAAGAAGGAGATATACAT-3′
3′-CCACGGTGCGCCATCAGTAGTGGTGGTAGTG...ATTGAAATTCTTCCTCTATATGTA-5′
```

The RT-PCR ('15' and '43') and inverse PCR ('44' and '45') products may be purified and combined together in a third, cross-priming PCR that includes only the forward caspase I cDNA oligonucleotide '15' and the reverse pET-50b(+) oligonucleotide primer '45'. Initially these are not linked together in any template; however, a cross-priming between the cDNA and vector may occur at the 16-nt overlapping region (underlined) that they share:

```
       caspase I 'reverse' end
 ...TCTACCTCTTCCCGGGACATCTGGTGCCACGCGGTAGT-3′
                 3′-CCACGGTGCGCCATCAGTAGTGGTGGTAGTG...
                         pET-50b(+) 'forward' end
```

The products may act as primers against each other, and be extended to yield a fused product that has the oligonucleotide '15' primer sequence at one end, and the oligonucleotide '45' primer at the other (see *Figure 7.7B*). The 1.2-kbp of caspase I sequence and 6.7-kbp of pET-50b(+) sequence would become a single long PCR product of 7.9 kbp, which should be completed using a thermostable DNA polymerase such as *Pfu*, which does not leave an overhanging 3′ adenylate nucleotide (see *Figure 7.7C*).

Upon completion of this long PCR, the 5′ ends can be phosphorylated using the enzyme T4 polynucleotide kinase and ATP as a substrate. This is an important step because the 5′ ends of oligonucleotides typically have a 5′-OH that would not be suitable for use in a ligation reaction. The final step is blunt-end ligation of the 7.9-kbp linear fragment so that it closes upon itself as a circular DNA (see *Figure 7.7D*). The ligation joins the ATG start codon of the caspase I gene to the ribosome binding site (RBS) region

Figure 7.7

Insertion of a PCR product into a vector by cross-priming. (A) A PCR product is prepared, on the left, using a pair of oligonucleotides, 'forward 1' and 'reverse 1'. The 5′ end of the 'reverse 1' oligonucleotide is complementary to the 5′ end of the 'forward 2' oligonucleotide. The recipient vector is prepared as a linear fragment by inside-out PCR, on the right, using 'forward 2' and 'reverse 2' oligonucleotides (this may require the use of a 'long PCR' method). (B) The products of the two previous reactions are mixed and allowed to cross-prime during PCR, based on the complementary region included in the oligonucleotides. (C) The PCR with the mixed template is conducted using 'forward 1' and 'reverse 2' oligonucleotides, generating a double-stranded product from the cross-primed template. (D) The linear product is treated with T4 polynucleotide kinase and T4 DNA ligase and closed into a circular plasmid containing the inserted sequence.

from the pET-50b(+) (the ends below are shown opposed to each other, ready for ligation):

```
        pET-50b(+)                     caspase I
           RBS                    MetAlaAspLysIleLeu...
...TAAGAAGGAGATATACAT-3'          5'-ATGGCTGACAAGATCCTG...
...ATTCTTCCTCTATATGTA-5'          3'-TACCGACTGTTCTAGGAC...
```

This can be performed in a dilute reaction because the ends of the linear fragment occupy a limited volume in solution and are less subject to diffusion.

Plasmid mutagenesis by inverse PCR

In the previous example, a plasmid was prepared for mutagenesis by inverse PCR, and the mutagenesis was conducted by cross-priming, a general engineering tool discussed in more detail below. Inverse PCR allows us to control how a circular sequence is linearized, and gives us control over the structure of the ends. For example, if we have cloned caspase I into pET-50b(+), or any other plasmid for that matter, we might use inverse PCR to generate site-specific mutations. Suppose that we wish to change amino acid 204 from an Asn residue into an Asp residue. The Asn codon in question is AAT, and it is underlined in this reference sequence:

```
    LeuGlyTyrThrValLysValLysGluAsnLeuThrAlaLeuGluMetValLysGlu
781 ATCTGGGGTATACCGTGAAAGTGAAAGAAAATCTCACAGCTCTGGAGATGGTGAAAGAGG
```

We only need to change the codon from AAT to GAT to make it encode an Asp residue, and this can easily be done with a pair of oligonucleotides for inverse PCR. The forward oligonucleotide could start with the mutated nucleotide at its 5' end (lowercase), and copy an additional 18 nt of the sense strand of the gene:

```
46. Forward (caspase I mutation): 5'-gATCTCACAGCTCTGGAGA-3'
```

The reverse oligonucleotide would be

```
47. Reverse (caspase I): 5'-TTCTTTCACTTTCACGGT-3'
```

which is the reverse complement sequence of the preceding six codons Thr Val Lys Val Lys Glu. Upon completion of the inverse PCR of the caspase I clone, using '46' and '47' and an error-correcting polymerase such as *Pfu* to leave blunt ends, the product would have the following general structure:

```
5'-gATCTCACAGCTCTGGAGATGGTGAAAGAGG...ATCTGGGGTATACCGTGAAAGTGAAAGAA-3'
3'-cTAGAGTGTCGAGACCTCTACCACTTTCTCC...TAGACCCCATATGGCACTTTCACTTTCTT-5'
```

(where the mutation is indicated in lowercase). The fragment could be treated with T4 polynucleotide kinase to phosphorylate the 5' ends, and then treated with T4 DNA ligase to close the DNA as a circle. This would rejoin the codons exactly as before, except for the fact that the 204th amino acid would now be an Asp (underlined) instead of an Asn:

```
...   LeuGlyTyrThrValLysValLysGluAspLeuThrAlaLeuGluMetValLysGlu...
...ATCTGGGGTATACCGTGAAAGTGAAAGAAGATCTCACAGCTCTGGAGATGGTGAAAGAGG...
...TAGACCCCATATGGCACTTTCACTTTCTTCTAGAGTGTCGAGACCTCTACCACTTTCTCC...
```

We could also apply this method to making deletions or insertions of any size; for example, if we wished to delete amino acid 204 entirely, we could just use an alternative forward oligonucleotide that starts with the codon for amino acid 205. That oligonucleotide would be:

```
48. Forward (caspase I deletion) 5'-CTCACAGCTCTGGAGATGG-3'
```

and could be paired in inverse PCR with the reverse oligonucleotide '47' used for the previous mutation. The PCR product would have the general structure:

```
5'-CTCACAGCTCTGGAGATGGTGAAAGAGG...ATCTGGGGTATACCGTGAAAGTGAAAGAA-3'
3'-GAGTGTCGAGACCTCTACCACTTTCTCC...TAGACCCCATATGGCACTTTCACTTTCTT-5'
```

Upon treatment with T4 polynucleotide kinase and T4 DNA ligase the circularized DNA would have lost the three nucleotides encoding the original Asn codon:

```
...   LeuGlyTyrThrValLysValLysGluLeuThrAlaLeuGluMetValLysGlu...
...ATCTGGGGTATACCGTGAAAGTGAAAGAACTCACAGCTCTGGAGATGGTGAAAGAGG...
...TAGACCCCATATGGCACTTTCACTTTCTTGAGTGTCGAGACCTCTACCACTTTCTCC...
```

(the point of ligation is shown by underlined nucleotides).

Mutational insertion can be performed by extending the forward or reverse oligonucleotides at their 5' ends. For example, the forward oligonucleotide

```
49. Forward (caspase I insertion): 5'-cccAATCTCACAGCTCTGGAGATGG-3'
```

is extended by 3 nt at its 5' end, and used in conjunction with the same reverse oligonucleotide '47', would yield a PCR product with the general structure:

```
5'-cccAATCTCACAGCTCTGGAGATGGTGAAAGAG...TCTGGGGTATACCGTGAAAGTGAAAGAA-3'
3'-gggTTAGAGTGTCGAGACCTCTACCACTTTCTC...AGACCCCATATGGCACTTTCACTTTCTT-5'
```

Upon treatment with T4 polynucleotide kinase and T4 DNA ligase, the circularized DNA would have a gain of three nucleotides (a CCC proline codon) between the Glu-203 and Asn-204 codons:

```
...   LeuGlyTyrThrValLysValLysGluProAsnLeuThrAlaLeuGluMetValLysGlu...
...ATCTGGGGTATACCGTGAAAGTGAAAGAACCCAATCTCACAGCTCTGGAGATGGTGAAAGAGG...
...TAGACCCCATATGGCACTTTCACTTTCTTGGGTTAGAGTGTCGAGACCTCTACCACTTTCTCC...
```

(the point of ligation is shown by underlined nucleotides).

Synthetic oligonucleotide insertions

Long insertions into a plasmid can be engineered by cross-priming, as discussed previously for the insertion of the caspase I cDNA, or by dropping a synthetic oligonucleotide into the cohesive ends of a plasmid. For example, suppose we wished to make a tandem duplication of the 10-amino-acid sequence Val Lys Glu Asn Leu Thr Ala Leu Glu Met that lies in the middle of the reference line:

```
    LeuGlyTyrThrValLysValLysGluAsnLeuThrAlaLeuGluMetValLysGlu
781 ATCTGGGGTATACCGTGAAAGTGAAAGAAAATCTCACAGCTCTGGAGATGGTGAAAGAGG
```

We could design a forward oligonucleotide that contained the sense coding sequence of the first six codons Val Lys Glu Asn Leu Thr, or

```
50. Forward: 5'-GTGAAAGAAAATCTCACA-3'
```

and a reverse oligonucleotide that is complementary to nt 783-800, the sequence encoding the preceding six codons Leu Gly Tyr Thr Val Lys, or

```
51. Reverse: 5'-TTTCACGGTATACCCCAG-3'.
```

Then the product of inverse PCR using '50' and '51' would have the general structure:

```
5'-GTGAAAGAAAATCTCACAGCTCTGGAGATGGTGAAA...GGAGGATCTGGGGTATACCGTGAAA-3'
3'-CACTTTCTTTTAGAGTGTCGAGACCTCTACCACTTT...CCTCCTAGACCCCATATGGCACTTT-5'
```

It could be treated with Klenow enzyme or T4 DNA polymerase and the substrate dGTP to generate overhanging ends of 12 nt and 3 nt:

```
5'-GTGAAAGAAAATCTCACAGCTCTGGAGATGGTGAAA...GGAGGATCTGGGGTATACCGTG    -3'
3'-            GAGTGTCGAGACCTCTACCACTTT...CCTCCTAGACCCCATATGGCACTTT-5'
```

Into these cohesive ends we could ligate a pair of annealed, phosphorylated oligonucleotides

```
52. Top (insert): 5'-aaaGTGAAAGAAAATCTCACAGCTCTGGAGATG-3'
53. Bottom (insert): 5'-attttcttTcacCATCTCCAGAGCTGTGAGATTTTCTTTCAC-3'
```

that fill and expand the gap, adding 10 additional amino acids (the overlapping nucleotides are in lowercase):

```
        ValLysGluAsnLeuThrAlaLeuGluMet
  5'-aaaGTGAAAGAAAATCTCACAGCTCTGGAGATG
        CACTTTCTTTTAGAGTGTCGAGACCTCTACCcactttctttta-5'
```

The four 5' ends would be phosphorylated using T4 polynucleotide kinase, prior to treatment with T4 DNA ligase.

The use of oligonucleotides to build up sequence in engineering can be generalized; a long sequence can be constructed using purely synthetic DNA segments that are overlapping, and the ends can be designed for insertion into a cohesive site in a plasmid vector. Assembly of overlapping oligonucleotides into a hydrogen-bonded duplex can be very rapid when the T_M is optimized, because the concentrations of oligonucleotides in solution can be very high. The mixture of annealed oligonucleotides may need to be diluted substantially before ligation into a plasmid site, so that the molar concentrations of oligonucleotide-annealed DNA and plasmid target are similar. For example, two oligonucleotides might be annealed at a concentration of about 5 pmol per 10 µl, which is 500 nM, but the plasmid target would very likely be at a much lower concentration of about 11 nM; or 50 µg ml^{-1} for a 6.7-kbp plasmid such as pET-50b(+). The annealed oligonucleotides would consequently be diluted about 50-fold before use in a ligation reaction with the plasmid.

Cross-priming in PCR engineering

Using cross-priming for assembly

In the previous example of bringing together the caspase I cDNA and pET-

50b(+) vector by cross-priming of templates, we have seen how two overlapping DNA molecules can be joined to make a contiguous duplex. This can be applied as a general principle for assembly of DNA segments; for example, constructing a coding sequence from individual exons when a cDNA is not available. The mouse caspase I coding sequence is distributed among 8 exons, spread over 7.5 kbp of genomic DNA (GenBank locus NC_000075.3), and we will illustrate the method by designing an approach to bring together the first three exons by PCR.

```
Exon 1 coding sequence (274 nt):
ATGGCTGACAAGATCCTGAGGGCAA...GCAGGAATTCTGGAGCTTCAATCAG
TACCGACTGTTCTAGGACTCCCGTT...CGTCCTTAAGACCTCGAAGTTAGTC

Exon 2 (63 nt):
CTCCATCAGCTGAAACATTTGTTGC...AAAGGAGGACATCCTTCATCCTCAG
GAGGTAGTCGACTTTGTAAACAACG...TTTCCTCCTGTAGGAAGTAGGAGTC

Exon 3 (113 nt):
AAACAAAGGAAGAACAGAACAAAGA...GTTATGGAAAGAAAATCCTTCAGAG
TTTGTTTCCTTCTTGTCTTGTTTCT...CAATACCTTTCTTTTAGGAAGTCTC
```

The oligonucleotides used to generate exons 1, 2, and 3 individually will be the pairs '54' and '55', '56' and '57', and '58' and '59', respectively:

```
54. Forward 1: 5'-ATGGCTGACAAGATCCTG-3'
55. Reverse 1: 5'-tgatggagCTGATTGAAGCTCCAGAA-3'
56. Forward 2: 5'-tcaatcagCTCCATCAGCTGAAACAT-3'
57. Reverse 2: 5'-ctttgtttCTGAGGATGAAGGATGTC-3'
58. Forward 3: 5'-atcctcagAAACAAAGGAAGAACAGAACAAAGA-3'
59. Reverse 3: 5'-CTCTGAAGGATTTTCTTTCCATA-3'
```

(the exon-to-exon junctions are shown by a transition from lower- to uppercase letters). The principle of design is that '55' and '56' will overlap by 16 nt, and '57' and '58' will overlap by 16 nt. Had we extended the example to include additional exons, then similar overlaps between 'reverse' and 'forward' oligonucleotides would be established.

The two oligonucleotides for exon 3, '58' and '59', were extended at their 3' ends beyond 18 nt of perfect matching sequence to compensate for a lack of sequence complexity (a problem discussed in Section 7.2). Exon 1 would be generated in a reaction using a mouse genomic DNA template, and '54' and '55'. Exon 2 would be generated in a separate reaction using the same template, and '56' and '57'. Then the products of the two reactions would be purified and brought together in a third reaction that makes use of oligonucleotides '54' and '57', with cross-priming occurring along the 16 nt of overlapping sequence (underlined) between the two exons:

```
                '55' end
...TTCTGGAGCTTCAATCAGctccatca-3'
          3'-agttagtcGAGGTAGTCGACTTTGTA...
                '56' end
```

Exon 3 would be generated using a mouse genomic DNA template, and '58' and '59', then purified and brought together in a cross-priming reaction with the combined exon 1 + 2 product. The oligonucleotides '54' and '59' would be used to generate the combined exon 1 + 2 + 3 product, with cross-priming between the 16 nt of overlapping sequence (under-lined) between exons 2 and 3:

```
                  '57' end
... GACATCCTTCATCCTCAGaaacaaag-3'
          3'-taggagtcTTTGTTTCCTTCTTGTCTTG...
                  '58' end
```

Assembling a complete coding sequence exon by exon could be a tedious job, particularly if the gene has many exons, but it can be done and it is an alternative to cDNA cloning.

Using cross-priming for site-specific mutagenesis

Cross-priming during PCR is a method that is broadly applicable to site-specific mutagenesis. The region of DNA that is to be subjected to mutage-nesis is also the location of oligonucleotide primers that are used in PCR, and the mutations are built into the product by being incorporated into the primer sequences.

Let us return to an example used in the discussion of inverse PCR, in which we planned to change amino acid 204 in the mouse caspase I gene from an Asn residue into an Asp residue. The Asn codon is underlined in this reference sequence, and we needed to change it from an AAT codon to a GAT codon, an A to G transition at nt 810:

```
    LeuGlyTyrThrValLysValLysGluAsnLeuThrAlaLeuGluMetValLysGlu
781 ATCTGGGGTATACCGTGAAAGTGAAAGAAAATCTCACAGCTCTGGAGATGGTGAAAGAGG
```

To define the boundaries of the caspase I cDNA, we could make use of any of the forward and reverse oligonucleotides that were devised for the RT-PCR; for example:

```
15. Forward (caspase I): 5'-ATGGCTGACAAGATCCTG-3'
14. Reverse (caspase I): 5'-TTAATGTCCCGGGAAGAG-3'
```

The pair of new oligonucleotides we would devise for mutagenesis might be:

```
60. Reverse (nt 820): 5'-GCTGTGAGATcTTCTTTCACTTTCACGGT-3'
```

which would be used in conjunction with '15' to generate a product of 590 bp; and

```
61. Forward (nt 801): 5'-GTGAAAGAAgATCTCACAGCTCTGGAGA-3'
```

which would be used in conjunction with '14' to generate a product of 609 bp. The lowercase letters in '60' and '61' indicate the point of mutation an A to G transition in the forward oligonucleotide and a T to C transition in the reverse oligonucleotide. These 590- and 609-bp products could be added to a third PCR and they anneal over 20 nt of complementary sequence:

```
...GGGGTATACCGTGAAAGTGAAAGAAgATCTCACAGC-3'
           3'-CACTTTCTTcTAGAGTGTCGAGACCTCTACCACTTT...
```

In a PCR with '15' and '14', the mutated 1409-bp template would be established during the first few cycles and would be amplified geometrically thereafter.

Combinatorial mutagenesis with cross-priming

The previous example involved changing a specific nucleotide base to a different nucleotide base, but the method could be broadened to allow for combinatorial mutagenesis as well. Suppose that we wished to prepare 20 different versions of the caspase I protein, in which Asn204 was allowed to be any of the 20 amino acids. We would need to prepare degenerate oligonucleotides that changed the AAT codon to an NNN codon, where the 'N' could be any of the four nucleotide bases. These might be:

```
62. Reverse (nt. 827): 5'-CTCCAGAGCTGTGAGnnnTTCTTTCACTTTCACGGT-3'
63. Forward (nt. 794): 5'-CGTGAAAGTGAAAGAAnnnCTCACAGCTCTGGAGATG-3'
```

Oligonucleotide '63' extends from nt 794 to nt 830, and oligonucleotide '62' extends on the complementary strand from nt 827 to nt 792. Both contain a degenerate triplet NNN codon that replaces nt 810–812 (lowercase letters). Once again, '15' and '62' would be used in a first PCR and '63' and '14' would be used in a second PCR, and the products of those two reactions would be purified and combined for cross-priming in a third PCR that uses '15' and '14'. The degenerate sequences from any two products have a 1 in 64 chance of being complementary, and so the design of the oligonucleotides has been changed slightly; they are lengthened at their 5' ends to allow the complementary 3' ends to cross-prime efficiently.

```
...GGGGTATACCGTGAAAGTGAAAGAAnnnCTCACAGCTCTGGAG-3'
        3'-GCACTTTCACTTTCTTnnnGAGTGTCGAGACCTCTACCACTTT...
```

It is important to note that although the cross-priming may occur between products that do not have complementary NNN codons, there is no mismatch among individual PCR products of the third PCR. The NNN sequence is copied into a complementary sequence by the polymerase, and is amplified precisely by the '15' and '14' oligonucleotides thereafter.

The use of the 'NNN' degenerate sequence in '62' and '63' allows all 64 codons to be represented, including all 20 amino acids and the three stop codons. However, the 20 amino acids will not be represented in equal proportions, because of the nature of the genetic code. For example, the amino acids Ser, Leu, and Arg would be present three times as often as Asp, Phe, and Cys, because Ser, Leu, and Arg each have six codons specifying them and Asp, Phe, and Cys each have two. A slightly more proportionate representation of all 20 amino acids can be arranged by using the degenerate codon 5'-NNK-3', in which the 'K' is either a 'G' or 'T' nucleotide. The anticodon sequence in the reverse oligonucleotide would be 5'-MNN-3', where the 'M' is either a 'C' or 'A' nucleotide.

Finally, PCR can be used in other ways to generate diversity in sequence. It is possible to 'breed' chimeric sequences by a process called domain shuffling, in which overlapping sequence between alleles allows for cross-priming during PCR. We allow this mixed template to generate a combinatorial library of sequences of mixed parentage, and we may screen among the elements of the library for desirable characteristics or phenotypes.

Problems

The answers to these problems are provided at the end of the chapter.

(viii) Suppose that you decide to clone the mouse caspase I coding sequence (underlined or boxed sequence in *Figure 7.5*) using the *Eco*RI site of pET-50b(+) (nt 1686 in *Figure 7.6*). To prepare unique ends you digest the pET-50b(+) with *Eco*RI and treat the DNA with Klenow enzyme and only the substrate dCTP.
A. What do the ends of the plasmid look like after this treatment?
B. Design a forward oligonucleotide for capture of the caspase I 5′ end that can be treated with Klenow enzyme to make an end compatible with the plasmid, and that maintains the reading frame of the fusion protein upon ligation.
C. Design a reverse oligonucleotide for capture of the caspase I 3′ end that can be treated with Klenow enzyme under the same conditions used to treat the 5′ end. This oligonucleotide should introduce a stop codon at the end of the caspase I sequence.

(ix) Suppose that you want to take the plasmid you generated in the previous problem and delete most of the fusion protein sequences, so that the ATG start codon of the caspase I gene follows the His·Tag® sequence at nt 115 of pET-50b(+) (see *Figure 7.6*). Design a pair of oligonucleotides that can be used to create this deletion by inverse PCR, so that the inverse PCR product can be re-closed by ligation.

(x) The sequence shown below is the *Aequorea victoria* coding sequence for green fluorescent protein (GFP) (GenBank® locus E17099):

```
  1 ATGAGTAAAG GAGAAGAACT TTTCACTGGA GTTGTCCCAA TTCTTGTTGA ATTAGATGGT
 61 GATGTTAATG GGCACAAATT TTCTGTCAGT GGAGAGGGTG AAGGTGATGC AACATACGGA
121 AAACTTACCC TTAAATTTAT TTGCACTACT GGAAAACTAC CTGTTCCATG GCCAACACTT
181 GTCACTACTT TCTCTTATGG TGTTCAATGC TTTTCAAGAT ACCCAGATCA TATGAAACGG
241 CATGACTTTT TCAAGAGTGC CATGCCCGAA GGTTATGTAC AGGAAAGAAC TATATTTTTC
301 AAAGATGACG GGAACTACAA GACACGTGCT GAAGTCAAGT TTGAAGGTGA TACCCTTGTT
361 AATAGAATCG AGTTAAAAGG TATTGATTTT AAAGAAGATG GAAACATTCT TGGACACAAA
421 TTGGAATACA ACTATAACTC ACACAATGTA TACATCATGG CAGACAAACA AAAGAATGGA
481 ATCAAAGTTA ACTTCAAAAT TAGACACAAC ATTGAAGATG GAAGCGTTCA ACTAGCAGAC
541 CATTATCAAC AAAATACTCC AATTGGCGAT GGCCCTGTCC TTTTACCAGA CAACCATTAC
601 CTGTCCACAC AATCTGCCCT TTCGAAAGAT CCCAACGAAA AGAGAGACCA CATGGTCCTT
661 CTTGAGTTTG TAACAGCTGC TGGGATTACA CATGGCATGG ATGAACTATA CAAATAA
```

Suppose that you want to prepare a pET-50b(+) clone that expresses a fusion protein of GFP and caspase I. You will introduce the GFP sequence (minus the 'TAA' stop codon at the end) into your pET-50b(+) clone of caspase I (from the previous problem). The GFP coding sequence will follow the ribosome binding site in pET-50b(+), and the caspase I coding sequence will follow and be in frame with the GFP coding sequence.

You should plan one pair of oligonucleotides to generate the GFP sequence from a cDNA clone, and a second pair of oligonucleotides to generate the vector sequence (including the caspase I). Your plan will be to introduce the GFP by cross-priming during a third PCR (as outlined in *Figure 7.7*). Design the four oligonucleotides you will need for this job, and explain how they will be used.

7.6 Diagnostic methods

When we consider the role of PCR in diagnostic methods, we will divide the discussion into two parts. PCR may be part of a direct test of a sample, for example when the presence or absence of annealing of primers distinguishes between alleles, or it may be used to generate material for a different test, such as restriction analysis or molecular beacon binding. In this section we will be primarily concerned with the first type of application, in which the PCR is directly generating data.

The polymerase chain reaction is a powerful tool in molecular diagnostics, because it can be used with very small starting samples and made highly specific. The PCR method may be used in conjunction with a wide variety of detection systems, including fluorescent imaging dyes, such as ethidium bromide or SYBR® Green I, and fluorescent or radiolabeled probes. For most applications, the information that is gathered by PCR answers questions about the amount, size, or structure of the template, or whether oligonucleotides match the template sequence and can be used as primers.

Can the primers be extended?

The 3′ end of an oligonucleotide must be base-paired to a template in order for it to serve as a primer for DNA synthesis. This general principle is the basis of many diagnostic tests, because it allows the detection of mismatched sequence underlying the putative primer. For example, we can design an oligonucleotide that will detect the presence or absence of key nucleotides, as in this example taken from the human CFTR gene (GenBank loci # NM_000492, and # S64640):

```
  64. Reverse (WT CFTR): 5′-TTCATCATAGGAAACACCAAA-3′
```

anneals to the template as follows:

```
Wild type:
1621 ATTATGCCTGGCACCATTAAAGAAAATATCAtctTTGGTGTTTCCTATGATGAATATAGA
                                 3′-AAACCACAAAGGATACTACTT-5′
```

Oligonucleotide '64' will bind to the wild-type template carrying the three lowercase nucleotides; however, the 3′-adenylate nucleotide of '64' is mismatched when used in conjunction with the ΔF508 template and cannot be extended.

```
ΔF508 deletion:
1621 ATTATGCCTGGCACCATTAAAGAAAATATCATTGGTGTTTCCTATGATGAATATAGA
                               3′-AAACCACAAAGGATACTACTT-5′
```

We may design a second oligonucleotide

```
  65. Reverse (ΔF508): 5′-TTCATCATAGGAAACACCAAT-3′
```

that anneals to the ΔF508 sequence but not the wild-type sequence:

```
Wild type:
1621 ATTATGCCTGGCACCATTAAAGAAAATATCAtctTTGGTGTTTCCTATGATGAATATAGA
                                  3'-TAACCACAAAGGATACTACTT-5'
ΔF508 deletion:
1621 ATTATGCCTGGCACCATTAAAGAAAATATCATTGGTGTTTCCTATGATGAATATAGA
                                3'-TAACCACAAAGGATACTACTT-5'
```

The two oligonucleotides '64' and '65' can be used individually in PCR, in conjunction with a common forward oligonucleotide that may lie several hundred nucleotides upstream in the genomic DNA. When the oligonucleotides are used in paired experiments, the genetic makeup of a diploid genome can be uncovered: The DNA from an individual who is homozygous for the ΔF508 allele would generate, during PCR, a product with '65' but not '64'. The DNA from a heterozygote (a carrier) would generate bands with either oligonucleotide, and the DNA from a wild-type individual would generate a band with only '64' and not '65'.

Taq DNA polymerase would be an appropriate enzyme to use in this type of test for mismatch, because it lacks a 3' to 5' exonuclease activity. An error-correcting polymerase such as *Pfu* DNA polymerase would be capable of removing the mismatched 3' nucleotide and extending the oligonucleotide from the shortened sequence, interfering with the discrimination between matched and mismatched primers.

How much template is in the sample?

The polymerase chain reaction can be used as a diagnostic tool, to detect the presence of a specific template in a sample. The test may answer a simple 'yes or no' question: is the template detectable in the sample? A more complicated application is trying to determine exactly how much is present, or to make relative comparisons between amounts of two different templates, and this matter is discussed in Section 7.7.

The limit of detection of a PCR-based assay depends in part on the number of cycles used in the method, because an increase in the number of cycles will allow more opportunity for a rare template in a population of DNA molecules to be amplified. Different templates may be amplified with different efficiencies, so the simplistic calculation of productivity used in Section 7.3 may fall short of being predictive of the behavior of a complex PCR with multiple products. A researcher developing such a method may wish to start with known 'positive' and 'negative' control samples, and test them in different mixtures; for example, if the samples are mixed in a 1:10, 1:100, or 1:1000 ratio of 'positive' to 'negative' control, but using the same overall amount of DNA in each reaction, is the positive signal still detectable among the PCR products?

Notwithstanding the problem of template contamination, which was discussed in Section 7.4, it is easier to prove that a sample contains a template than to prove that it does not. A sample could appear to be 'negative' in an assay for reasons that are unrelated to the amount of template; for example, a sample could contain an inhibitor of the polymerase chain reaction, perhaps a polysaccharide that co-purifies with a plant DNA, or hemoglobin from a tissue-derived sample, or a small molecule such as urea. In these cases a 'positive' sample may only appear 'negative' because the polymerase efficiency was reduced, and the amount of product produced at the end point of the assay was below the limit of detection. A researcher may test for diffusible inhibitors by using mixed samples prepared as 1:10, 1:100, or 1:1000 ratios of 'positive' control to 'negative' sample. If the 'negative' sample contains a template that is not detected because of diffusible inhibitors, then a trace amount of positive template added to the sample should also be inhibited, and the mixed samples should remain 'negative'. On the other hand, if the 1:10, 1:100, or 1:1000 ratio samples remain positive, then the researcher can have greater certainty that the 'negative' sample is indeed template-negative and not inhibitor-positive.

The output of a PCR usually includes a mixture of expected DNA products of known size, and unexpected background bands. We may easily overlook these background bands when we are admiring one of our gel pictures, because we focus our attention on the one 'positive' band in the picture. Indeed, it is common for researchers to crop the images of their gels for publications so that only the positive 'band' is displayed, and the readers are allowed to assume that the rest of the gel lane was blank. A problem with this practice of overlooking 'the rest of the data' is that we may miss some important clues. Two samples may have equal amounts of a positive signal; however, if one of the samples generates an extra band on a gel, that might change the experimental interpretation, then cropping the image to only show the one canonical band gets in the way of discovering the truth.

It is difficult to perform an experiment to test for the presence or absence of a template if the size of the product is not established, because any emerging background band on a gel must be considered as a possible 'positive' result. For example, if a researcher is using a pair of primers that may generate a product in the range of 500 bp to 5000 bp, depending on the sample, then any spurious background band in that range may be misinterpreted. If the method of detection of PCR products does not include sizing information, for example if the products are detected by overall fluorescence in a qPCR method (discussed in Section 7.7), then some caution needs to be taken to ensure that 'background' or side reactions are not misconstrued to be positive results.

There are several ways to establish the validity of PCR products, and it may be important for a researcher to spend some time doing so. A legitimate PCR product should only be amplified from a sample if both primers are present in the reaction, and background bands may be excluded from consideration if they can be generated with only one of the two oligonucleotide primers. Where a researcher has enough information about the template to design nested PCR primers, a situation in which one set of primers might generate a PCR product of 1000 bp and a second nested set of primers might be able to generate a 950 bp sub-fragment of the first product, the two reactions can be run sequentially to help establish the identity of a product. Finally, a band on a gel may be excised, and the DNA eluted, and if there is a sufficient yield the DNA can be subjected to DNA sequencing using one of the two oligonucleotide primers. If there is an insufficient yield for direct DNA sequencing, there may still be a sufficient amount for plasmid cloning, followed by sequencing of the recombinant plasmid insert.

What is the size of the product?

The size of a PCR product is a measure of the distance between the primers, and this distance can be the basis for a diagnostic test. A deletion or insertion in DNA can be detected from the size of a PCR product generated using flanking oligonucleotides. For example, a natural polymorphism of the human chemokine receptor gene CCR5 involves a deletion of 32 nt from the coding sequence, as indicated by the strikethrough text (GenBank loci # AY221093, HSU66285):

```
501 CAGATCTCAA AAAGAAGGTC TTCATTACAC CTGCAGCTCT CATTTTCCAT
551 ACAGTCAGTA TCAATTCTGG AAGAATTTCC AGACATTAAA GATAGTCATC
601 TTGGGGCTGG TCCTGCCGCT GCTTGTCATG GTCATCTGCT ACTCGGGAAT
651 CCTAAAAACT CTGCTTCGGT GTCGAAATGA GAAGAAGAGG CACAGGGCTG
701 TGAGGCTTAT CTTCACCATC ATGATTGTTT ATTTTCTCTT CTGGGCTCCC
751 TACAACATTG TCCTTCTCCT GAACACCTTC CAGGAATTCT TTGGCCTGAA
```

Two PCR oligonucleotide primers may be used to capture the region:

```
66. Forward (CCR5): 5'-TCATTACACCTGCAGCTC-3'
67. Reverse (CCR5): 5'-CCAAAGAATTCCTGGAAG-3'
```

Primer '66' has a 5' end at nt 522 (underlined) and '67' has a 5' end complementary to nt 794 (underlined), and together they would generate a PCR product of 794 − 522 + 1 = 273 bp from the wild-type allele, and a product of 241 bp from the allele with the deletion. In PCR of a template taken from a heterozygote, both the 273-bp and the 241-bp bands are detected, and occasionally a third band of lower mobility that represents a heteroduplex between a single strand of 241 bases and a single strand of 273 bases. Heteroduplex molecules form by annealing of products during the last PCR cycle, and the lower mobility is due to the aberrant shape of the heteroduplex.

Differences in product size may also help a researcher identify variants in a splicing pattern when the product is generated by RT-PCR. For example, a skipped exon might be noticed as a decrease in the size of a product, whereas an incompletely spliced product might generate a larger product. However, a word of caution is in order: When the products of a single reaction are heterogeneous, there may be heteroduplex formation between reaction products in the last PCR cycle, complicating the pattern of bands on a gel. Cross-priming during synthesis may also lead to reaction artifacts; for example, if there are two alleles of a 1-kbp gene in a sample from a heterozygote, one template version having a polymorphism at nt 300 and the other having a polymorphism at nt 700, then a doubly polymorphic version might also be recovered among the PCR products.

7.7 Real time or quantitative PCR

Real time or quantitative PCR (qPCR) is a method that allows continual monitoring of product accumulation during the polymerase chain reaction (6).

The method requires a special instrument, a thermocycler that has a built-in fluorometer, so that a fluorescently labeled product can be detected. The qPCR method is an improvement over regular PCR because it allows a researcher to monitor the progress of the amplification and take data points during the cycles when the reaction is optimal. In regular PCR, a researcher may need to perform many 'end point' trials to establish the efficiency or productivity of the reaction, as discussed in Section 7.3.

One simple approach to qPCR makes use of a fluorescent dye such as SYBR® Green I that fluoresces when it is bound to double-stranded DNA. As a PCR cycle starts at 94°C to 98°C, the fluorescence drops as the templates are denatured to yield single strands. After annealing of oligonucleotides and during the synthetic phase of the cycle at 72°C to 74°C, the fluorescence increases again, as double-stranded product is formed and binds to the dye. The fluorescence is not high enough to be detected during the early cycles of a method; however, as the amplification proceeds the signal becomes detectable and the progress of the reaction can be monitored. The data gathered from a qPCR instrument is the amount of fluorescence emission from the product mixture, as a function of time, and the fluorescence is proportional to the mass of double-stranded DNA in solution.

There is no direct indication of the size distribution of products, and the qPCR method may need to be combined with gel electrophoresis to make sure that the fluorescent products are not artifactual. Primer dimer artifacts may be distinguishable from large synthetic products on some qPCR instruments, by analysis of the fluorescence profile during melting of the DNA products. As the temperature increases from the 72°C to 75°C synthetic step of one cycle to the 94°C to 98°C denaturation step of the next, the rate at which the SYBR® Green I fluorescence decreases is an indication of the T_M, and therefore indirectly the size of the products.

An alternative approach to DNA detection by fluorescence is to use a molecular beacon probe that contains one fluorescent dye (for example FAM, ROX, or TAMRA) at one end of an oligonucleotide, and a fluorescence quenching dye (for example DABCYL) at the other end. The beacon probe is designed so that it will anneal to itself, forming a short duplex hairpin between the 5′ and 3′ ends of the oligonucleotide but otherwise remaining as a large single-stranded loop. The apposition of the 5′ and 3′ ends in this stem structure brings the fluorescent dye and the quencher together, and the quencher prevents fluorescence by an effect called fluorescence resonance energy transfer (FRET). A beacon probe might be designed so that the loop region will anneal to a PCR product during the annealing step; this hybridization between probe and target separates the ends of the hairpin and the fluorescent dye and quencher molecule are separated as well. If a beacon probe is added to a PCR, it will fluoresce and report on the molar accumulation of product in the reaction. The beacon probe is displaced during synthesis, and not consumed in the reaction, so it is available as a fluorescent reporter in each cycle.

A third approach to qPCR is called TaqMan® (Roche Molecular Systems, Inc.), which involves annealing of a short probe molecule with a 5′ end fluorescent dye and a 3′ end quencher dye (see *Figure 7.8A*). The TaqMan® probe does not contain a hairpin loop, and is sufficiently small that FRET prevents fluorescence emission whether the probe is annealed to a PCR product target or not. The key point about the TaqMan® method is that the probe is degraded during the reaction, if it is annealed to the template. Upon encountering the annealed probe during primer extension (see *Figure 7.8B*), the 5′ to 3′ exonuclease activity of *Taq* DNA polymerase releases the fluorescent reporter dye into solution (see *Figure 7.8C*). This separates the fluorescent dye from the quencher at the 3′ end and the fluorescent dye becomes fluorescent in solution. As product accumulates during the cycles of PCR, the rate of release of fluorescent dye during the synthetic step will increase, and this can be used to determine the molar amount of product. However, if the TaqMan® probe does not anneal to the template, either because a template is not being generated by the forward and reverse primers or because the template being generated does not base-pair to the probe, then the probe is not degraded and the fluorescent dye is not released.

In all qPCR approaches, fluorescence is used to track the accumulation of product, whether at a mass level as in the case of SYBR® dyes (Molecular Probes, Inc.) or a molar level as in the case of molecular beacon or TaqMan® probes. Fluorescence intensity can be detected over a wide range in many qPCR instruments, and once fluorescence is detectable above background in a reaction, the progress and efficiency of the cycles can be measured. The PCR efficiency will typically reach a maximum value, a point of inflection in the accumulation curve, and then decrease as reaction substrates become limiting.

With molecular beacon or TaqMan® methods of qPCR, a researcher may conduct multiplexed (multiple) assays in the same reaction tube. This capability allows for the development of multiplexed internal controls; for example the amplification of a sequence from an endogenous transcript in the cell such as 18S rRNA or actin, using reverse transcription as an initial step. If the endogenous control transcript is expressed at the same steady-state levels in all cells, the qPCR results of that control template may be used as a normalizing factor to determine the relative steady-state levels of other transcripts. A researcher may also add exogenous control templates in known amounts, for example *in vitro* transcribed RNA templates or cloned DNA sequences from a plasmid.

There are several mathematical approaches used to gather information from the fluorescence profiles generated during qPCR. For example, if one sample contains less template than another, then the first sample should require more cycles of qPCR to generate the same level of fluorescence as the second. The number of cycles that it takes for each sample to achieve a fixed level of fluorescence can be recorded, and used to infer the relative amounts of starting template in each reaction. This approach is valid if the DNA in the two samples was amplified with the same efficiency during PCR, or if the efficiencies can be reasonably determined and included in the calculation.

For example, if a template in one tube is amplified to yield a fixed level of fluorescence in 28 cycles, and a template in a second tube reaches the same point after only 25 cycles, then we could apply the type of mathematical model introduced in Section 7.3, and write $A_1P^{28} = A_2P^{25}$, where A_1 is the starting amount of template in the first tube and A_2 is the starting

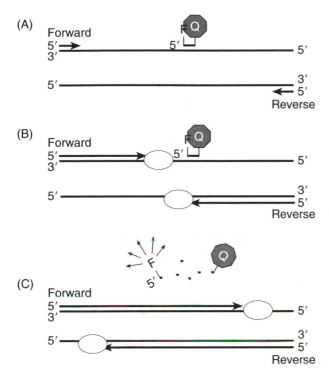

Figure 7.8

The TaqMan® method of quantitative PCR. (A) PCR of a template using a forward and a reverse oligonucleotide. Also included is a third 'probe' oligonucleotide that anneals to the template and carries a fluorescent dye molecule 'F' conjugated to the 5' nucleotide, and a quencher dye molecule 'Q' conjugated to the 3' nucleotide. The 'Q' dye prevents fluorescence from the 'F' dye because of their proximity. (B) During the course of PCR, an annealed 'probe' will stand in the way of the *Taq* DNA polymerase (ovals) on one strand. (C) The *Taq* DNA polymerase has a 5' to 3' exonuclease that will degrade the annealed 'probe' oligonucleotide, separating the dyes. Once separated from 'Q', the 'F' dye is actively fluorescent. Fluorescence is an indication of probe-template binding, and is used to measure template production during PCR.

amount of template in the second tube. The relative amounts of starting template (the ratio $A_1:A_2$) is therefore P^{25}/P^{28}.

A difficulty with this mathematical treatment is that the value of P may not be the same in the two tubes, or even the same between two different templates being amplified in a multiplex reaction in a single tube. There is some danger that a lower efficiency with one template might not be noticed, and might be misconstrued as a difference in the amount of starting material. For example, 10^4 starting copies of a template can be amplified to 1.4×10^{11} product molecules over 37 PCR cycles, with a geometric mean value for P of 1.56; however, the same amount of amplification of 10^4 templates could occur in just 36 cycles if the value for P were 1.58. If we take the cycle numbers as an indication of the starting copies, without correction for the different efficiencies, we will reach an incorrect conclusion.

The qPCR instruments allow us to measure the efficiency of a reaction during the last cycles of a method, before the amount of product reaches a plateau; however, a value of P based on these last few cycles may be a poor indicator of the geometric mean value of P that must be applied in an expression such as P^{25}/P^{28}. The same concern might be expressed about mathematical treatments in which the rate of accumulation of product and its overall amount are used to extrapolate back to the first cycle, to estimate the amount of starting template. The slope of the product accumulation line at the inflection point is a snapshot of the efficiency, but may be an overestimate of the mean efficiency during the previous cycles in the method.

7.8 Troubleshooting

At times it can be difficult to get a PCR working reproducibly. The reason for this is that minor changes in reaction efficiency are compounded in each cycle, sometimes resulting in too much or too little amplification of a product. Unwanted side reactions with the primers, cross-priming between products, and heteroduplex annealing artifacts may also result in a mixture of bands or a smear on a gel.

The blank gel

When we perform a PCR and see no product of any kind on a gel, there are a few things that need to be considered. We need to ask some difficult questions, such as 'did we order the right primers?', and review our calculations. However, a blank gel lane is not automatically a devastating result, and should be taken in your stride. The hoped-for product might actually be present on the gel, but just below the limit of detection of the DNA stain. With a bit of optimization, the yield may improve to the point that the experiment is successful.

Did the gel and the staining work?

If the gel and stain worked, then we expect to see a molecular-weight marker on the gel that appears to be normally stained. The marker lane can help us decide if the gel was subjected to electrophoresis for the right amount of time; a product band that is run off the end of a gel will not be visible. The marker lane also carries information about the quality of the gel, the ability of the gel to resolve single species of DNA into tight bands. If a product is not resolved into a tight band, for example if the gel matrix is not the right percentage of agarose or acrylamide, then its signal will be spread over a larger area and may be undetectable. For example, it is much easier to detect a 300-bp product on a 10% polyacrylamide gel than on an agarose gel, because the bands are resolved much better.

We can use information about the amount of marker loaded on the gel to tell us something about the limit of detection of the gel. For example, if we are using γ DNA digested with *Hind*III as a marker, our marker has eight DNA species: 23 130, 9416, 6557, 4361, 2322, 2027, 564, and 125 bp.

Suppose that we load 2 μl of a 0.5-mg ml^{-1} sample, which would be 1 μg of DNA, and after running and staining the gel we can only barely discern the band at 564 bp on the gel, and we can't see the band at 125 bp at all. We need to do a little calculation. The eight species of DNA are present in equimolar amounts, and when added together their total length is 48.5 kbp. The 564-bp and 125-bp bands are therefore 0.564/48.5 = 1.16% and 0.125/48.5 = 0.258% of the 1 μg that we loaded, or respectively 11.6 ng and 2.58 ng. We now have the information that our limit of detection lies somewhere in this range, and that if we had succeeded in generating 2 ng of PCR product, it would probably not be visible on this stained gel.

Are the primers working?

When we are facing a PCR that is not working, we have to consider the possibility that the primers are not annealing to the target template. Were the primers designed correctly so that they anneal to complementary strands and can be extended in the right direction (see Section 7.2)? The 3' end of the oligonucleotide is the most critical part, because if it does not anneal then it cannot serve as a primer. Were the primers ordered from a reputable supplier? Are the oligonucleotide primers being added in sufficient amounts (see Section 7.4)? If the PCR products are analyzed on a high-percentage agarose gel, or a polyacrylamide gel, the primers may be visible at the bottom of the stained gel as a 'primer glow' with a stain such as ethidium bromide, which provides visual confirmation that the primers are present. Sometimes it is possible to determine whether the primers are being consumed in side reactions such as primer dimer formation.

Is the template pure and correct?

We must consider the possibility that the template does not have the expected sequence to match the oligonucleotides, or that the template may not have been added to the reaction in sufficient quantity, or with adequate purity (see Section 7.4). If the template is derived from cultured cells, are we certain that the cells are what they are purported to be? Has the culture become contaminated and been overtaken by a different type of cell line? A reaction may fail if the template does not base-pair with the 3' ends of both oligonucleotide primers.

If the template is a genomic DNA, has it been adequately denatured during the initial stage of the PCR? A period of denaturation at a higher temperature (e.g., 98°C) may help improve the availability of single-stranded template during the first cycles, particularly if the template is a high molecular weight genomic DNA (e.g., >20 kbp) and not extensively nicked.

Does the sample of DNA contain inhibitors of the polymerase, for example polysaccharides or salts, that might co-purify with the DNA? For example, a DNA template that is prepared by gel electrophoresis may contain contaminants derived from the agarose, or may be contaminated with borate salts if the gel running buffer was TBE-based. Alcohol precipitation will not help to separate these types of contaminants from a DNA sample, and a researcher may need to use an affinity binding matrix or a spin column to purify the template.

Is the magnesium ion concentration optimal?

When a researcher develops a new PCR method, one of the first tasks is to optimize the concentration of magnesium salts. As explained in Section 7.3, the magnesium salts have an important effect on the efficiency of the reaction, and the difference between success and failure may rest on a change of only 0.5 to 1.0 mM Mg^{2+} concentration. If the stock of $MgCl_2$ or $MgSO_4$ is switched, or the concentrations of oligonucleotides or dNTP substrates are changed, the magnesium salts may need to be optimized again.

Are the other reaction components present in adequate amounts?

As discussed in Section 7.3, the buffer and salt conditions, the dNTP substrates, and the thermostable DNA polymerase are all critical components of the reaction. A researcher trying to troubleshoot a PCR may want to review the amounts of these components that are added, and the age and quality of the materials. For example, a thermostable DNA polymerase that has not been stored properly may have lost activity, so that the number of enzyme units per μl marked on the tube is incorrect. Similarly, a mixture of dNTP substrates that has been prepared or stored improperly may have a short shelf life in the freezer, and may need to be replaced in due time.

Many researchers use additives such as betaine or DMSO to enhance the efficiency of PCR. These components require optimization when they are used, but might be considered for difficult methods that are not yielding satisfactory results. Nucleoside analogues such as deoxyinosine triphosphate or 7-deaza-2'-deoxyguanosine are often used to improve the efficiency of PCR with GC-rich templates; however, the latter analogue diminishes the staining of products with ethidium bromide.

Is the number of cycles appropriate?

A PCR protocol that works well with one set of primers and templates may not work well with another; in particular, the number of PCR cycles may need to be adjusted up or down, depending on the efficiency of the reaction. The simple predictive calculations presented in Section 7.3 can help a researcher determine whether the number of cycles is appropriate for the needed amplification. For example, if a laboratory uses 35 cycles for every PCR, never allowing for the possibility of differences in efficiency, they may struggle to obtain a result if the efficiency suddenly drops from P = 1.58 to P = 1.42, in a new experiment. The reason is that the lower efficiency will cause more than a 40-fold drop in yield over 35 cycles, resulting in a gel lane that carries too little product to detect after staining. Approximately 45 to 46 cycles would be needed to restore the yield with the lower efficiency reaction, because 1.42^{45} is close to 1.58^{35}.

Is the annealing temperature correct?

If the PCR is not working, it may be because the annealing temperature is too high, and the oligonucleotides are not efficiently priming synthesis from the template. The T_M of an oligonucleotide may be estimated using several different mathematical approaches, as discussed in Section 7.2; however, the optimal annealing temperature may need to be determined empirically. This is especially true when the oligonucleotide is degenerate, and the exact T_M is indeterminate.

A researcher may set up trial reactions in which the annealing temperature is allowed to vary, for example from 65°C down to 50°C in steps of 2°C. Some thermocycler instruments allow a researcher to establish a temperature gradient in a block of tubes, and test multiple annealing conditions simultaneously. An alternative is to use a thermocycler that allows for timed ramping of temperatures. For example, an annealing step might include a gradual ramp from 50°C up to 65°C, over a 2- to 3-min period, to allow for the annealing of a range of degenerate oligonucleotides to their targets.

Is the timing of the synthetic step appropriate?

The 72°C to 74°C synthetic step is when most of the actual work of DNA polymerization takes place. For DNA products shorter than approximately 500 bp, a synthetic step of 30 s is usually adequate. However, some consideration may need to be given to the timing of the synthetic step if longer products are being planned. Approximately 1 to 2 min may be needed for each kbp of product, depending on the type of DNA polymerase and the advice from the manufacturer. Long PCR methods may require the use of proofreading DNA polymerases such as *Pfu*, or formulations of several different polymerases.

The smear

Finding a smear on a gel lane of PCR products may be more disheartening than finding a blank lane. A smear brings with it all the implications of a messy result; the possibility that locating the product may be like finding a needle in a haystack. Many of the pieces of advice given for troubleshooting a blank gel lane may be equally applicable to the case of a smeary gel lane, because a poorly optimized reaction may spawn side reactions that are responsible for the smear. The concentration of Mg^{2+} may need to be optimized, for example, if a researcher has not yet performed that important task.

What is a smear?

One thing to recognize is that a broad smear on a stained gel represents a lot of DNA. A smear that is distributed over 5 cm of gel lane may represent 1 to 10 µg (or more) of DNA. We are used to seeing bands on a gel in which 10 to 100 ng of DNA are collected into a 5- to 10-mm² area; however, if we are considering a broad smear that spans 50 times the area of a typical band then we must integrate the signal over the length of the smear to estimate the total amount of DNA product.

Where did the smeary DNA come from? It is likely to be new DNA synthesis, unless an inordinate amount of genomic DNA starting template was added to the reaction and is being detected as part of the final product. The new DNA synthesis must have been primed either by the oligonucleotides that were added, or by an artifactual priming from the template molecules. Unless the template has very low sequence complexity, it is unlikely to participate in many of the priming reactions because of the limitations of kinetics. The most likely culprit for the priming of the DNA synthesis is therefore the oligonucleotide primers that we added.

We must bear in mind that a smear may represent two different effects: a variation in the size of product or a variation in the conformation of product. A smear that spans a range of 1 to 10 kbp on a gel is unlikely to include duplex DNA that is actually 5 to 10 kbp, unless we have used a long 72°C to 74°C synthetic step of 5 to 20 min. For example, if we see a 1- to 10-kbp smear with a synthetic step of only 30 s, we should conclude that most of the smear is due to oddly shaped heteroduplex molecules, perhaps cruciform-shaped products that have annealed. In other words, we have generated so much product that it is annealing to itself. A smear may also represent partially completed products that accumulate when a PCR is driven to the point of reaction failure. The job of completing the synthesis of a large amount of product in the last few cycles, in the allotted time at 72°C to 74°C, may be too great for the depleted activity of the enzyme or the remaining concentrations of substrates.

When a smear is too much of a good thing

The first thing to suspect in the case of a smear is that the number of PCR cycles is too high. A higher estimate of efficiency may be appropriate (see Section 7.3), and reducing the number of cycles or the amount of starting template may resolve the problem. In troubleshooting the problem, it may be helpful for a researcher to set up a series of identical side-by-side reaction tubes, and to sit by the thermocycler during its operation so that tubes may be pulled out at intervals (e.g., upon completion of 25, 27, 29, 31, 33, and 35 cycles).

When primers behave badly

The second thing that should be suspected about a smear on a gel is that it may represent nonspecific side reactions of the primer, particularly if the sequence of the primer has low complexity or if the T_M is low. One or both of the oligonucleotide primers may be finding nonspecific sites in the template at which a reaction can be initiated, perhaps a sequence that is repetitive. It may be helpful to perform a control PCR with only one primer in the reaction at a time; if the smear is evident with only one primer and not the other, then the result is certainly nonspecific and this one primer may need to be redesigned.

Any primer may yield nonspecific side reactions if it is used at too low an annealing temperature, because a low temperature of annealing is less stringent than a higher temperature, and favors pairings between oligonucleotide and template that are based on fewer base pairs. The T_M of an oligonucleotide can be predicted, as discussed in Section 7.2, but the optimal annealing temperature may need to be determined empirically.

In cases where primers generate a mixture of specific and nonspecific reaction products, it is sometimes possible to favor the specific reactions by performing a touchdown PCR (see *Figure 7.9*) (7). In this approach, a thermocycler is programmed so that the annealing temperature is initially set high for the first cycle, and is gradually reduced by 0.5°C to 1°C in each subsequent cycle (see *Figure 7.9B*), until a minimal annealing temperature is reached. That minimal annealing temperature is maintained in all remaining cycles (see *Figure 7.9C*). The more specific annealing reactions will be favored at the higher temperatures, and this gives the specific reactions a 'head start' of several cycles over the nonspecific reactions. That head start will be reflected in the final composition of products.

For example, the first 10 cycles of a touchdown program might have annealing temperatures of 65°C, 64°C, 63°C, 62°C, 61°C, 60°C, 59°C, 58°C, 57°C, and 56°C, and the remaining 30 cycles of the program might use an annealing temperature of 55°C. Let us suppose that 55°C is the optimal annealing temperature for generating the specific product, but that nonspecific products are also made in abundance at 55°C because they have only a slightly lower optimal annealing temperature. By bringing the annealing temperature down gradually from a much higher temperature, the specific products may accumulate and be better represented in the final PCR results.

When primers jump the gun

An oligonucleotide primer may anneal to nonspecific sites in a genome while the components of a reaction are being assembled on ice or at room temperature. This would not be a problem, except for the fact that the enzyme may be active as the temperature of the tube is being raised for the first time (see left section of *Figure 7.4A* or *Figure 7.9A*). In other words, some primers may 'jump the gun' and start an extension reaction before we have set an adequately stringent temperature condition. Once that happens, the nonspecific products are replicated in subsequent cycles because the oligonucleotide primers are incorporated into the 5′ ends. We may see a smear on the gel if these nonspecific products accumulate.

We can prevent these low-temperature annealing events from generating nonspecific products by preventing polymerization from taking place at the lower temperatures. This is called a hot start PCR, because the reaction is not started until the tube is hot enough to prevent nonspecific annealing. There are several ways of arranging for a hot start reaction. Some thermostable enzymes are sold as 'hot start' enzymes, and these are bound to an antibody molecule that blocks function. The enzyme is not functional as the reaction is being assembled, but becomes functional at high temperature when the antibody denatures.

An alternative approach is to delay adding the polymerase until the temperature of the tube is higher than the proper annealing temperature. Without the polymerase being present, the spurious annealing reactions cannot be extended and don't become fixed into the nonspecific products. For example, the thermocycler might be programmed to raise the tempera-

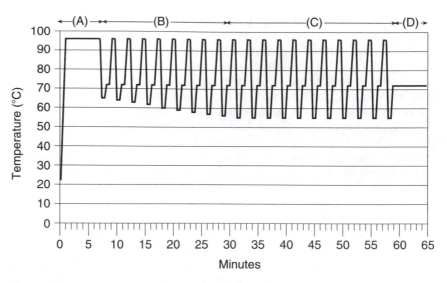

Figure 7.9

A profile of 'touchdown' PCR. The temperature of the reaction is shown as a function of time. (A) An initial temperature rise to 96°C for complete denaturation of the template. (B) Repetitive cycling in which the annealing temperature is 64°C in the first cycle, and reduced by 1°C in each subsequent cycle until an annealing temperature of 55°C is used in the tenth cycle. (C) Repetitive cycling between 96°C, 55°C, and 72°C. Only 10 additional cycles are shown. (D) A segment during which the temperature is held at 72°C to allow for completion of products being synthesized in the last cycle.

ture to 96°C for 5 min to denature the template, drop the temperature to 70°C so that the enzyme can be added, then enter into the repetitive cycling part of the program. The brief stop at the lower temperature of 70°C allows for addition of enzyme when the reaction is at a lower vapor pressure, and reduces the risk of burning the fingers on a hot temperature block.

A disadvantage of adding polymerase to the tubes after denaturation is that it presents an opportunity for contamination of the reactions. Some researchers use wax beads that are loaded with a preset amount of $MgCl_2$; these beads melt and add their contents to the reaction after the tube has reached a high enough temperature to prevent the unwanted nonspecific annealing. This is a hot start method because the polymerization cannot be initiated without the $MgCl_2$ salts, and the reaction is delayed until the solution is hot. The beads can be added to the tubes under sterile conditions, so that the tubes do not need to be reopened after they are inserted into the thermocycler.

Multiple bands

Many of the comments that were made about smears on a gel will also apply to the problem of multiple unwanted bands. Whereas a smear suggests a wide variety of different products or conformations, the presence of a few extra bands on a gel suggests a more limited problem.

Multiple unwanted bands can be a sign of a poorly optimized reaction in which too much or too little Mg^{2+} is added, or it might be an indication that the annealing temperature is too low, or that too many PCR cycles are being employed. If the goal is to generate a product for genetic engineering or construction work, the one desired band might be excised from the gel and purified away from the unwanted bands. On the other hand if the goal is to use the PCR as a diagnostic tool then it might be best to improve the performance of the method to prevent the synthesis of the extra bands.

Extra bands on a gel could represent side reactions that originated from nonspecific annealing reactions between the primer and template, or they could be a consequence of cross-priming reactions between templates during the course of PCR. In those cases, a higher annealing temperature or a set of newly designed primers may help to resolve the problem. Alternatively, the extra bands could be due to aberrant migration of heteroduplex molecules, as discussed in Section 7.6, from templates that are not homogeneous.

Further reading

Ausubel FW, Brent R, Kingston RE, Moore DD, Seidman JG, Smith JA and Struhl K (eds) (1987) *Current Protocols in Molecular Biology*, John Wiley & Sons, New York.

Sambrook J, Russell D (2001) *Molecular Cloning: A Laboratory Manual*, 3rd Edn. Cold Spring Harbor Laboratory Press. Cold Spring Harbor, NY.

References

1. Mullis K, Faloona F, Scharf S, Saiki R, Horn G, Erlich H (1986) Specific enzymatic amplification of DNA in vitro: the polymerase chain reaction. *Cold Spring Harb. Symp. Quant. Biol.* **51**(1): 263–273.
2. Saiki RK, Gelfand DH, Stoffel S, Scharf SJ, Higuchi R, Horn GT, Mullis KB, Erlich HA (1988) Primer-directed enzymatic amplification of DNA with a thermostable *DNA polymerase. Science* **239**: 487–491.
3. McPherson M, Møller S (2006) PCR *2e: The Basics*, Taylor & Francis, Oxford.
4. Novagen® EMD Biosciences (2004) *pET-50b(+) Manual* (TB418 0804 Rev. A), Merck KGaA, Darmstadt.
5. Aslanidis C, de Jong PJ (1990) Ligation-independent cloning of PCR products (LIC-PCR). *Nucleic Acids Res.* **18**: 6069–6074.
6. Higuchi R, Dollinger G, Walsh PS, Griffith R (1992) Simultaneous amplification and detection of specific DNA sequences. *Biotechnology (NY)* **10**: 413–417.
7. Don RH, Cox PT, Wainwright BJ, Baker K, Mattick JS (1991) 'Touchdown' PCR to circumvent spurious priming during gene amplification. *Nucleic Acids Res.* **19**: 4008.

Solutions to problems

Solutions to problems posed in Section 7.2

(i)   A. The 5′ ends of the oligonucleotides define the boundaries of the product. We want to have the product start at nt 1 and end at nt 200, so the forward oligonucleotide will have a 5′ end with nt 1 and the reverse oligonucleotide will have a 5′ end complementary to nt 200. The forward and reverse oligonucleotides will have these sequences:

```
Forward: 5′-CTTCCAAGTGTTGAAGAAGAATCAT-3′
Reverse: 5′-GGCGTTTCTTTTCTACACCGCAGAG-3′
```

We can check that these two oligonucleotides anneal to complementary strands and that the 3′ ends 'point towards' each other (i.e., the DNA polymerase should extend the 3′ ends to fill in the sequence between the primer sites).

B. The forward oligonucleotide has 9 G or C nucleotides out of 25, so its estimated T_M will be:

$$T_M = 64.9°C + 41°C \times (9 - 16.4)/25 = 53°C$$

The reverse oligonucleotide has 13 G or C nucleotides out of 25, so its estimated T_M will be:

$$T_M = 64.9°C + 41°C \times (13 - 16.4)/25 = 59°C$$

C. The reverse oligonucleotide would anneal to the mRNA, and could be used to prime the synthesis of a first-strand cDNA. The forward oligonucleotide could be added as well, when PCR is commenced.

(ii) The primers will be identical to those described above, except that our previously labeled 'forward' primer will anneal to the single strand that is shown and the 'reverse' primer will anneal to its complement. The important point is that DNA doesn't intrinsically have a 'top' and 'bottom' strand. Our labeling of oligonucleotides as 'forward' or 'reverse' only has meaning after we have picked one of the strands and labeled it as 'top'. Only one of the oligonucleotides will anneal to the mRNA, but that is not a concern in this problem.

(iii) To address this problem we need to look at how these oligonucleotides might base-pair at their 3′ ends.
A. The second oligonucleotide on the list, the one ending '...AGCT' could base-pair to itself over those four nucleotides (underlined):

```
5′- GGCGTTTCTTTTCTACACCGCAGAGCT
                    TCGAGACGCCACATCTTTTCTTTGCGG-5′
```

This pairing is the most extensive among the oligonucleotide sequences on the list, and could lead to the generation of a primer dimer of 50 nt:

```
5′-GGCGTTTCTTTTCTACACCGCAGAGCTCTGCGGTGTAGAAAAGAAACGCC-3′
3′-CCGCAAAGAAAAGATGTGGCGTCTCGAGACGCCACATCTTTTCTTTGCGG-5′
```

B. There are several possible pairings, but the most significant may be the last oligonucleotide on the list, the one ending '...TACT', which can form six base pairs with the new oligonucleotide, as follows:

```
5′-CAGGACTACGATTTAGTAGCGGTAGT-3′
                3′-TCATCGAGACGCCACATCTTTTCTTTGCGG-5′
```

Extension of the lower strand could lead to formation of this 44-mer primer dimer:

3′-GTCCTGATGCTAAATCATCGAGACGCCACATCTTTTCTTTGCGG-5′

C. The 44-mer would not be amplified geometrically, because the 3′ end of the new oligonucleotide does not base-pair with the 44-mer. However, there are several ways in which the primer dimer could be extended in additional reactions, so that it would be amplified geometrically. The observant reader may have also noticed that the new oligonucleotide can form a primer dimer with itself!

5′-CAGGACTACGATTTAGTAGCGGTAGT-3′
    3′-TGATGGCGATGATTTAGCATCAGGAC-5′

Solutions to problems posed in Section 7.3

(iv) We need to figure out how many moles of template and primer have been added to the reaction. In the case of the template, we first need to determine the formula weight of the genome:

$(4 \times 10^8 \text{ bp})(660 \text{ Da bp}^{-1}) = 2.6 \times 10^{11} \text{ Da}$

Now we can calculate how many moles 20 ng represents:

$(2 \times 10^{-8} \text{ g}) \div (2.6 \times 10^{11} \text{ g mol}^{-1}) = 7.6 \times 10^{-20} \text{ mol}$

On the other hand, we've added 1 μl of a 5-μM stock of oligonucleotide to the reaction, and that is 5 pmol:

$(1 \times 10^{-6} \text{ l})(5 \times 10^{-6} \text{ mol l}^{-1}) = 5 \times 10^{-12} \text{ mol}$

The molar ratio between oligonucleotide and template at the very start of the PCR is therefore:

$5 \times 10^{-12} \text{ mol} : 7.6 \times 10^{-20} \text{ mol} = 6.6 \times 10^7 : 1$

(v) We consume one oligonucleotide primer each time we make a product molecule, so we are limited to making 5 pmol of product. The product is 2 kbp, so its formula weight is:

$(2 \times 10^3 \text{ bp})(660 \text{ Da bp}^{-1}) = 1.3 \times 10^6 \text{ Da}$

Now we may calculate the theoretical yield in mass units:

$(5 \times 10^{-12} \text{ mol})(1.3 \times 10^6 \text{ g mol}^{-1}) = 6.6 \times 10^{-6} \text{ g} = 6.6 \text{ μg}$

(vi) If you have obtained 200 ng of a 2-kbp product, then the molar yield is:

$(2 \times 10^{-7} \text{ g}) \div (1.3 \times 10^6 \text{ g mol}^{-1}) = 1.5 \times 10^{-13} \text{ mol}$

The amount of template amplification (A/A_0) is therefore:

1.5×10^{-13} mol product $\div 7.6 \times 10^{-20}$ mol genome $= 2 \times 10^6$-fold

If this amplification occurred during the course of 30 PCR cycles, the productivity factor P can be determined from the equation $A = A_0 P^n$, where $A = 1.5 \times 10^{-13}$ mol product, $A_0 = 7.6 \times 10^{-20}$ mol genome, and $n = 30$ cycles. P^n is the amount of amplification, or 2×10^6, and we may determine n by using logarithms:

$$\log_{10}P^{30} = \log_{10}(2 \times 10^6)$$
$$30 \log_{10}P = 6.30$$
$$\log_{10}P = 0.21$$
$$P = 10^{0.21} = 1.62$$

(vii) We've taken 1/1000 of our 200-ng yield, or 200 pg, and want to use that to generate 100 ng of product by PCR. We assume P = 1.62. Since the template and product have the same formula weight, we need not bother with converting from grams to moles. We can observe that 100 ng is 500 times more than 200 pg, and so we need 500-fold amplification, and therefore,

$$1.62^n = 500$$
$$n \log_{10}1.62 = \log_{10}500 = 2.699$$
$$n = 2.699/0.21 = 12.8$$

Rounding up our result to the next integer, we will need 13 cycles to obtain 100 ng.

Solutions to problems posed in Section 7.5

(viii)A. After digestion of the plasmid with *Eco*RI, the ends have the following structure:

```
...TACCAGGATCCG              AATTCTGTACAGGCC...
...ATGGTCCTAGGCTTAA              GACATGTCCGG...
```

Upon treatment with Klenow enzyme and dCTP, the 3′ ends are digested to yield a 5-nt and a 6-nt overhanging end:

```
...TACCAGGATCC              AATTCTGTACAGGCC...
...ATGGTCCTAGGCTTAA              CATGTCCGG...
```

B. We need to design the forward oligonucleotide so that it starts with a sequence that is compatible with the overhanging end in the plasmid (GAATT). Following that, we need to adjust the number of nucleotides so that the ATG of the caspase I will be in frame for the generation of a fusion protein. The 'AAT' in the 'GAATT' overhanging end is an Asn codon, and the 'T' that follows is the first nucleotide of the next codon. We need to fill in two more nucleotides to finish the codon. Furthermore, the nucleotide that follows the last 'T' should be the point where the 3′ to 5′ exonuclease is halted on the opposite strand.

We don't have much choice here, because three nucleotides have already been used in the overhanging end. If we imagine which dNTP substrates have to go into the Klenow enzyme digestion, it is clear that dCTP, dTTP, and dATP will be left out. If these three substrates were present they would halt the exposure of the 'GAATT' overhanging end. Therefore the nucleotide following 'GAATT' must be a 'C', and we will leave dGTP out of the Klenow enzyme reaction to generate the 5-nt overhanging end. This also forces the codon to be a Ser codon, and the choice of the last nucleotide (following 'GAATTC') is arbitrary (we'll select 'C', again). We will finish the design by adding on the first 18 nt of the caspase I coding sequence (underlined):

68. Forward: 5'-GAATTCC<u>ATGGCTGACAAGATCCTG</u>-3'

Following PCR of the caspase I sequence, using oligonucleotide '68', the 5' end will have this structure:

```
5'-GAATTCCATGGCTGACAAGATCCTGAGGGCA...
3'-CTTAAGGTACCGACTGTTCTAGGACTCCCGT...
```

Upon treatment with Klenow enzyme and only the substrate dGTP, a 5-nt overhanging end will be generated:

```
5'-GAATTCCATGGCTGACAAGATCCTGAGGGCA...
3'-     GGTACCGACTGTTCTAGGACTCCCGT...
```

This end is compatible with the pET-50b(+) end generated in the previous part of the problem, and upon ligation the reading frame is preserved (the caspase I-derived sequence is underlined):

```
      TyrGlnAspProAsnSer MetAlaAspLysIleLeuArgAla...
...TACCAGGATCCGAATTCCATGGCTGACAAGATCCTGAGGGCA...
...ATGGTCCTAGGCTTAAGGTACCGACTGTTCTAGGACTCCCGT...
```

C. At the 3' end of the caspase I gene, we need to design an oligonucleotide for PCR that will generate a compatible overhanging end upon treatment with Klenow enzyme and dGTP (our choice of substrate was established in the previous part of the problem). The 5' end of the oligonucleotide (5'-AGAATT) will copy the sequence compatible with this overhanging end from pET-50b(+):

```
5'-AATTCTGTACAGGCC...
3'-     CATGTCCGG...
```

Following the 'AGAATT' we introduce a 'C' nucleotide, so that the 3' to 5' exonuclease will halt on a 'G' nucleotide. Finally, we add 18 nt complementary to the end of the caspase I sequence (underlined below):

69. Reverse: 5'-AGAATTC<u>TTAATGTCCCGGGAAGAG</u>

Following PCR of the caspase I sequence, using oligonucleotide '69', the 3' end will have this structure:

```
...TTCTACCTCTTCCCGGGACATTAAGAATTCT-3'
...AAGATGGAGAAGGGCCCTGTAATTCTTAAGA-5'
```

Upon treatment with Klenow enzyme and only the substrate dGTP, a 6-nt overhanging end will be generated:

```
...TTCTACCTCTTCCCGGGACATTAAG      -3'
...AAGATGGAGAAGGGCCCTGTAATTCTTAAGA-5'
```

This end is compatible with the pET-50b(+) end generated in the previous part of the problem, and upon ligation the segment would have the following structure (the caspase I-derived sequence is underlined):

```
...TTCTACCTCTTCCCGGGACATTAAGAATTCTGTACAGGCC...
...AAGATGGAGAAGGGCCCTGTAATTCTTAAGACATGTCCGG...-5'
...PheTyrLeuPheProGlyHisStop
```

(ix) The 'reverse' oligonucleotide that should be used in the inverse PCR will start at the end of the His·Tag®.

70. Reverse: 5'-ATGGTGATGGTGGTGATG-3'

This oligonucleotide '70' will anneal to the sequence between nt 114 and nt 97 (see *Figure 7.6*).

```
   MetGlySerSerHisHisHisHisHisHisSerSerMetAsnLysGluIleLeuAla
...ATGGGCAGCAGCCATCACCACCATCACCATTCTAGTATGAACAAAGAAATTTTGGCT...
           3'-GTAGTGGTGGTAGTGGTA-5'
```

The 'forward' oligonucleotide will start with the ATG start codon of the caspase I gene:

15. Forward (caspase I): 5'-ATGGCTGACAAGATCCTG-3'

Following inverse PCR using '70' and '15', the ends of the plasmid will have the following structure:

```
  MetGlySerSerHisHisHisHisHisHis              MetAlaAspLysIleLeuArgAla...
...ATGGGCAGCAGCCATCACCACCATCACCAT-3',    5'-ATGGCTGACAAGATCCTGAGGGCA...
...TACCCGTCGTCGGTAGTGGTGGTAGTGGTA-5'     3'-TACCGACTGTTCTAGGACTCCCGT...
```

Upon treatment of the ends with T4 polynucleotide kinase and T4 DNA ligase, the ends can be fused to generate a contiguous open reading frame (the caspase I sequence is underlined):

```
  MetGlySerSerHisHisHisHisHisHisMetAlaAspLysIleLeuArgAla...
...ATGGGCAGCAGCCATCACCACCATCACCATATGGCTGACAAGATCCTGAGGGCA...
...TACCCGTCGTCGGTAGTGGTGGTAGTGGTATACCGACTGTTCTAGGACTCCCGT...
```

(x) The GFP sequence can be generated by PCR using this pair of oligonucleotides:

71. Forward (GFP): 5'-ATGAGTAAAGGAGAAGAA-3'
72. Reverse (GFP): 5'-gtcagccatTTTGTATAGTTCATCCAT-3'

Oligonucleotide '71' is simply nt 1-18 of the GFP sequence. Oligonucleotide '72' is complementary to nt 697-714 of GFP (in uppercase) and the first 9 nt of the caspase I coding sequence (in lowercase). The PCR product will have the following structure (the caspase I sequence is underlined):

```
5'-ATGAGTAAAGGAGAAGAACTTTT...CATGGCATGGATGAACTATACAAAATGGCTGAC-3'
3'-TACTCATTTCCTCTTCTTGAAAA...GTACCGTACCTACTTGATATGTTTTACCGACTG-5'
```

The pET-50b(+) vector containing caspase I can be generated by inverse PCR, using these primers:

73. Forward (pET): 5'-ctatacaaaATGGCTGACAAGATCCTG
45. Reverse (pET): 5'-ATGTATATCTCCTTCTTA-3'

Oligonucleotide '73' contains the first 18 nt of the caspase I coding sequence (in uppercase) and the final 9 nt of the GFP sequence (in lowercase). Oligonucleotide '45' has been discussed earlier, and is complementary to the pET-50b(+) sequence upstream of the ATG start codon (see the first site marked '7' in *Figure 7.6*). The PCR product generated using '73' and '45', and a caspase I clone as a template, would have the following structure (the caspase I sequence is underlined):

```
5'-CTATACAAAATGGCTGACAAGATCCTGAGGGCA...TAACTTTAAGAAGGAGATATACAT-3'
3'-GATATGTTTTACCGACTGTTCTAGGACTCCCGT...ATTGAAATTCTTCCTCTATATGTA-5'
```

This product would be combined with the GFP product, and could cross-prime in a third PCR (the GFP product is shown on top, and the paired region is underlined):

```
...CATGGCATGGATGAACTATACAAAATGGCTGAC-3'
             3'-GATATGTTTTACCGACTGTTCTAGGACTCCCGT...
```

The reading frame would be preserved through this cross-priming of GFP and caspase I (caspase I is underlined):

```
...HisGlyMetAspGluLeuTyrLysMetAlaAspLysIleLeuArgAla...
...CATGGCATGGATGAACTATACAAAATGGCTGACAAGATCCTGAGGGCA...
...GTACCGTACCTACTTGATATGTTTTACCGACTGTTCTAGGACTCCCGT...
```

This third PCR would be conducted using the two distal oligonu-cleotides, '71' and '45', and would generate ends with the following structure:

```
                         MetSerLysGlyGluGluLeu
...TAACTTTAAGAAGGAGATATACAT-3' 5'-ATGAGTAAAGGAGAAGAACTTTT...
...ATTGAAATTCTTCCTCTATATGTA-5' 3'-TACTCATTTCCTCTTCTTGAAAA...
```

After treatment with T4 polynucleotide kinase and T4 DNA ligase, the plasmid would be circularized and the GFP gene (underlined below) would be adjacent to the ribosome binding site of the vector.

```
                    MetSerLysGlyGluGluLeu
...TAACTTTAAGAAGGAGATATACATATGAGTAAAGGAGAAGAACTTTT...
...ATTGAAATTCTTCCTCTATATGTATACTCATTTCCTCTTCTTGAAAA...
```

Working with DNA: the reasons

8

8.1 Engineering DNA sequences

When we work with DNA in the laboratory, it is usually for one or more of the following reasons: we want to engineer a new sequence, identify or measure the amount of a nucleic acid, or determine the biological activity of a DNA molecule. In this chapter we will consider some of these reasons in more detail, and look at applications that depend on the materials and methods discussed in the previous chapters.

Using restriction enzymes to manipulate DNA sequences

Genetic engineering usually commences with work done *in vitro*, using DNA-modifying enzymes such as restriction endonucleases, DNA polymerases, and DNA ligase. Some of the earliest work in recombinant DNA involved joining fragments of plasmids that had been digested with restriction endonucleases (1). The cohesive ends left by some restriction endonucleases, such as *Eco*RI, allowed specificity in ligations. DNA from the eukaryote *Xenopus laevis* was inserted into a bacterial plasmid using restriction endonucleases, and the foreign DNA was transcribed in the *E. coli* host (2). Restriction endonucleases are now widely used for genetic engineering, and the list of commercially available restriction enzymes grows yearly (see Chapter 5). Plasmids and bacteriophage vectors with multiple cloning sites have been developed to aid researchers in assembling DNA sequences and checking their organization by restriction mapping.

Combining restriction endonucleases and DNA polymerases

When restriction enzymes are used to prepare fragments for cloning, the critical concern during assembly of fragments is whether the ends are compatible (see Chapter 4). Blunt ends are always mutually compatible, regardless of how they were generated, so cohesive ends that are incompatible can be brought together if the overhanging ends are first treated with a DNA polymerase such as Klenow enzyme or T4 DNA polymerase and the dNTP substrates (see Chapter 6). This technique expands the cloning strategies using restriction endonucleases, and an example is provided in *Figure 8.1*. Suppose the goal is to isolate a fragment of a gene using the enzymes *Eco*RI and *Hind*III (see right side of *Figure 8.1A*) and introduce it into a plasmid vector (left side of *Figure 8.1A*). The cloning vector has a multiple cloning site with both *Eco*RI and *Hind*III sites, so the cloning could be accomplished by ligating *Eco*RI to *Eco*RI ends and *Hind*III to *Hind*III ends. However, doing it this way would leave no remaining sites in the multiple cloning site of the plasmid for insertion of additional pieces of DNA, for example a transcriptional regulatory element.

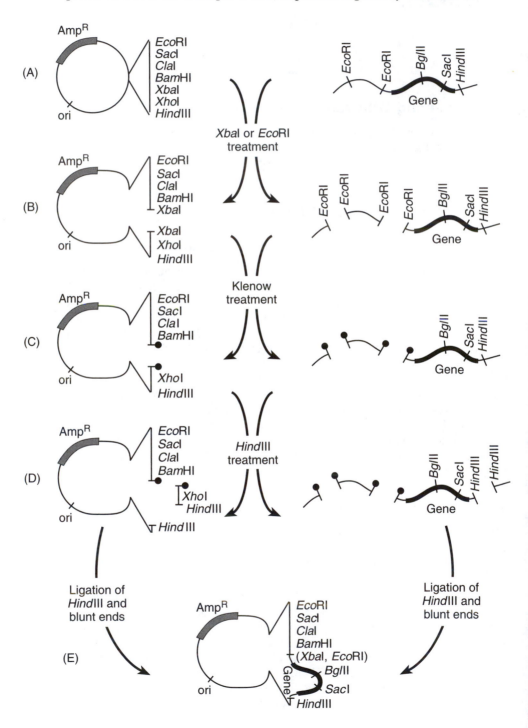

One solution is to treat the target DNA with *Eco*RI (G/AATTC) and then make the overhanging ends blunt by treatment with Klenow enzyme and the dNTP substrates (see right side of *Figure 8.1B, C*). The plasmid could be linearized with another enzyme that digests the DNA in the multiple cloning site, for example *Xba*I (T/CTAGA) and similarly made blunt (see left side of *Figure 8.1B, C*). With those reactions finished, the plasmid and target DNA could be treated with *Hind*III (A/AGCTT) (see *Figure 8.1D*). The ligation between plasmid and target DNA would involve matching two cohesive ends (*Hind*III) and two blunt ends (*Eco*RI and *Xba*I, filled in) (see *Figure 8.1E*).

The joining of the two *Hind*III-generated ends preserves the sequence AAGCTT in the ligation products, so the junction between plasmid and target DNA has a *Hind*III site. However, the joining of the filled-in *Xba*I and *Eco*RI sites has the interesting effect of regenerating both of those sites. The *Xba*I end has the following structure before and after treatment with Klenow enzyme and the dNTP substrates (the nucleotides added by Klenow enzyme are underlined):

```
      XbaI end                     XbaI end after Klenow treatment
...NNNNT    -3'                   ...NNNNTCTAG-3'
...NNNNAGATC-5'                   ...NNNNAGATC-5'
```

The *Eco*RI end has the following structure before and after treatment with Klenow enzyme and the dNTP substrates (the nucleotides added by Klenow enzyme are underlined):

```
      EcoRI end                    EcoRI end after Klenow treatment
5'-AATTCNNNN...                   5'-AATTCNNNN...
3'-    GNNNN...                   3'-TTAAGNNNN...
```

When these different ends are joined by blunt end ligation, the structure is as follows:

```
...NNNNTCTAGAATTCNNNN...
...NNNNAGATCTTAAGNNNN...
```

The *Xba*I (T/CTAGA) and *Eco*RI site (G/AATTC) are both regenerated.

The regeneration of the *Xba*I site is fortuitous, because *Xba*I can be used a second time to insert foreign DNA sequences adjacent to the gene. However, it is a bit of a nuisance that the *Eco*RI site was regenerated because now we have two *Eco*RI sites in close proximity. If we treat the plasmid with *Eco*RI during cloning we will not be able to make use of other enzymes in the multiple cloning site.

Figure 8.1

A cloning strategy using restriction endonucleases and DNA polymerase. (A) A plasmid vector is shown on the left, with a multiple cloning site. A genomic DNA is shown on the right, with a gene sequence (thick line). (B) The plasmid vector is linearized with *Xba*I, and the genomic DNA is digested with *Eco*RI. (C) The plasmid and genomic DNA ends are made blunt, using Klenow enzyme and dNTP substrates. (D) The plasmid and genomic DNA are treated with *Hind*III. (E) The plasmid and genomic DNA are ligated at their *Hind*III ends, and at their blunt ends.

Destroying restriction endonuclease sites

One way to handle this type of problem is to destroy one of the restriction enzyme sites. Suppose that the plasmid shown in *Figure 8.1A* were first treated with *Eco*RI (G/AATTC) in order to linearize it, then treated with Klenow enzyme and the dNTP substrates. The *Eco*RI ends would be made blunt, as shown below (the nucleotides added by Klenow enzyme are underlined):

```
...NNNNGAATT    AATTCNNNN...
...NNNNCTTAA    TTAAGNNNN...
```

If these ends were rejoined by ligation the sequence at the end of the polylinker would have the following structure:

```
...NNNNGAATTAATTCNNNN...
...NNNNCTTAATTAAGNNNN...
```

The GAATTC sequence would not be present after this treatment. The products of ligation could be transformed into *E. coli*, and a plasmid clone isolated that lacked the *Eco*RI site. If this alternative plasmid were used to initiate the cloning work outlined in *Figure 8.1*, the result of ligation would be the clone in *Figure 8.1E*, with only a single *Eco*RI site instead of two.

Treating overhanging ends left by a restriction enzyme with a DNA polymerase, such as Klenow enzyme or T4 DNA polymerase, will usually destroy the recognition sequence for the restriction enzyme when the blunt ends are rejoined. We can use that technique to destroy recognition sequences we don't want, and more generally to make specific insertion or deletion mutations in DNA. In the example shown in *Figure 8.1*, the gene of interest has an internal *Bgl*II (A/GATCT) site. We can make a mutation in the gene by linearizing the clone (shown in *Figure 8.1E*) at the *Bgl*II site, filling in the 5′ overhanging end with Klenow enzyme and the dNTP substrates, and re-closing the plasmid by ligation of blunt ends. The sequence AGATCT will be changed to AGATCGATCT, where the underlined nucleotides represent new synthesis by Klenow enzyme, and the net result is a four-nucleotide insertion. If the *Bgl*II recognition sequence is in a coding sequence, the four extra nucleotides will cause a frameshift mutation. Such a mutation may be useful to us if we want an interrupted gene as a negative control for an experiment.

We can also introduce a mutation at the *Sac*I (GAGCT/C) site of the gene, but here it is a little more complicated because the original plasmid has a *Sac*I site in the multiple cloning site. We could treat the plasmid carrying the cloned gene (see *Figure 8.1E*) with a limiting amount of *Sac*I to cause a partial digestion, so that some of the products of digestion are cleaved at only one of the two *Sac*I sites. Then we could treat the products with Klenow enzyme and the dNTP substrates, which would remove the four nucleotides in the 3′ overhanging end. The NNNGAGCTCNNN sequence would be changed to NNNGCNNN. If we gel-purified the mixture, selecting the DNA fragments that are the full size of the plasmid (i.e., digested exactly once with *Sac*I), these could be re-closed by ligation and used to transform *E. coli*. We could then confirm that the plasmid clone candidates are digested only once with *Sac*I, and search for a candidate that has lost the *Sac*I site in the inserted gene rather than in the multiple cloning site. If the loss of the four nucleotides were within a coding sequence, it would constitute a frameshift mutation.

Using exonucleases to make deletions

When DNA deletions are engineered using an exonuclease, the extent of the deletion is controlled by the number of units of enzyme added to the DNA or the amount of time allowed for digestion. For example, the exonuclease *Bal*31 degrades double-stranded DNA and leaves a mixture of blunt and overhanging ends. A DNA can be linearized with a restriction endonuclease (see use of *Eco*RI site in *Figure 8.2A*), and samples treated with *Bal*31 for

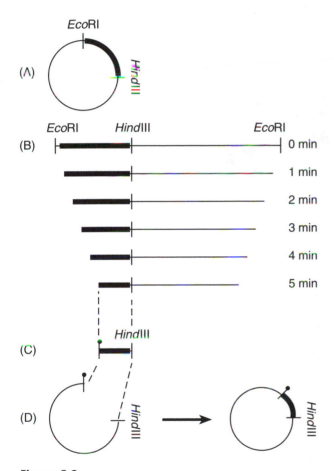

Figure 8.2

Generation of DNA deletions using *Bal*31 exonuclease. (A) The intact plasmid vector has an *Eco*RI site and a *Hind*III site. (B) The plasmid is linearized at the *Eco*RI site and treated with *Bal*31 for 1 to 5 min. The DNA becomes progressively shorter at each end. (C) The digested DNA is treated with Klenow enzyme and dNTP substrates to make a blunt end (black dot), and with *Hind*III to generate an internal cohesive end. (D) The fragment is ligated into the plasmid vector prepared with one blunt end (black dot) and one end digested with *Hind*III.

varying periods of time. The digestion will proceed in both directions away from the starting point (see progression from both ends in *Figure 8.2B*). After inactivating *Bal*31 the DNA ends can be made uniformly blunt with a DNA polymerase such as Klenow enzyme or T4 DNA polymerase, and the dNTP substrates. A segment of DNA that is reduced in size can then be excised using another restriction endonuclease site (see use of *Hind*III site in *Figure 8.2C*), and cloned into a plasmid vector (see *Figure 8.2D*). The plasmid clones generated after different amounts of digestion with *Bal*31 are a set of deletion mutations that can be used to map functional characteristics of a DNA sequence (as discussed later, in *Figure 8.8*).

Using PCR to isolate a fragment of a gene

A specific DNA sequence may be generated by PCR from a genomic DNA, or by RT-PCR from RNA, if enough information is known about the target

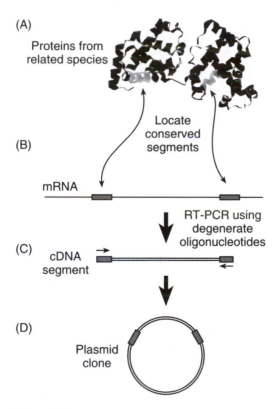

Figure 8.3

PCR cloning of a gene segment by RT-PCR, using sequence information from related species. (A) Conserved segments (light gray) of amino acid sequence can be identified by analysis of proteins from a group of related species. (B) The conserved amino acid sequences, if maintained in the species of interest, will be encoded by the mRNA. (C) RT-PCR is conducted using degenerate oligonucleotides matching the conserved segments. (D) The RT-PCR product is cloned into a plasmid vector.

sequence so that oligonucleotides can be designed. If a researcher is working with an organism that has a sequenced genome or a large collection of expressed sequence tags (ESTs, which are collections of cDNA sequences), the information for designing oligonucleotide primers is easily collected from GenBank® or other public databases of DNA sequences.

Even if specific sequence information is not available for an organism of interest, many times the same gene will have been cloned and sequenced from a closely related organism. A researcher can study an alignment of the amino acid sequences for the gene from related organisms to determine which domains are highly conserved (see *Figure 8.3A*). These protein sequences can be used to design degenerate oligonucleotides for PCR (see *Figure 8.3B*), or hybridization probes. The amino acid sequences are usually more highly conserved than the underlying DNA sequences, because of the degeneracy of the genetic code. However, once the amino acid sequence information is used to design DNA oligonucleotides, it is the specific base-pairing between oligonucleotide and target DNA that is important. The PCR oligonucleotide binds to a specific sequence and is extended to make a product, as discussed in Chapter 7, while a hybridization probe binds to a sequence and tags it with a radioactive or chemical label, as discussed in Chapter 6. A small region of a gene sequence may be isolated using PCR (see *Figure 8.3C*), and cloned into a plasmid vector (see *Figure 8.3D*).

If a researcher wants to clone a gene encoding a protein but has no examples of the gene sequence from related organisms, then an alternative strategy is to purify a sufficient amount of the protein to obtain amino acid sequence by Edman degradation. As discussed in Chapter 7, degenerate oligonucleotides can be prepared using the amino acid sequence as a guide, and a segment of the coding region may be isolated by RT-PCR. This captured segment of a gene can be used as a hybridization probe to identify a cDNA or genomic DNA clone that contains a larger segment of the gene (as discussed later).

8.2 Identifying DNA sequences

Base-pairing and fragment length

Many methods of identifying DNA or RNA sequences are based on binding of a chemically or radioactively labeled probe to a target. If the hybridization is conducted under stringent conditions, the binding of probe to target indicates a good match in base-pairing. The size of the target may also be a factor in identification; for example, restriction fragment lengths on a Southern blot and mRNA sizes on a Northern blot help to identify the target. In other methods, such as microarray analysis, the lengths of the probes and targets are not used in the identification.

Base-pairing is also involved in methods used to identify DNA molecules by their conformation, such as denaturing-gradient gel electrophoresis (DGGE) and single-strand conformational polymorphism (SSCP). DGGE involves running a double-stranded DNA fragment on a gel that includes a gradient of denaturant, such as urea. As the DNA fragment migrates during electrophoresis, it eventually reaches a point in the gel where the urea concentration causes partial denaturation of the base-pairing; that is, a change in DNA conformation that causes the migration to be slowed. Different DNA sequences tend to have different patterns of denaturation in urea, so DGGE is used as a way of determining whether two sequences are likely to be the same.

SSCP is also based on detecting conformational differences; however, in this case a DNA fragment is denatured completely and the single-stranded DNA is allowed to form a secondary structure through intramolecular base-pairing. The resulting single-stranded DNA molecules adopt secondary structures in solution and are separated on a non-denaturing gel. If two DNA fragments differ in sequence, even at a single nucleotide, they are likely to have differences in secondary structure and show differences in fragment migration.

Identification through enzymatic sites

We can identify DNA fragments by creating a map of their restriction endonuclease sites (as discussed in Chapter 5). Most restriction enzymes recognize four to six nucleotides of sequence, and the enzyme-DNA interactions are highly specific. A single nucleotide change can cause a restriction enzyme to no longer digest the DNA at a site, and we take advantage of that specificity when we use restriction endonucleases for diagnostic purposes.

One diagnostic technique involves characterizing DNA samples by similarities and differences in their restriction enzyme maps. This is called restriction fragment length polymorphism (RFLP) analysis, and it is often done in conjunction with PCR preparation of a fragment from a genomic sample. RFLP analysis is performed on the PCR product, and gel electrophoresis and DNA staining are the method of detection. Alternatively, RFLP analysis can be performed on genomic samples digested with restriction endonucleases, using Southern analysis as a means of focusing on one small region. A collection of RFLP markers can be used to identify a sample genetically, much as a fingerprint can be used to identify a person, except that the genetic identification carries much more statistical weight.

A pattern of restriction endonuclease digestion sites provides a snapshot of a DNA fragment, but it does not provide the same level of identification or information as DNA sequencing. There are also several challenges in using restriction enzymes to characterize DNA, as discussed in Chapter 5. Enzymes can lose activity and fail to cut the DNA to completion, leaving a confusing array of partially digested fragments. Furthermore, DNA fragments may not be noticed on a gel if they co-migrate with other fragments, and small DNA fragments may be lost when they run off the bottom of a gel.

Southern and Northern blotting

Southern and Northern blotting are methods used to identify DNA and RNA molecules, following transfer of nucleic acids from a gel to a solid support (e.g., nitrocellulose or nylon) and hybridization of a labeled probe (3,4). Northern blots are a way of determining the native sizes of RNA molecules, for example a specific mRNA that is part of a pool of RNA extracted from a cell. In Southern blotting the DNA fragments are usually the products of restriction endonuclease digestions, so the fragment sizes provide information on the spacing of restriction endonuclease recognition sequences (as discussed in Chapter 5). Gathering this information requires that molecular weight standards be included in the gel. The gel may be stained and photographed with a ruler in the picture, so that the migration of the standards is recorded for later use.

There are some important differences between the techniques used in Southern and Northern blotting. Northern blots are generated after running denatured RNA samples on a gel under RNase-free conditions. The RNA may be glyoxalated prior to being run on a gel, or the gel may be prepared with formaldehyde or other denaturants to keep the RNA free of secondary structures as the electrophoresis is being conducted. If the RNA molecules are larger than 3 kb, the transfer to a solid support may be inefficient; some protocols call for random nicking of the RNA *in situ* by a 15-min incubation of the gel in a mildly alkaline solution (50 mM NaOH), to reduce the sizes of the fragments. Since this nicking occurs after electrophoresis is complete, it does not affect the locations of the bands on the blot.

Southern blots are typically generated from a non-denaturing gel on which DNA restriction fragments are loaded. The double-stranded DNA fragments are then separated by size through electrophoresis, after which they are transferred to a nylon or nitrocellulose membrane (see *Figure 8.4A, B*). If the DNA fragments are larger than approximately 10 kbp, the transfer efficiency may be reduced, which is why some methods call for a 15-min incubation of the gel in an acidic solution (0.20 M HCl) to partially depurinate the DNA. The DNA is denatured prior to, or during the process of transfer to a solid support (see *Figure 8.4C, D*), which allows the probe to base-pair to the target at a later stage, during hybridization (see *Figure 8.4G*). The composition of solutions used to prepare the gel for DNA transfer varies between protocols and the type of solid support used. In the case of nitrocellulose as a solid support, the transfer is conducted under high salt conditions and a pH of 7 to 8, while in the case of nylon as a solid support the transfer may be conducted in a high pH solution such as 0.4 M NaOH.

The actual 'blotting' step of the Southern and Northern blot technique is performed by placing the gel on a chromatography paper wick that is in contact with a transfer solution, placing a pre-wetted sheet of nitrocellulose or nylon on the gel (removing all air bubbles separating the gel and sheet during the process), and adding layers of dry blotting paper on top of the nitrocellulose or nylon (see *Figure 8.4C, D*). In horizontal gel electrophoresis methods the nucleic acids tend to run along the bottom face of the gel, the side of the gel closest to the electrodes, so it is best to flip the gel over and apply the nitrocellulose or nylon sheet to the bottom face during blot assembly. The blotting paper gradually soaks up the transfer solution, a process that may take overnight. The solution passes from the wick through the gel, and in doing so washes the nucleic acid onto the nitrocellulose or nylon sheet where it sticks. A vacuum blot apparatus may be used in place of the blotting paper, to draw liquid out of the gel. Alternatively, the transfer can be conducted by electrophoresis

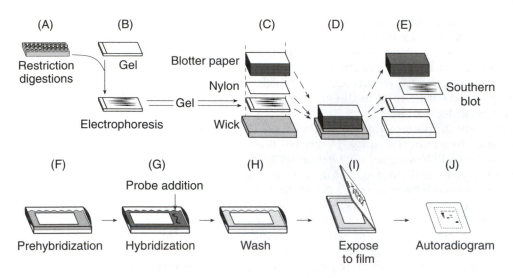

(A) Restriction digestions

(B) Gel

Electrophoresis

Blotter paper

Nylon

Gel

Wick

(C)

(D)

(E)

Southern blot

(F) Prehybridization

(G) Probe addition

Hybridization

(H) Wash

(I) Expose to film

(J) Autoradiogram

Figure 8.4

Southern blotting. (A) Samples of DNA treated with restriction endonucleases are loaded on a gel. (B) Electrophoresis of the gel and staining. (C) The gel is placed on a wet wick and layered with a nylon or nitrocellulose membrane and dry blotter paper. (D) The solution flows from the wick through the gel, washing the DNA to be transferred to the membrane. (E) The membrane 'blot' with the transferred DNA is removed. (F) After the DNA is bound, the membrane is subjected to prehybridization to reduce background. (G) Hybridization with a labeled probe. (H) Washing of the blot to remove loosely bound probe. (I) Exposure of the blot to film (or other methods of signal development). (J) Film exposure of the blot. The autoradiogram indicates the positions of DNA bands that hybridized to the probe.

perpendicular to the face of the gel, using an 'electroblot' transfer apparatus.

Following disassembly of the transfer apparatus or blotting paper (see *Figure 8.4E*), the DNA should be bound to the nylon or nitrocellulose sheet. The location of the wells of the gel should be marked during disassembly so that the migration of bands on the blot can be measured later from a consistent starting point, and compared to the photograph of the standards visible on the stained gel. It is also important to mark the orientation of the gel on the solid support so that the location of the first lane of the gel is known. Some protocols call for baking the solid support for 2 hours at 80°C in a vacuum oven (the vacuum is to prevent ignition of the nitrocellulose). In the case of nitrocellulose this baking step is important for cross-linking of the nucleic acid to the solid support; however, the method varies depending on the type of solid support being used and the manufacturer's recommendations. Cross-linking can also be conducted by UV irradiation, or simply by air-drying. The gel may be equilibrated in a buffer and re-stained to determine whether the DNA was successfully transferred out. It is often the case that high-molecular-weight DNA remains in the gel unless a depurination step was included in the method earlier.

Prior to hybridization of a Southern or Northern blot to a labeled probe, the solid support must be pre-hybridized in a blocking solution to cover up any nonspecific binding sites (see *Figure 8.4F*). Pre-hybridization solutions have different formulations, and some commercial solutions are proprietary, but they may include a carrier RNA (such as *E. coli* or yeast RNA) or carrier DNA (such as denatured salmon sperm or calf thymus DNA) that are unrelated to the probe sequence. Many pre-hybridization solutions also include protein carriers such as reconstituted nonfat dry milk, or bovine serum albumin. The solution is usually based on a high-salt buffer, for example $6\times$ standard saline citrate ($6\times$ SSC), although many researchers favor a formamide-containing buffer. A $1\times$ SSC solution is 0.15 M NaCl, and 0.015 M sodium citrate, pH 7.0, and a $6\times$ SSC solution is a six-times concentrated stock. The pre-hybridization may be conducted at 65°C for several hours, with constant agitation, or 42°C if the pre-hybridization includes formamide. During this process the blot can be sealed in a plastic bag to prevent air exposure, or placed in a hybridization chamber that rotates constantly and keeps the blot wet.

The hybridization solution typically has the same composition as the pre-hybridization solution, except that a denatured probe is added (see *Figure 8.4G*). The probe must be single-stranded so that it can base-pair to the single-stranded target that is bound to the solid support. In the case of DNA probes, the probe solution may be heated in a boiling water bath for 5 min to denature the double-stranded DNA, or treated with 0.12 M NaOH for 5 min before being neutralized with 0.1 M Tris-HCl, pH 8, and an equivalent amount of 0.12 M HCl to titrate out the NaOH. RNA probes are heat-denatured to remove secondary structures, often in the presence of formamide as a denaturant. The hybridization is conducted in a high-salt buffer such as $6\times$ SSC, or a formamide-containing buffer, but in either case the salt ions shield the nucleic acids so that probe and target are less affected by electrostatic repulsion during hybridization. The temperature (65°C without formamide, or 42°C with formamide) also affects the stringency of hybridization: higher temperatures prevent the nonspecific binding of probe. The length of time needed for hybridization depends on the concentration of probe, but for most applications an overnight incubation is sufficient.

Following the hybridization step, the blot is washed in a lower-ionic-strength solution to dislodge probe molecules that are nonspecifically bound. For example, the blot might be washed several times in a $2\times$ SSC solution at 65°C for 15 min, then several more times in a $1\times$ SSC solution under the same conditions. In addition to SSC, these wash solutions also contain 0.1% (w/v) sodium dodecyl sulfate (SDS), which helps to reduce the amount of nonspecific binding of probe. If the probe is radiolabeled, the hybridization and wash solutions must be disposed of properly, and in accordance with laboratory or institutional radiation safety policies.

The blot may be wrapped and placed in an X-ray film cassette for exposure to X-ray film (see *Figure 8.4I*), or exposed to a phosphorimager plate. If the signal is being generated by phospholuminescence, a substrate is added to the wrapped blot in advance of the exposure and there may be a time lag before light production is steady and maximal. If the probe is radiolabeled and the blot is exposed to X-ray film, the cassette may be placed at −70°C to increase the effect of the exposure, then allowed to thaw completely before the film is removed for development. The film image is called an autoradiogram (see *Figure 8.4J*), and the signals indicate where probe was bound to the nucleic acid on the blot.

Identification of library elements by hybridization

DNA in a bacterial colony or bacteriophage plaque can be identified by hybridization to a probe (5,6). In this application a dry nitrocellulose filter is applied to a bacteriological plate containing colonies or plaques, and some of the bacteria or bacteriophage adhere to the filter. They are lysed *in situ*, and the DNA is cross-linked to the nitrocellulose. The filter is then processed in much the same way as a Southern blot, and the hybridization of a probe to a specific spot on the filter indicates that complementary DNA from the colony or plaque is present.

For example, a bacteriophage lambda library carrying genomic DNA fragments could be grown on bacteriological plates so that each clone in the library forms a plaque (see *Figure 8.5A–C*). Libraries are usually plated at high density so that thousands of clones are represented on each plate; the low density of plaques represented on the plate in *Figure 8.5C* is for the purposes of illustration only. A nitrocellulose filter or similar type of solid support is used to make an imprint of the bacteriophage λ on the plate. The filter is allowed to rest on the surface of the plate for several minutes, during which time some of the bacteriophage λ adhere. It is then marked with alignment markings so that it can be re-oriented later, and carefully peeled off of the surface of the plate (see *Figure 8.5E*). The bacteriophage are lysed on the filter and their DNA is covalently bound to the filter. The filter is then pre-hybridized with a blocking solution to cover all sites on the filter to which the probe might bind nonspecifically. The filter is hybridized with a labeled probe (see *Figure 8.5D*), in the same way described for Southern blotting (see *Figure 8.5F*). During hybridization the probe base-pairs to target sequences that are part of the cloned DNA in the bacteriophage λ vector; each λ clone in the library may carry a different sequence and only some of them will base-pair to the probe.

Following the period of hybridization the excess probe is washed away from the filter, and the filter is exposed to X-ray film or a phosphorimaging plate. Plaque-sized spots on the autoradiogram are indicative of probe binding to the filter (see *Figure 8.5G*). The plate is aligned to the spot using alignment markings, and the area corresponding to the spot is cut from the plate so that residual λ bacteriophage can be recovered.

This sample of λ bacteriophage taken from the plate is a mixture of clones, because several different plaques are usually excised in the area taken from the plate. Furthermore, the bacteriophage have had an opportunity to spread by diffusion in the agar medium during the time between initial plating (*Figure 8.5C*) and collection of hybridization results (*Figure 8.5G*). The λ bacteriophage are eluted from the material taken from the plate and re-plated at low density so that the plaques are separated. The filter-hybridization test is repeated (see *Figure 8.5D–G*), and isolated plaques that contain phage-clone DNA hybridizing to the probe are identified and

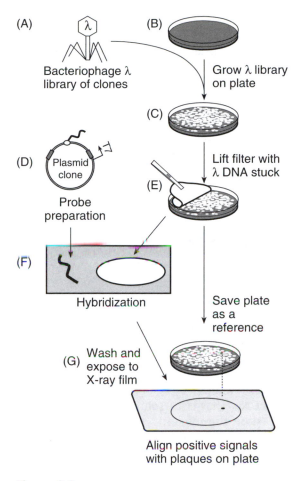

(A) Bacteriophage λ library of clones

(B) Grow λ library on plate

(C) Lift filter with λ DNA stuck

(D) Plasmid clone — Probe preparation

(E)

(F) Hybridization

Save plate as a reference

(G) Wash and expose to X-ray film

Align positive signals with plaques on plate

Figure 8.5

Bacteriophage plaque hybridization. (A) A library of bacteriophage is combined with bacteria to initiate an infection. (B) The infected bacteria are combined with a low percentage agar medium and distributed on a fresh bacteriological plate. (C) After overnight growth, plaques appear on the plate. (D) A plasmid clone carrying a DNA insert is used to generate a labeled probe for hybridization. (E) A nitrocellulose filter is used to adsorb bacteriophage from the plaques, making an imprint of the library clones. The bacteriophage are lysed *in situ* and the DNA is bound to the filter. (F) The filter is prehybridized, then hybridized to the labeled probe. (G) The washed filter is exposed to film (or another method of signal development). Plaques on the original plate that correspond to spots on the film are candidate clones; they may carry DNA overlapping with the probe sequence.

excised from the plate. The re-plating process is repeated several times until 100% of the plaques on the plate are hybridization-positive, and the clone is said to be 'plaque purified'.

This method of identification of clones by hybridization can generally be applied to bacterial colony hybridization as well as bacteriophage hybridization, but there are some important differences. The nitrocellulose filter that is used to make an imprint of the bacteriological plate may leave few cells behind as a reference colony for isolation. This is not a problem with phage plaques because some of the phage are left behind on the plate and can be extracted and isolated later. In conducting experiments with bacterial colonies it is important to make a separate reference plate, so that when a positive colony is identified by hybridization there will be live cells available on a reference plate that carry the DNA. Preparation of a reference plate is a process called 'replica plating' in which the agar plate surface is pressed onto a sterile velvet cloth surface to make an imprint of the colonies, prior to application of the nitrocellulose filter that may strip away the remaining cells. A new sterile plate is pressed onto the velvet cloth to capture the imprint of colonies on a reference plate, and the plate is placed in an incubator for re-growth.

Expression profiling by microarray analysis

Microarrays are slides that carry a large number of DNA samples, suitable for testing by hybridization, and many of the principles of hybridization used in Southern and Northern blotting also apply to microarray hybridization (7). The ways of generating these slides vary from robotic printing using a set of dip pens to more complicated methods of photolithography, but the basic similarity is that the slide will have thousands of individual spots of DNA. These spots of DNA are a set of reference sequences fixed to the slide, and they may be cloned DNA samples or oligonucleotides. These reference sequences hybridize to a fluorescently labeled sample of cDNA that is in solution, by base-pairing to it, making that spot on the slide fluorescently labeled. Each spot of DNA on the slide serves as a probe for measuring the amount of one type of cDNA in solution. This use of the word 'probe' may be initially confusing because the probe is fixed to a solid support; in the case of Southern and Northern blotting the probe was in solution and the targets were fixed. The fundamental techniques of hybridization that were discussed earlier for Southern and Northern blotting still apply to microarrays; the hybridization is conducted at an optimal temperature and salt condition for annealing of the targets and probes.

The targets of these probes in an expression profiling experiment are cDNA prepared from cellular mRNA. For example, a collection of cDNA from a eukaryotic cell line might be generated by reverse-transcription synthesis using an oligo(dT) primer labeled with the red fluorescent dye Cy™5 (Amersham Biosciences, Ltd.). A second collection of cDNA might be prepared from a different cell line, or the same cell line under different growth conditions, and labeled with the yellow fluorescent dye Cy™3 (Amersham Biosciences, Ltd.). The Cy™5- (red) and Cy™3- (yellow) labeled cDNA could be mixed together and used in hybridization against the microarray. If a microarray probe sequence matches an mRNA that is expressed in only one of the two cell lines, the spot will hybridize with only one of the two cDNA target collections and be dyed with either Cy™5 (red) or Cy™3 (yellow). If the mRNA is expressed in both cell line samples then the spot will give a mixture of Cy™5 and Cy™3 colors, and if the mRNA is not expressed in either cell line then the spot will not be fluorescent.

The dye emission colors that are used to record the amounts of bound cDNA may not bear any relationship to the colors used in publication, when the data are reported. In figures for publication the intensity and colors of the spots may redrawn to represent the amount of change in expression between an experimental and reference sample, rather than the overall amount of expression. For example, if the Cy™5-cDNA signal on a specific probe spot is less than the Cy™3-cDNA signal, that spot may be drawn with a red color in a figure for publication; the red color in the figure indicates to the reader that the mRNA was up-regulated in the second cell line compared to the first. Furthermore, the intensity of the red color in the figure may be linked to how much up-regulation exists, or the magnitude of the $\log_2$ of Cy™3:Cy™5. Conversely the spot in the figure may be drawn with a green color to indicate down-regulation of the mRNA (Cy™3 < Cy™5), and again the intensity may be related to the magnitude of the $\log_2$ of Cy™5:Cy™3. A black spot may indicate a case where the expression levels were similar in the two cell lines (Cy™5 ~ Cy™3), after normalization.

Primer binding and extension

PCR (as discussed in Chapter 7) and primer extension analysis (as discussed in Chapter 6) are used to identify DNA and RNA sequences through base-pairing of an extendable oligonucleotide primer. The extension of an oligonucleotide during an experiment of this type allows a researcher to infer that the base-pairing was sufficiently high at the 3′ end of the oligonucleotide for a DNA polymerase to extend it as a primer. This provides a limited amount of sequence information, and a single nucleotide polymorphism at the 3′ end of the primer is all that is needed to prevent the extension. However, two oligonucleotides are involved in PCR so there are two opportunities to use primer base-pairing to identify a sequence. The spacing between the primer binding sites on the complementary strands, which is inferred from the length of the PCR product, is an additional piece of information that is also used to identify a target.

Sanger DNA sequencing is one of the most important ways of identifying DNA sequences, and it is also fundamentally a primer extension reaction (see Chapter 6). It is based on binding of an oligonucleotide primer to a target, and then on extension of the primer to multiple points of chain termination. The DNA sequence is inferred from the lengths of chain-terminated products; however, the lengths of these products can only be interpreted meaningfully if they have the same 5′ end (and different 3′ ends). The primer used in DNA synthesis establishes this fixed 5′ reference point on the target, and contributes to the specificity of the reaction.

Specific sequence pull-down

Southern and Northern blotting, phage plaque hybridization, and microarray analysis all involve the base-pairing of one nucleic acid fixed to a solid support to a second nucleic acid that is pulled out of solution. In blotting or plaque hybridization methods, the probe in solution is a known reference sequence and the information gathered in such an experiment helps to characterize the targets fixed to the nitrocellulose or nylon membrane. In microarray analysis each fixed spot of DNA is a reference sequence, and it is the labeled nucleic acid in solution that is being characterized by the probes. What is similar among these techniques is that either the probe or the target is immobilized on a solid support. However, it is not critical that probes or targets be immobilized for diagnostic techniques; for example, molecular beacon probes (discussed in Chapter 7), allow the detection of a sequence by hybridization between two molecules that are in solution. The hybridization of the molecular beacon probe physically separates a fluorescent dye from a quenching dye, permitting fluorescence emission.

Hybridization or primer-binding experiments are often improved by capturing a target or probe sequence on a solid support, a method that is sometimes referred to as a 'pull-down' assay. For example, a PCR could be conducted in which the forward oligonucleotide was labeled with a fluorescent dye such as Cy™3 and the reverse oligonucleotide was conjugated to biotin. A bead coated with avidin could be used to capture the reverse oligonucleotides, by binding of biotin to avidin. Co-localization of Cy™3 with the avidin beads would indicate that the forward and reverse oligonucleotides were linked by a successful PCR. The point is that the immobilization and the detection 'events' can both be made specific and informative.

The specific capture of nucleic acids on a solid support allows them to be concentrated for signal detection. For instance, an array of fiber optic sensors can be used as solid supports, with each glass fiber tip in the array carrying a different probe oligonucleotide. The binding of fluorescently labeled targets to the fiber tips is detected by fluorescence emission, with a laser used to 'interrogate' the array in such a way that each fiber provides information on a different target (8).

Alternatively, collections of very small polystyrene beads coated with probes can capture fluorescently labeled targets in solution (9). Each bead detects a specific target or collection of targets by fluorescence, and can report on the levels of a molecular target in much the same way as a microarray. However, the beads are not in a fixed position and are identified through other specific dye signatures. The dye combinations on the beads can be read by their patterns of fluorescence in flow cytometry.

Pull-down assays are also used to separate reactants from products. For example, DNA ligation can be used as a diagnostic test for the presence of a DNA sequence, followed by a pull-down assay to isolate the ligation products. Two single-stranded DNA oligonucleotides can be brought together for ligation by a sample DNA to which both anneal. T4 DNA ligase will form a covalent connection between the oligonucleotides if the 5' phosphate and 3' hydroxyl groups are next to each other. For example, one of the DNA oligonucleotides may be tagged with a biotin group (represented as a 'B' in *Figure 8.6A, E*), and the others with fluorescent dyes (for example Cy™3 or Cy™5 in *Figure 8.6A, E*).The point of the assay is to see whether the biotin-labeled oligonucleotide can be linked to one of the dye-labeled oligonucleotides; ligation can only happen if the two oligonucleotides are properly annealed to the sample DNA.

Suppose that we have two different DNA samples, 'allele 1' and 'allele 2', and that the Cy™3 oligonucleotide base-pairs to allele 1 while the

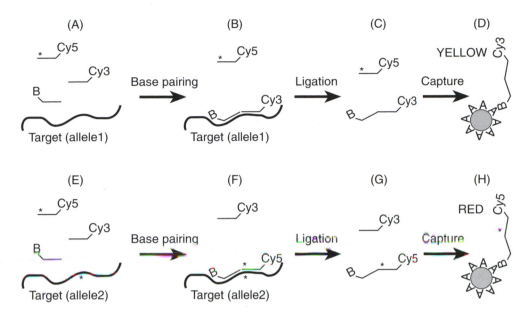

Figure 8.6

Pull-down of a fluorescent oligonucleotide in a diagnostic test. (A) The assay includes two fluorescently labeled oligonucleotides (Cy3 and Cy5), a biotin-tagged oligonucleotide 'B', and a DNA sample (allele 1). (B) The Cy3 oligonucleotide and the biotin-tagged oligonucleotide base-pair to the allele 1 DNA adjacent to each other. The Cy5 oligonucleotide does not anneal because it contains a different sequence (*). (C) The Cy3- and biotin-tagged oligonucleotides are joined by ligation. (D) The biotin-tagged DNA is captured by avidin-conjugated beads. The fluorescent oligonucleotide is also pulled down if it is ligated to the biotin-tagged DNA. In this example, the Cy3 oligonucleotide is pulled down rather than the Cy5 oligonucleotide because it is complementary to the allele 1 sequence, and the Cy5 oligonucleotide is not. (E) The same assay as above, but with a different DNA sample (allele 2). (F) The Cy5 oligonucleotide base-pairs to the allele 2 DNA because they have complementary sequences (*), but the Cy3 oligonucleotide does not base-pair. (G) The Cy5 oligonucleotide and biotin-tagged oligonucleotide are joined by ligation. (H) The biotin-tagged DNA is captured by avidin-conjugated beads. In this example, the Cy5 oligonucleotide is pulled down and the Cy3 oligonucleotide is not.

Cy™5 oligonucleotide base-pairs to allele 2 (see *Figure 8.6B, F*). The difference between the two alleles could be as minor as a single base change at the 3′ end of the dye-labeled oligonucleotides. The biotin-labeled oligonucleotide base-pairs to both allele 1 and allele 2 DNA, so it can be ligated to whichever fluorescently labeled oligonucleotide is base-paired next to it (see *Figure 8.6C, G*). Finally, avidin-coated beads can be used to pull down the biotin-labeled oligonucleotides and the amount of fluorescence assessed (see *Figure 8.6D, H*). The fluorescent dyes can be detected independently, if the original sample is from an individual heterozygous for the two alleles and both Cy™3 (yellow) and Cy™5 (red) dyes are pulled down.

Indirect identification through an encoded product

Immunological screening

If a DNA sequence encodes a gene product, the DNA may be identified indirectly if the gene product can be expressed. For example, suppose that a researcher has developed a polyclonal antiserum against a purified protein and wishes to identify a DNA clone that encodes the protein. He or she might use a cDNA library prepared from the organism of interest, and cloned into Lambda ZAP II® or λgt11 (10). As explained in Chapter 4, these vectors have a cloning site within a β-galactosidase gene and expression of the cloned sequence is induced by addition of IPTG (isopropyl-β-D-thiogalactopyranoside). The foreign sequence inserted into the bacteriophage could be transcribed and translated as a fusion protein with β-galactosidase, and some of the protein would be released into the bacteriophage plaque as the *E. coli* cells lyse and release progeny bacteriophage. A nitrocellulose filter applied to a bacteriological plate with these plaques would pick up some bacteriophage, as discussed earlier, but would also absorb some of the fusion protein in each plaque. The nitrocellulose filter could be blocked with nonspecific proteins and developed as a Western blot using the polyclonal antiserum. Plaques that are scored as positive in this assay may carry fusion proteins that are cross-reactive with the antiserum, which is to say that the bacteriophage in the plaque may have a foreign DNA insert that encodes the cross-reactive protein. The cDNA clone is therefore identified indirectly by the amino acid sequence it encodes. A partial cDNA clone may only carry a few hundred nucleotides of foreign sequence, but this may be used to prepare a labeled probe for detection of a longer cDNA clone or a genomic DNA clone by hybridization methods, as discussed earlier.

For immunological screening to work, a cDNA must be cloned into the vector in the correct orientation so that the sense strand of the gene is transcribed. In an unbiased cloning strategy only half of the clones may meet this criterion. Furthermore, the coding sequence must be cloned in frame with the β-galactosidase gene, only one of three reading frames being concordant, or inserted in such a way that a ribosome can re-initiate synthesis. A commonly stated idea is that one out of every six unbiased insertions of a cDNA will be in the correct orientation and reading frame, so the size of the cDNA library may need to be expanded accordingly.

The cDNA does not need to be a full-length copy of the RNA, and indeed the technique may be more successful if the 5' non-coding sequences of the cDNA are not included among the cloned sequences. The amount of foreign sequence expressed must be sufficient to generate a protein epitope recognized by the antiserum, and that will vary between different cloning projects. Polyclonal antisera are more likely to be successful in immunological screenings than monoclonal antibodies, because the latter will only work in the unlikely event that the monoclonal target epitope is formed during cDNA expression in *E. coli*.

Polyclonal antisera can be generated from synthetic polypeptides, so the initial steps in cloning a gene by this method do not necessarily require the biochemical purification of a protein in quantities sufficient for injection as an immunogen. A protein might be purified in small amounts and subjected to Edman degradation so that a partial amino acid sequence could be established. Alternatively, a conserved segment of a protein might be identified by a multiple alignment of sequences from related species, without any 'wet work' performed in the lab. Rather than make degenerate oligonucleotides for a PCR cloning attempt (as discussed in Section 8.1), a researcher might prepare (or order from a commercial source) short synthetic peptides with desired amino acid sequences that are conjugated to a hapten-carrier such as keyhole limpet hemocyanin (KLH). These conjugates would be injected into an animal such as a rabbit for preparation of a polyclonal antiserum.

Display libraries

Bacterial and bacteriophage display libraries (discussed in Chapter 4) are a different way to identify DNA sequences indirectly. The researcher screens or selects the library for specific binding between a target ligand and a polypeptide epitope, expressed on the surface of a bacterium or bacteriophage. The foreign DNA carried in the bacterium or bacteriophage library element encodes the polypeptide epitope that is expressed on the surface, and can be recovered by re-growth of a clonal population of cells or bacteriophage.

8.3 Functioning of DNA sequences

One of the most important reasons for working with DNA is to find out how biological systems work. DNA molecules have some functions that are integral, for example the ability to serve as a binding site for transcriptional regulatory elements, and other functions that are carried by transcribed or translated gene products. We can learn about these functions by making mutated versions of sequences, and seeing whether they work differently. Sometimes the function of a sequence can be tested *in vitro*; for example, an *in vitro* transcribed RNA may be biologically active in a cellular extract. Other DNA sequences may have functions that are only revealed *in vivo*. For example, the intracellular trafficking of a protein may be studied by introducing recombinant genes with mutated sequences, some of which may encode proteins that are targeted differently.

In vitro function

Gel shift assay

The purpose of a gel shift assay, also known as an electrophoretic mobility shift assay (EMSA), is to determine whether a specific fragment of DNA possesses a binding site for a protein (11,12). If a DNA fragment is bound by protein, the DNA-protein complex will exhibit reduced mobility during gel electrophoresis and will run on the gel more slowly than the non-complexed DNA. In a typical experiment, a researcher might be interested in characterizing binding sites in the DNA surrounding a transcriptional promoter. A double-stranded DNA representing part of the region of inter-est might be assembled from two complementary oligonucleotides, and radiolabeled at the 5′ ends using γ-[^{32}P]-ATP and polynucleotide kinase. The source of protein might be a nuclear extract that is expected to contain DNA binding proteins.

In the absence of any nuclear extract, the labeled DNA fragment will have the form of a duplex DNA and will display an electrophoretic mobil-ity that is predictable from its size (see lane 1, *Figure 8.7*). The gel is exposed to X-ray film so that the migration of the labeled DNA is tracked in each experiment. If the labeled DNA is first incubated with the nuclear extract, some of the DNA may migrate more slowly because the DNA-protein complex is bulkier and has a different conformation than a straight chain duplex (see lane 2, *Figure 8.7*). The amount of mobility shift is not easily predictable, and there is no way to determine the sizes of the

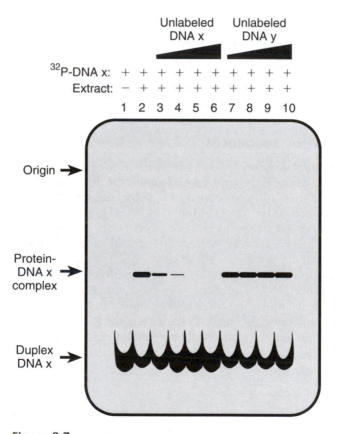

Figure 8.7

Electrophoretic mobility shift assay. A schematic diagram of an autoradiogram of a gel is shown, in which electrophoresis was conducted from the origin (arrow at the top). The ^{32}P-DNA is 'DNA x' and its migration as a native duplex is indicated (arrow at the bottom). The protein-DNA x complex has reduced mobility (arrow in the middle). The constituents of each assay are indicated above the autoradiogram. Lane 1, no extract; lane 2, complete assay; lanes 3–6, competition with DNA x; lanes 7–10, competition with DNA y.

binding proteins from molecular weight standards. Furthermore, the shift may depend on the concentration of polyacrylamide used to make the gel, and other experimental conditions.

If the binding between DNA and protein is specific, then addition of an excess of unlabeled duplex DNA with the same sequence should compete for binding to the protein factors in the nuclear extract (see lanes 3–6, *Figure 8.7*). The binding still occurs, but it is predominantly between the protein factor and the unlabeled DNA, which does not generate a signal during autoradiography. The labeled and unlabeled DNA will migrate predominantly as duplex DNA, without gel shift.

However, a nonspecific unlabeled duplex oligonucleotide, perhaps a DNA with the same nucleotide composition but a different sequence, should not compete for binding to a specific protein factor. If such an oligonucleotide is added in excess, the protein factors that have specific binding sites on the labeled sequence should still bind, and there should still be a gel shift (see lanes 7–10, *Figure 8.7*).

Once a gel shift assay is shown to be specific, a researcher may decide to make mutations in the DNA fragment of interest to map the boundaries and determine the consensus sequence of the protein-binding site. Alternatively, a researcher may decide that biochemical purification of the protein factors involved in specific binding is an important research direction. If the genes for the binding proteins can be cloned and expressed in a recombinant host, it may be possible to prepare mutations in the genes to determine which amino acid residues in the proteins are critical for binding activity.

Depletion and replacement of components in an extract

An *in vitro* system for studying the functions of biological macromolecules can be more elaborate than just a binding assay. Many cellular processes can be conducted *in vitro* using cell-free extracts, and these extracts can sometimes be studied in great detail by reconstitution of the extract after biochemical fractionation. A historical example of this is the reconstitution of *in vitro* translation extracts, which allowed the identification of initiation factors and elongation factors for protein synthesis. Another example is the development of *in vitro* splicing extracts that permitted rapid progress in the discovery of the components of the RNA splicing machinery of cells. RNA substrates for splicing could be generated from cloned DNA sequences, by use of bacteriophage RNA polymerase promoters (as discussed in Chapter 6), and the molecular details of the splicing reactions were uncovered in this way.

Differential centrifugation of a cell lysate and dialysis are some of the initial steps used to prepare a cellular extract. Individual components of the extract may be purified by biochemical techniques such as chromatography, and a biologically active extract reconstituted from individual fractions. The composition of an extract can be changed further by immunoprecipitation of specific components to deplete the extract. For example, if the extract loses biological function after a specific factor is pulled out by a monoclonal antibody, then that factor may be critical for activity. If activity can be replaced by reconstitution of the extract with the purified factor, or better still a cloned factor expressed in a recombinant host, then the involvement of the factor in the activity can be shown unambiguously.

In vivo function

Finding out how cells work

We can learn a lot about how cells work by introducing foreign DNA into them. For example, if we are interested in the signals that cause targeting of a specific protein to a cellular compartment, we can make recombinant genes that fuse the specific protein coding sequence with a green fluorescent protein (GFP) gene. If we introduce this recombinant gene into the cells, with the regulatory sequences needed for expression, we might be able to track the distribution of the gene product in the cell by fluorescence microscopy. By making changes in the sequence, we can learn what amino acid sequences are responsible for targeting the protein.

We can also change the activity of a cell by introducing genes that encode new enzymes, adding biochemical pathways or changing the regulation of existing ones. This is sometimes called 'metabolic engineering' and it has many applications in biotechnology. New enzymatic activities in a cell can sometimes expand the diversity of secondary metabolites that are generated, for example leading to new antibiotics produced from a microorganism or new pigments in a flower.

An interesting discovery is that many functions of proteins are modular, and can be separated into polypeptide domains. A protein that acts as a restriction endonuclease may have a DNA-binding domain and a catalytic domain that conducts the enzymatic step. If the DNA-binding domain were placed in a different protein, it could target the protein to the specific DNA-binding site. In a similar way, a protein that is a growth factor may have isolated regions that have defined functions: a signal sequence that causes the protein to be secreted, a domain that is responsible for binding to a cell surface receptor, and another domain that functions in cellular activation. If the receptor-binding domain were placed in a new protein, the new protein could have receptor-binding activity. The important point is that domains in proteins may often be dissected and made part of different proteins, sometimes without loss of domain function.

Much can be learned about biological systems by making genetic mutations that lead to a loss of function in a cell. For example, a gene may be 'knocked out' by homologous recombination between an introduced recombinant DNA molecule and the host cell genome (as discussed in Chapter 4). The second copy of the gene in a diploid organism may be similarly disrupted to engineer a double knockout, or if the cell has a sexual life cycle then classical genetics can be used to engineer a homozygous knockout. Any loss or alteration of cell function in the genetic knockout is traceable to changes made in the genome. On the other hand if no changes in cell function are detected in a genetic knockout, then the gene may not be required, at least in the cells as they are being grown. The gene may have a function that is not revealed by the differentiated cell being studied or the culture conditions. Cells often have levels of redundancy in their functions that are not easily discerned without molecular genetic research.

Short interfering RNA molecules (siRNA) are now widely used to 'knock down' the expression of genes in some types of cells, by using an effect called RNA interference (RNAi) (13). The siRNA, which can be made *in vitro* or *in vivo*, base-pairs to a specific mRNA in a cell and targets that mRNA for destruction through the RNA-induced silencing complex (RISC). In many species, RNAi is a rapid way of generating a loss of function of a specific gene without engineering a genetic change in the genome of a cell. The technique is an important contribution towards research into cellular functions.

Bait and prey

Interactions between proteins can be revealed *in vivo* through the use of bait and prey (or 'fish') systems, such as the yeast two-hybrid libraries described in Chapter 4. This technique takes advantage of the modular nature of proteins: the ability to separate the DNA-binding domain of a yeast transcriptional factor from the activation domain, and have those be functional in separate fusion proteins (the 'bait' and 'fish', respectively). There are ongoing efforts to catalog all of the interactions between the proteins in different types of cells, that is, to characterize the cellular system holistically. Some master regulatory proteins in a cell may have many different interaction partners, like a politician who get things done by shaking a lot of different hands. Other regulatory proteins might have a small number of interaction partners. What those protein–protein interactions may lead to is not known; however, the whole 'system' of the cell depends on these interactions, and a systematic analysis may prove informative.

Wreck and check

When we suspect that a specific segment of DNA or an encoded product has a defined function, we can study the relationship between structure and function by systematic mutagenesis. That is, we try to find ways to 'wreck' the function by genetic mutation, then we 'check' to see whether we have succeeded. If we alter a sequence and find that the function is not lost, then we have some certainty that the original sequence was not required. If a sequence is changed and the function is lost, then we have an idea that the sequence is required, in some way. The interpretation is complicated because nucleotides may be required as placeholders, in which case any sequence of the same length may substitute. For example, if two DNA-binding domains must be separated by 55 bp for function of a promoter, it is possible that there are few constraints on the sequence of those 55 bp. Furthermore, binding sites may also allow for heterogeneity. If a consensus sequence for a transcription-factor-binding site is 'TATGARAT', then both 'TATGAAAT' and 'TATGAGAT' meet the requirements.

Amino acid sequences are similarly both restricted and flexible, in that proteins have some amino acid residues that must be conserved for function and others that have an allowable range of side-chain variation, such as a serine substituting for a threonine, or an aspartic acid substituting for a glutamic acid. Uncovering functional domains by mutagenesis is complicated because some changes may cause a protein to mis-fold, to become insoluble, or to have a shorter half-life in a cell. There may be pleiotropic effects on the activity being studied, though the changed amino acids are not directly related to the functional domain.

A schematic representation of a 'wreck and check' analysis of a transcriptional promoter region is shown in *Figure 8.8*, where a 420-nt segment of DNA is being tested for its effect on the expression of a β-galactosidase reporter gene. Each of the DNA constructions shown in *Figure 8.8* would be part of a recombinant plasmid that was introduced into cells by transformation or transfection. After time was provided for gene expression in the cells, the amount of β-galactosidase expression would be measured as an indirect indication of promoter activity. The negative index numbers in the figure indicate nucleotides upstream of the transcriptional start site, and the entire 420-nt piece is shown in *Figure 8.8A*. The β-galactosidase expression is reported as '+++' with the full 420 nt, but progressive deletion of the nt −180 to −420 region exposes a sequence element that is required (see *Figure 8.8B–E*) since β-galactosidase is not expressed (level drops to '−'). An internal deletion removing nt −250 to −280 also eliminates expression (see *Figure 8.8F*), suggesting that this may be a critical part of the promoter region. If the nt −240 to −300 region is re-inserted in the opposite orientation the β-galactosidase expression is not restored,

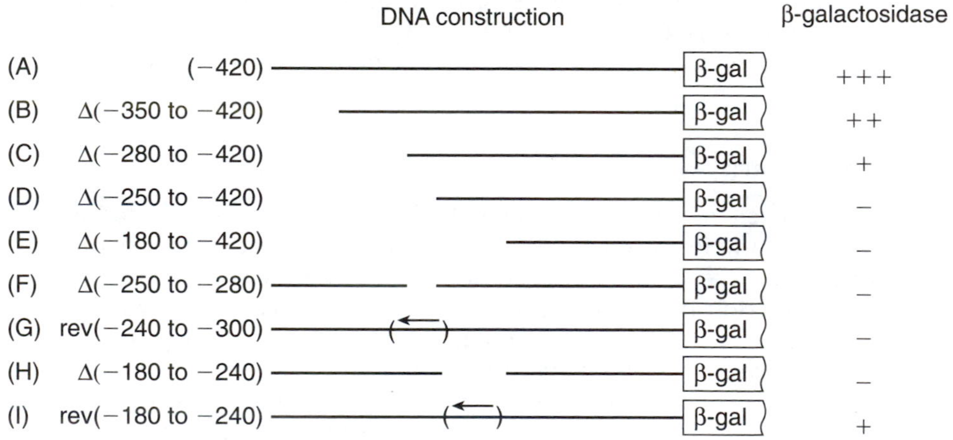

Figure 8.8

Deletion analysis of a promoter region. The DNA construction is indicated schematically, and the levels of β-galactosidase reporter expression are indicated on the right. (A) A 420-nt sequence upstream of a gene of interest is placed in front of a β-galactosidase reporter gene. (B) Deletion of the −350 to −420 region. (C) Deletion of the −280 to −420 region. (D) Deletion of the −250 to −420 region. (E) Deletion of the −180 to −420 region. (F) Deletion of the −250 to −280 region. (G) Inversion or reversal of the −240 to −300 region. (H) Deletion of the −180 to −240 region. (I) Inversion or reversal of the −180 to −240 region.

suggesting that the required sequence in the region is more than just a placeholder (see *Figure 8.8G*). The deletion of sequence between -180 and -240 results in no β-galactosidase expression (see *Figure 8.8H*); however, re-insertion of the sequence in the opposite orientation partially restores the expression of β-galactosidase (see *Figure 8.8I*). This shows how, with a series of examples of DNA constructions and *in vivo* expression, the function of a specific sequence element can be elucidated.

Complementation

The idea of making a cell or organism 'whole' again is the basis for studies of genetic complementation. A loss of function in a cell or organism that is caused by genetic change can help to identify the gene products that are responsible for the function, but an important tool in genetics is showing how that function may be restored. For example, if we start with a cell that has lost the ability to make a metabolite 'M', we might be able to discover how 'M' is made by genetic complementation using a library (as discussed in Chapter 4). If the 'M'-negative cells are transformed or transfected with the library and 'M'-positive cells are discovered, it is possible that the introduced DNA restored a gene that was mutated in the original cell. It is also possible that the introduced DNA provides an entirely new biochemical function that bypasses the defective pathway. In both cases the biochemistry underlying metabolite 'M' is uncovered through the application of genetic research.

Transgenesis

Many organisms can be made transgenic by introducing DNA into a progenitor cell that has the developmental potential to regenerate the whole organism or to populate the germ line of a mosaic organism. The genetic engineering work that leads to a transgenic organism typically starts with plasmid vectors grown in *E. coli*, and the cloning techniques described in Chapters 4–7. Depending on the organism that is to be genetically manipulated, the DNA may be introduced into cells by electroporation, biolistics, microinjection, or other methods (see Chapter 4). DNA can be introduced into embryonic stem cells of many animals, and then clonal populations of genetically modified cells isolated. These cells can be cultured and injected into an early stage embryo, and become part of a mosaic individual. These individuals are bred, and if their germ line is mosaic for the transgene then some of their progeny may be heterozygous for the transgene. These heterozygotes are bred to generate homozygous progeny. In the case of higher plants, many somatic cells are pluripotent and a whole plant may be established from a single genetically modified cell in culture.

Foreign DNA may be introduced into a genome by homologous or non-homologous (ectopic) integration, as discussed in Chapter 4. A selectable marker gene can be inserted in place of a normal allele by homologous recombination, and a transgenic knockout organism may exhibit a loss of function that is related to the genetic defect. Research of this kind sometimes reveals the relationship between genes and physiology in the whole organism. For example, many gene knockouts in mice lead to changes in development or effects on multiple organs. These effects could not be detected in a cell culture because the functions of the DNA sequence and its encoded gene products are only fully elucidated in the whole organism.

Further reading

Ausubel FW, Brent R, Kingston RE, Moore DD, Seidman JG, Smith JA and Struhl K (eds) (1987) *Current Protocols in Molecular Biology*, John Wiley & Sons, New York.

Sambrook J, Russell D (2001) *Molecular Cloning: A Laboratory Manual*, 3rd Edn. Cold Spring Harbor Laboratory Press, Cold Spring Harbor, NY.

References

1. Cohen SN, Chang AC, Boyer HW, Helling RB (1973) Construction of biologically functional bacterial plasmids in vitro. *Proc. Natl Acad. Sci. USA* **70**: 3240–3244.
2. Morrow JF, Cohen SN, Chang AC, Boyer HW, Goodman HM, Helling RB (1974) Replication and transcription of eukaryotic DNA in *Escherichia coli*. *Proc. Natl Acad. Sci. USA* **71**: 1743–1747.
3. Southern EM (1975) Detection of specific sequences among DNA fragments separated by gel electrophoresis. *J. Mol. Biol.* **98**: 503–517.
4. Alwine JC, Kemp DJ, Stark GR (1977) Method for detection of specific RNAs in agarose gels by transfer to diazobenzyloxymethyl-paper and hybridization with DNA probes. *Proc. Natl Acad. Sci. USA* **74**: 5350–5354.
5. Grunstein M, Hogness DS (1975) Colony hybridization: a method for the isolation of cloned DNAs that contain a specific gene. *Proc. Natl Acad. Sci. USA* **72**: 3961–3965.
6. Benton WD, Davis RW (1977) Screening lambda gt recombinant clones by hybridization to single plaques in situ. *Science* **196**: 180–182.
7. Schena M, Shalon D, Davis RW, Brown PO (1995) Quantitative monitoring of gene expression patterns with a complementary DNA microarray. *Science* **270**: 467–470.
8. Graham CR, Leslie D, Squirrell DJ (1992) Gene probe assays on a fibre-optic evanescent wave biosensor. *Biosens Bioelectron.* **7**: 487–493.
9. Spiro A, Lowe M, Brown D (2000) A bead-based method for multiplexed identification and quantitation of DNA sequences using flow cytometry. *Appl. Environ. Microbiol.* **66**: 4258–4265.
10. Young RA, Davis RW (1983) Efficient isolation of genes by using antibody probes. *Proc. Natl Acad. Sci. USA* **80**: 1194–1198.
11. Garner MM, Revzin A (1981) A gel electrophoresis method for quantifying the binding of proteins to specific DNA regions: application to components of the *Escherichia coli* lactose operon regulatory system. *Nucleic Acids Res.* **9**: 3047–3060.
12. Fried M, Crothers DM (1981) Equilibria and kinetics of *lac* repressor-operator interactions by polyacrylamide gel electrophoresis. *Nucleic Acids Res.* **9**: 6505–6525.
13. Fire A, Xu S, Montgomery MK, Kostas SA, Driver SE, Mello CC (1998) Potent and specific genetic interference by double-stranded RNA in *Caenorhabditis elegans*. *Nature* **391**: 806–811.

Index

Bold type is used to indicate the main entry where there are several.